Nature of Fish Growth Hormones

Nature of Fish Growth Hormones

Edited by:

Dr. Sekar Megarajan and Dr. Prerna Pandey

Delve PUBLISHING

www.delvepublishing.com

Nature of Fish Growth Hormones

Dr. Sekar Megarajan and Dr. Prerna Pandey

Delve Publishing
2010 Winston Park Drive,
2nd Floor
Oakville, ON L6H 5R7
Canada
www.delvepublishing.com
Tel: 001-289-291-7705
001-905-616-2116
Fax: 001-289-291-7601
Email: orders@arclereducation.com

ISBN: 978-1-77361-199-0 (Hardcover)

DECLARATION

Some content or chapters in this book are open access copyright free published research work, which is published under Creative Commons License and are indicated with the citation. We are thankful to the publishers and authors of the content and chapters as without them this book wouldn't have been possible.

ABOUT THE EDITORS

Dr. Sekar Megarajan

Dr Sekar Megarajan graduated in fisheries science, and then obtained master degree in Fish Genetics and Biotechnology and PhD in Fish Nutrition and Biochemistry with specialization in fish molecular endocrinology from Central Institute of Fisheries Education (CIFE), India. He worked for two years with Transasia Bio-Medicals Ltd, Mumbai in the area of immunology for development of diagnostic kits for human infectious disease and joined Indian Agriculture Research service in 2013 as a scientist. He currently works as scientist in mariculture division at Central Marine Fisheries Research Institute, Cochin, India. Dr Megarajan has been working on breeding, seed production and culture of high value marine finfishes. He has published more than 10 research papers in peer valued journals in the area of fish population genetics, molecular endocrinology and fish culture.

Prerna Pandey

Dr. Prerna Pandey, a biotechnologist with several years of wet lab research experience, worked at International Center for Genetic Engineering and Biotechnology, New Delhi. Her field of research in PhD included isolation and molecular characterization of geminiviruses, genome sequencing, gene annotation, and gene silencing using the RNA interference technology. She has also worked at Transasia Biomedicals and Advance Enzyme Technologies as a scientist. Prerna has published papers in peer-reviewed journals, and has submitted a number of annotated Geminiviral genome sequences in the GenBank, including two novel ones. She has also completed her editing and proof-reading courses from Society for the promotion of Editors and Proof-readers, UK and now works as a freelance scientific editor and writer. When Prerna is not busy with her assignments, she loves to explore historical places.

TABLE OF CONTENTS

LIST OF CONTRIBUTORS

Sekar Megarajan

Dr Sekar Megarajan graduated in fisheries science, and then obtained master degree in Fish Genetics and Biotechnology and PhD in Fish Nutrition and Biochemistry with specialization in fish molecular endocrinology from Central Institute of Fisheries Education (CIFE), India. He worked for two years with Transasia Bio-Medicals Ltd, Mumbai in the area of immunology for development of diagnostic kits for human infectious disease and joined Indian Agriculture Research service in 2013 as a scientist. He currently works as scientist in mariculture division at Central Marine Fisheries Research Institute, Cochin, India. Dr Megarajan has been working on breeding, seed production and culture of high value marine finfishes. He has published more than 10 research papers in peer valued journals in the area of fish population genetics, molecular endocrinology and fish culture.

Prerna Pandey

Dr Prerna Pandey, a biotechnologist with several years of wet lab research experience, worked at International Center for Genetic Engineering and Biotechnology, New Delhi. Her field of research in PhD included isolation and molecular characterization of geminiviruses, genome sequencing, gene annotation, and gene silencing using the RNA interference technology. She has also worked at Transasia Biomedicals and Advance Enzyme Technologies as a scientist. Prerna has published papers in peer-reviewed journals, and has submitted a number of annotated Geminiviral genome sequences in the GenBank, including two novel ones. She has also completed her editing and proof-reading courses from Society for the promotion of Editors and Proof-readers, UK and now works as a freelance scientific editor and writer. When Prerna is not busy with her assignments, she loves to explore historical places.

D. Linga Prabu

D. Linga Prabu With Masters in Fish Nutrition and Biochemistry, he joined as Agricultural Research Service Scientist in Marine Biotechnology Division of ICAR-Central Marine Fisheries Research Institute, Kochi, India in 2014. Before joining into ARS, Mr Prabhu worked with Growel Feeds Pvt Ltd.,

Andhra Pradesh, India, a leading fish and shrimp feed producer, as Assistant Manager- Fish Nutrition and Formulation for a period of two years since 2012 and gained rich experience in commercial aspects of aquafeed formulation and manufacture. Currently he is working in marine food fish nutrition with nutrigenomic approach, and also involving in the development of cost effective best performance feed for marine food fishes. Mr Prabhu has great interest in aquatic disease management in mariculture species through nutraceuticals and its effects on immune development.

Sanal Ebeneezar

Mr Sanal Ebeneezar joined as Scientist in the division of Marine Biotechnology at ICAR-Central Marine Fisheries Research Institute, Kochi, India. He has completing Masters in Fisheries Science (MFSc) from Central Institute of Fisheries Education, Mumbai, India and qualified the Agricultural Research Service Examination (ARS). Mr Ebeneezar has been working in the area of Fish Nutrition and Biotechnology for the last two years. Currently he is working on the aspects of enhancement of colour and growth in ornamental fish and also on the nutrient requirement studies of marine fin fish.

S. Chandrasekar

Mr S Chandrasekar is Masters in Fish Physiology and Biochemistry from Central Institute of Fisheries Education, Mumbai, India. He joined as Agricultural Research Service Scientist in Marine Biotechnology Division of ICAR-Central Marine Fisheries Research Institute, Kochi, India in 2014. Currently he works in marine food fish nutrition mainly focusing on fishmeal replacement and immunomodulation. He is also working on the nutrition of marine ornamental fish with the aim of colour enhancement.

Shubhadeep Ghosh

Completed my Master of Fishery Science (M F Sc) in Fishery Pathology and Microbiology in 2002 and my Ph. D. on the role of probiotics in enhancing growth and survival in live-bearing ornamental fishes in 2007. I was inducted in Agriculture Research Service (ARS) in 2005 and have since been working in Pelagic Fisheries Division of CMFRI. I am currently acting as Scientist-in-charge of Visakhapatnam Regional Centre of CMFRI. I have studied in detail the biology, population parameters and stock structure of most commercially important pelagic resources from the west and east coast of India. I was actively involved in pop-up satellite tagging of yellowfin tunas from the Indian seas. I have contributed to the expansion of cage mariculture in the country. Recently, under my guidance, Visakhapatnam RC has bred and produced seed of Indian pompano (Trachinotus mookalee) for the first time in the world.

Ritesh Ranjan

I graduated from College of Fisheries, Mangalore in 2003. I completed my post-graduation in 2005 and Ph. D. in 2012 in Fish Pathology and Microbiology from Central Institute of Fisheries Education, Mumbai. I joined CMFRI in 2008 and since then I have been working in the field of mariculture, particularly on marine finfish breeding. I have worked on the sex reversal of grouper and it's broodstock development. Presently, I am working on breeding and seed production of grouper and Indian Pompano. I am interested in live feed culture and its application in marine finfish larval rearing.

Suresh. E

Dr. E. Suresh is working as Assistant Professor in the Dept. of Aquaculture in Fisheries College and Research Institute constituent of Tamil Nadu Fisheries University since 2015. Earlier he was working as Technical Assistant in National Bureau of Fish Genetic Resources, Lucknow for 18 months. He has obtained his PhD in Fish Genetics and MFSc in Inland Aquaculture from ICAR – Central Institute of Fisheries Education, Mumbai and BFSc from Tamil Nadu Veterinary and Animal Science University, Chennai. He has also gained research experience in working projects sponsored by DBT and ICAR-Norway collaborative project. In addition to research, he involved in the establishment of Shrimp brood bank in Kanyakumari Dist, Tamil Nadu funded by National Agricultural Development Project (NADP), Govt. of India. He has published more than 15 articles in National and International Journal. He was recipient of ICAR – CIFE fellowship for pursuing Master and Doctoral degree.

M. Kavitha

Ms M. Kavitha obtained her M.F.Sc in Aquatic Environment and Management, qualified Agricultural Research Service (ARS) in Aquaculture discipline and joined as a Scientist in ICAR-CMFRI during 2014 in Molluscan Fisheries Division. Currently, she works on molluscan resource management and molluscan breeding. Ms Kavitha is also involved in hatchery rearing of squid and cuttlefish from egg masses and sea ranching of young ones in the natural fishing grounds at Tuticorin coast, Gulf of Mannar.

J Raymond Jani Angel

Dr. J Raymond Jani Angel is a scientist in the Division of Genetics and Biotechnology, ICAR-Central Institute of Brackishwater Aquaculture, Chennai. He earned his PhD in Aquaculture from ICAR-Central Institute of Fisheries Education, Mumbai. His main research interests are applied to growth improvement of candidate brackishwater aquaculture species using molecular tools. Other research work includes molecular systematics of coastal decapod crustaceans especially shrimps of family Penaeidae and Palaemonidae. More recent research has focused on mangrove and ichthyo- faunal biodiversity of the remotely located Andaman and Nicobar group of Islands.

R. Panicz
Department of Aquaculture, West Pomeranian University of Technology, Szczecin, Kazimierza Krolewicza 4, 71-550 Szczecin, Poland

J. Sadowski
Department of Aquaculture, West Pomeranian University of Technology, Szczecin, Kazimierza Krolewicza 4, 71-550 Szczecin, Poland

R. Drozd
Department of Immunology, Microbiology and Physiological Chemistry, West Pomeranian University of Technology, Szczecin, Doktora Judyma 24, 71-466 Szczecin, Poland

Mare Lo˜hmus
Centre for Aquaculture and Environmental Research, Fisheries and Oceans Canada, West Vancouver, British Columbia, Canada
Section for Environment and Biosecurity, Department for Chemistry, Environment and Feed Hygiene, National Veterinary Institute, Uppsala, Sweden

L. Fredrik Sundstro¨ m
Centre for Aquaculture and Environmental Research, Fisheries and Oceans Canada, West Vancouver, British Columbia, Canada

Mats Bjo¨ rklund
Animal Ecology/Department of Ecology and Evolution, Evolutionary Biology Centre, Uppsala University, Uppsala, Sweden

Robert H. Devlin
Centre for Aquaculture and Environmental Research, Fisheries and Oceans Canada, West Vancouver, British Columbia, Canada

Stefanie M Hixson
Department of Ocean Sciences, Memorial University of Newfoundland, St. John's, Newfoundland and Labrador, Canada

Maysa Enayatmehr
Phd student in Marine Biology, Department of Marine Biology, Science and Research Branch, Islamic Azad University, Tehran, Iran

Shahla Jamili
Iranian Fisheries Science Research Institute, Agricultural Research Education & Extension Organization

CH. Swetha
Department of Biotechnology, Sri Venkateswara University, Tirupati-517 502, AP, India

S.B. Sainath
Department of Biotechnology, Sri Venkateswara University, Tirupati-517 502, AP, India

P. Ramachandra Reddy
Department of Biochemistry, Yogi Vemana University, Kadapa-516 003, AP, India

P. Sreenivasula Reddy
Department of Zoology, Sri Venkateswara University, Tirupati-517 502, AP, India

Anne-Constance Franz
Institute of Anatomy, University of Zurich, Winterthurerstrasse 190, Zurich 8057, Switzerland

Oliver Faass
Institute of Anatomy, University of Zurich, Winterthurerstrasse 190, Zurich 8057, Switzerland
Study Group "Magdeburger Arbeitsgemeinschaft für Forschung unter Raumfahrt- und Schwerelosigkeitsbedingungen" (MARS), Otto-von Guericke-University Magdeburg, Universitätsplatz 2, Magdeburg 39106, Germany

Bernd Köllner
Friedrich-Loeffler-Institute, Federal Research Institute for Animal Health, Greifswald-Insel Riems 17493, Germany

Natallia Shved
Institute of Anatomy, University of Zurich, Winterthurerstrasse 190, Zurich 8057, Switzerland

Karl Link
Institute of Anatomy, University of Zurich, Winterthurerstrasse 190, Zurich 8057, Switzerland
Institute of Evolutionary Medicine, University of Zurich, Winterthurerstrasse 190, Zurich 8057, Switzerland

Ayako Casanova
Institute for Fish and Wildlife Health, University of Bern, Länggassstrasse 122, Bern 3012, Switzerland

Michael Wenger
Institute of Anatomy, University of Zurich, Winterthurerstrasse 190, Zurich 8057, Switzerland
Institute for Fish and Wildlife Health, University of Bern, Länggassstrasse 122, Bern 3012, Switzerland

Helena D'Cotta
Fish Physiology and Genomics, UMR ISEM—Institut des Sciences de l'Evolution de Montpellier, Département Conservation et Domestication, CIRAD, Campus International de Baillarguet, TA B-116/A, Montpellier cedex 5, Montpellier 34398, France

Jean-François Baroiller
Fish Physiology and Genomics, UMR ISEM—Institut des Sciences de l'Evolution de Montpellier, Département Conservation et Domestication, CIRAD, Campus International de Baillarguet, TA B-116/A, Montpellier cedex 5, Montpellier 34398, France

Oliver Ullrich
Institute of Anatomy, University of Zurich, Winterthurerstrasse 190, Zurich 8057, Switzerland
Study Group "Magdeburger Arbeitsgemeinschaft für Forschung unter Raumfahrt- und Schwerelosigkeitsbedingungen" (MARS), Otto-von Guericke-University Magdeburg, Universitätsplatz 2, Magdeburg 39106, Germany

Manfred Reinecke
Institute of Anatomy, University of Zurich, Winterthurerstrasse 190, Zurich 8057, Switzerland

Elisabeth Eppler
Institute of Anatomy, University of Zurich, Winterthurerstrasse 190, Zurich 8057, Switzerland
Study Group "Magdeburger Arbeitsgemeinschaft für Forschung unter Raumfahrt- und Schwerelosigkeitsbedingungen" (MARS), Otto-von Guericke-University Magdeburg, Universitätsplatz 2, Magdeburg 39106, Germany
Institute of Anatomy, University of Basel, Pestalozzistrasse 20, Basel 4056, Switzerland
Institute of Anatomy II, University of Erlangen-Nürnberg, Universitätsstraße 19, Erlangen 91054, Germany

PREFACE

Growth rate is a primary trait of interest in most finfish selection programs and is intrinsically linked to productivity and profitability of finfish aquaculture industry in several parts of the world. Search for candidate genes influencing growth has been a major focus of research in aquaculture and several such genes have been isolated from the fish genome and their effects are quantified. Growth hormone genes have been studied extensively in many species because of its potential in enhancing the growth rate of fishes in aquaculture. The identified growth hormone gene in most of the fishes is roughly 1.5 kb long, with the protein coding region divided in to four to five blocks (exons) representing less than one third length of the genomic region. Growth hormone (GH) is a single chain polypeptide that plays a major role in the growth development of vertebrates. It is the main regulator of postnatal somatic growth and stimulates anabolic processes such as cell division, skeletal growth and protein synthesis. In addition, growth hormone is also involved in the regulation of fat oxidation, inhibition of glucose transport to peripheral tissues and regulation of ribosome translational activity, which in turn influences protein synthesis. Presently, the importance of GH as a candidate gene for studies of genetic variation in connection with growth traits and potential natural marker for studies of evolutionary relationship of various fishes has been demonstrated. Its potential application in the growth traits and fish evolution studies has been an active area of research. Natural growth rates of organisms have evolved to provide maximum fitness in wild environments and therefore, enhancement of natural growth rate of fish in aquaculture has been extensively explored, with gains arising from improvements in husbandry, nutrition, and genetical selection. Growth enhancement can be advantageous for aquaculture by shortening production time, and enhancing feed conversion efficiency. Therefore, endocrine approaches in controlling growth have been extensively explored. Several studies have demonstrated the efficacy of mammalian GH in the acceleration of growth of various species. In fish, GH administration by various modes of delivery like injection, and oral administration of hormones has been successfully studied. These studies have suggested that the exogenous growth hormone enhances fish growth by stimulating appetite and improving feed conversion. It is clearly understood that growth hormone plays major role in aquaculture industry by increasing the growth. Therefore, knowing the different aspects of growth hormone is very much essential for the researchers aim for growth enhancement in fishes.

Editor

CHAPTER 1

NEUROENDOCRINE HORMONE REGULATED GROWTH IN FISH

Dr Sekar Megarajan and Dr Prerna Pandey

Dr Sekar Megarajan graduated in fisheries science, and then obtained master degree in Fish Genetics and Biotechnology and PhD in Fish Nutrition and Biochemistry with specialization in fish molecular endocrinology from Central Institute of Fisheries Education (CIFE), India. He worked for two years with Transasia Bio-Medicals Ltd, Mumbai in the area of immunology for development of diagnostic kits for human infectious disease and joined Indian Agriculture Research service in 2013 as a scientist. He currently works as scientist in mariculture division at Central Marine Fisheries Research Institute, Cochin, India. Dr Megarajan has been working on breeding, seed production and culture of high value marine finfishes. He has published more than 10 research papers in peer valued journals in the area of fish population genetics, molecular endocrinology and fish culture.

Dr Prerna Pandey, a biotechnologist with several years of wet lab research experience, worked at International Center for Genetic Engineering and Biotechnology, New Delhi. Her field of research in PhD included isolation and molecular characterization of geminiviruses, genome sequencing, gene annotation, and gene silencing using the RNA interference technology. She has also worked at Transasia Biomedicals and Advance Enzyme Technologies as a scientist. Prerna has published papers in peer-reviewed journals, and has submitted a number of annotated Geminiviral genome sequences in the GenBank, including two novel ones. She has also completed her editing and proof-reading courses from Society for the promotion of Editors and Proof-readers, UK and now works as a freelance scientific editor and writer. When Prerna is not busy with her assignments, she loves to explore historical places.

INTRODUCTION

Growth is the one of the important phenotypic characters for a fish either cultured in a confined culture environment or available in the wild environment. This character is directly linked to the total production of a specific species in a particular environment. Eventually this supports for the food security and income generation for the continuously growing world population. Therefore, studying growth characters of different kind of fishes in various systems including freshwater, brackish-water and marine environment is considered an important in the biological study of a species. The growth in an animal is mainly characterised by the hypertrophy (increase in muscle fiber size) and hyperplasia (increase in muscle fiber numbers) of muscle fibers (Weatherley et al., 1988; Rowlerson and Veggetti, 2001). In nature, the growth in an animal is varies from continuous to non-continuous growth. Interestingly, fish grows continuously and this characteristic feature is unique, which has not been observed in mammals, birds, insects, etc (Hutchings, 2002). It indicates that the fish grows continuously even after reaching the sexual maturity. However, the growth may slowdown after the maturation because the energy resource is partially utilised for gonadal development and reproduction from the somatic growth, but still fish continue its growth till entire life span and this is called indeterminate growth. Whereas, in other animals, somatic growth is increases continuously at the initial stage of life and stops after attaining the sexual maturity and this kind of growth is called determinate growth. The biological nature of growth in fish is little complicated, exhibits different nature. Though the continuous growth has been observed in fish but the growth rate varies among the fish species and it is mostly genetically determined. It is evidenced from the nature that different sizes of the fishes, the growth ranges in size from 8-10 mm in length for mature female fish for one of the smallest identified fish the goby (*Trimmatom nanus)* to the giant whale shark (*Rhincodon typus*), which grows and attains a length of about 18 m (Nelson, 1984). It is also observed that the growth rate of the fish varies among the fish within the same living environment, the marine fish cobia (*Rachycentron canadum*) grows and attain an average weight of 4.0 kg within a year, whereas the different species of sardine fish lives in the same marine environment grows an average weight of 0.2 kg in a year. Similarly, *Wallago attu*, a fresh water catfish which lives in the freshwater environment grows up to 5 kg in a year, on the contrary most of the fishes belonging to carps reaches maximum of 1 to 1.5 kg in a year. Therefore, it is understood from different studies that there are different factors affects and regulate the growth of a fish in different pathways. It is well established fact that, in the most of the mammals and non-mammals, the important phenotypic characters such as growth is controlled by interaction of genetics and environmental factors. The genetic factors that influence growth are the endogenous factors, which include the gene responsible for the growth, i.e growth hormone gene and protein that encoded by the growth gene, i.e. growth hormone protein. However, the endogenous factors such as growth hormone gene expression and interactions of other growth-related hormones in the growth hormone gene axis are mainly determine and regulates the growth status of the fish. In addition, some other exogenous factors such as water temperature, seasonality, food availability and nutritional quality, photoperiod, population density and other factors are also influencing the growth of the fish (Mustapha et al., 2012). Though the endogenous factors are more responsible for the growth stimulation and enhancement, but this factor are influenced by some of the other endogenous factors present within an animal biological system. Thus, this chapter will explain more detailed about growth regulation in fish by growth hormones and other factors regulate the growth hormones in the fish.

NEUROENDOCRINE FACTORS IN GROWTH REGULATION IN FISH

Presence of growth promoting activity in the pituitary gland was observed in 1921, by observing the result of an experiment showed the fast growth rate in the rat administered with crude pituitary extract than the control rat. Thereafter, the factor present in the crude pituitary extract was identified as growth hormone after 20 years later the initial experiment and then the growth hormone was first purified from bovine pituitary glands. Growth hormone in fish is the polypeptide and it mainly stimulates the growth in fish. Subsequently it had been well understood that the growth in fish is controlled by endocrine regulatory system, mainly by the interaction of neuroendocrine - growth hormone (GH) - insulin-like growth factor (IGF) axis (Reinecke, 2005). Interactions of these hormones are well established and it is known that control of GH secretion is multi-factorial and it is balanced by stimulating and inhibiting effect of neuroendocrine hormones on the pituitary somatotrophs (Peng & Peter, 1997). This is similar to the system present in the mammals, where growth hormone secretion in the pituitary is under multifactorial control. Studies in few teleost fishes showed that the factors like growth-hormone releasing hormone (GHRH), dopamine (DA), neuropeptide Y (NPY), gonadotropin-releasing hormones (GnRHs), thyrotropin-releasing hormone (TRH), cholecystokinin (CCK), bombesin and activin shown stimulatory action on the growth hormone secretion and thereby increases growth. Whereas other factors like somatostatin (SRIF), noradrenaline and serotonin are the major inhibitors for release of growth hormone and indirectly affect the growth and other factors (Agustsson, 2001).

Growth Hormone

Growth hormone (GH) is a poly peptide of approximately 22 kDa produced by somatotrophs in the anterior pituitary gland which controls normal growth, development, and metabolism in all vertebrates (Kaplan, 1999). GH is a pleiotropic endocrine factor and has important metabolic effects such as feed utilization and mobilization (Mommsen, 2001). In vertebrates, GH also plays a role in reproduction and osmoregulation, as well as in appetite and behaviour actions (Canosa et al., 2007). Growth hormone (GH) secretion is regulated through a complex of neuroendocrine control system, which includes functional interplay among a number of hypothalamic factors and neurotransmitters (Muller et al., 1993). In particular, the secretion of GH in both mammals and fish has been found to be depending on the balance between growth hormone-releasing hormone (GHRH) and somatostatin (SST) acting on the pituitary somatotrophs (Peng and Peter, 1997). In mammals, the generally accepted model of the hypothalamic regulation of GH release is a dual control whereby GH-releasing factor (GRF) stimulates, and SRIF (somatotropin release-inhibiting factor) inhibits GH (Ghigo et al., 1997). The GH secretion in fish is also regulated by complex hypothalamic control. In fish, hypothalamic fibers terminate in the proximity of the GH-producing somatotrophs of the anterior pituitary, which play a major role in GH regulation (Agustsson et al., 2000). However, it does appear that although both SRIF and GRF can influence GH secretion in some teleostean species, they probably play a smaller role in GH regulation than other hypothalamic factors, such as gonadotrophin-releasing hormone (GnRH) and dopamine (DA). The cellular action of GH is through its widely distributed growth hormone receptors (GHR), especially in the liver where the GH-GHR interactions induce the expression of another secretory peptide molecule i.e, insulin like growth factor I (IGF-I) which in turn promotes growth in a variety of tissues. Other important functions of growth hormone in fish are to maintain osmoregulation and to increase the tolerance capacity to seawater, these are carried out by increase in the size of gill chloride cells, Na^+, K^+-ATPase, and Na^+, K^+, $2Cl^-$ co-transporter (NKCC) and ion transporters involved in salt secretion (McCormick, 2001).

The structure of the GH family includes GH, prolactin (PRL), chorionic somatomammotropin or placental lactogen (PL), and fish somatolactin (SL), all of which likely have arisen from a common ancestor (Walker et al., 1991). Members of the GH family have an approximate molecular mass of 22 kDa, with approximately 200 amino acids, including two disulfide internal bonds in GH, PL or three disulfide internal bonds in PRL. In teleost fish, GH is a single polypeptide protein of 21-23 kDa showing more variability in size than in other vertebrates. The cysteine residues involved in the disulfide bonds are highly conserved in all members of the family, suggesting an important role in biological actions. In this regard, it was found that the large disulfide loop is essential for growth-promoting activity (Chen et al., 1992). The fish hormone exhibits typical GH features, such as four cysteine residues, capable of forming two disulfide bonds that are assumed to contribute to the tertiary structure of the hormone molecule and a site for *N*-linked glycosylation (Clements et al., 2004; Sciara et al., 2006).

The importance of growth hormone as a potential growth promoting agent has long been recognized and GH administration has been shown to enhance growth rate in a number of species, including fish (McLean and Donaldson, 1993; Tsai et al., 1995). Thus, the synthesis, release and recombinant production of growth hormone have been the subject of extensive research. With advent of recombinant DNA technology in the past few decades, GH gene of many fishes have been identified and GH encoding cDNA has so far been cloned from more than one hundred fish species, including salmon, tuna, carp, flounders and catfish, etc. The cloned genes have been expressed in heterologous expression systems like *E. coli* and yeast (*Saccharomyces cerevisiae*, *Pichia pastrosis* and *Chlorella ellipsoidea,* etc) (Seeburg et al., 1983). Consequently, the large-scale production of hormone by recombinant techniques has enabled their commercial use for their pharmacological interventions and growth stimulation action in fishes. Initially, the use of recombinant growth hormones from animal source has been used to promote the growth and alter body composition in many of livestock animals. Subsequently, in animal and fisheries sciences the recombinant GHs were demonstrated to have a significant potential due to their role in growth enhancement, effects, increasing food conversion efficiency and milk production etc, without changing the amount of food consumption (Krick et al., 1993; Farmanfarmaian and Sun, 1999). In subsequent years, an advance in recombinant DNA technology has provided the means for producing large amounts of protein hormones including teleostean GHs (Bjornsson, 1997). As a result, the growth promoting effects of GH in fish have been demonstrated convincingly in different cultured fish species (McLean and Donalson, 1993).

Before use of recombinant hormone, the effect of administration of exogenous natural mammalian and piscine growth hormone has been studied in many fishes including, salmonids, tilapias, basses, eels, carps, mullets, and channel catfish (Donaldson et al., 1979). These studies suggest that the fish growth is enhanced through appetite stimulation by exogenous growth hormone and improved feed conversion (Matty, 1986). Exogenous growth hormone stimulate appetite in fish by a direct action on centers in the hypothalamus that influence feed intake or by inducing a number of metabolic changes that feed back on the hypothalamic centers to affect appetite (Markert et al., 1977). Earlier experiments suggested that exogenous growth hormone may improve feed conversion by one or more possible mechanisms 1) stimulation of lipid mobilization and oxidation 2) an action on the rate of protein synthesis or breakdown 3) stimulation of insulin synthesis and release. The experimental trials have shown that a more than double of individual weight relative to controls may be achieved depending upon the dose, treatment duration, and growth stage (McLean et al., 1996). Use of recombinant growth hormone of *Pangasianodon hypophthalmus* has been administered to the fish *Labeo Rohita* (Rohu), and it showed that the fish fed with GH protein of 1.0 μg/g body weight of fish (rohu) grew faster than the control fish as well as fish fed with the hormone dose of 0.25 and 0.5 μg/g body in a three months period of the study. Similarly, the biologically active purified recombinant fish growth hormones have resulted with the growth promoting effects of other fishes such as trout, tuna (Sato et al., 1988), eel (Sugimoto et al., 1990), common carp (Fine et al., 1993), flounder (Jeh et al., 1998) and giant catfish (Promdonkoy et al., 2004).

STIMULATORY FACTORS OF FISH GROWTH HORMONE

Insulin like Growth Factors

The insulin-like growth factors (IGFs) including IGF-I and IGF-II, are a family of polypeptide growth factors that structurally related to proinsulin. The name *insulin-like growth factor (s)* (IGFs) reflects that these substances have insulin-like structure with the some of it's domains. The IGF-I, also known as somatomedin C, mediates several growth-promoting effects of growth hormone (GH) in the fish postnatal life (Duan, 1997). In some of the studies conducted on mice have demonstrated that the both IGF-I & IGF-II are essential for embryonic growth (Baker *et al,* 1993). Understanding the importance in the growth regulation, several works has been initiated on insulin like growth factors in fish. Then it is well established that the basic structure and basic endocrine mechanisms underlying growth regulations has been highly conserved among vertebrates. It is proved that the human and fish IGF1 are equally potent in mammal and fish by mammalian and fish bioassay systems, respectively and with similar functions (Upton et al., 1998). In fish, at first the existence of IGF like molecules was identified in serum of eels by Duan and Iniu, which stimulated the cartilage sulfation and indicated the presence of IGF in fishes. Thereafter, with help of molecular biology, the IGF genes have been identified from several fishes including salmonids, catfish, gilthead seabream, carp, barramundi, goldfish, buffer fish, tilapia, orange spotted grouper and several other fishes (Moriyama et al., 2000). IGFs of two peptides also identified in fishes, namely IGF-I & IGF-II, and it has been suggested that these peptides may have evolved from a common ancestral gene. Similar to higher vertebrates, the mature IGF-I & IGF-II of fish is containing four different domains namely B,

C, A & D and these amino acid sequences in both IGF-I & IGF-II is conserved. The mature IGF-I & IGF-II proteins are formed from the pre-hormone (Pro IGF-I & IGF-II). The pre-hormone of both IGF-I & IGF-II contains domain E and signal peptides in addition to the domain B, C, A & D, present in the matured hormones. The signal peptides and E-domain are proteolytically removed from the pre-hormone and then yields mature IGF(s). The structures of IGF(s) are similar but the some of the domains in both IGFs are having greater divergence and it has been observed in several fishes that the amino acid in both C & D domains of both IGFs are less conserved than the B & A domains. In fishes, the first IGF-I & IGF-II sequence were determined in coho salmon and rainbow trout, respectively (Shamblott and Chen, 1992). The matured IGF-I in coho salmon contain 70 amino acids basic peptides with 3 intra-chain disulfide and in trout IGF-II is a 70 residues of neutral amino acid peptide that is similar in structure to IGF-I. The fish and mammalian IGF-I primary sequence comparisons indicate that IGF-I is highly conserved. The overall sequence identity of the mature IGF-I is 80% between mature salmon and human IGF-I (Wood et al., 2005).

Biological activity of IGFs in fish is similar to the mammals, the major production site of circulating IGF-I in adult fish is the liver, while all other tissues such as brain, muscle, kidney and guts produce IGF-I locally and in small quantities (Duan and Hirano, 1992). IGFs are involved in the regulation of proteins, lipid, carbohydrate, mineral metabolism in the cells, and eventually lead to body growth. The actions of IGF are mediated by IGF receptors. In general, the growth hormone produced in the pituitary gland is regulated by hypothalamic hormones. The produced GH binds to its receptors in different target organs especially in the liver and then stimulates synthesis and release of insulin like growth factors. The synthesised IGF(s) are involved in the lipid, carbohydrate and protein metabolism in the cell, and cell differentiation and proliferation and ultimately body growth. However, administration of GH also found to stimulate the secretion of IGF(s) in fishes, where the administration of GH significantly increased hepatic IGF-I mRNA levels in the liver. Several reports indicate that the growth-promoting effects of GH in fish is mediated primarily by IGF-I. In many fish species, blood levels of IGF-I or tissue levels of its mRNA positively correlate with dietary ration, dietary protein content, and body growth rate (Beckman et al., 2004).

IGF-I exerts its effects on cells through binding to the IGF-I receptor (IGF-IR), which binds IGF-II in a similar manner in zebrafish (Mendez et al., 2001; Pozios et al., 2001). The wide range of tissue distribution of the IGF-I hormone and the IGF-I receptors in fish, coupled with the varied functions associated with IGF-I, make fish attractive subjects for future study.

Gonadotropin Releasing Hormone

Gonadotropin releasing hormone (GnRH) is a brain neuropeptide, which helps for the production of gonadotropins by the pituitary. It is a decapeptide showing the structural similarities in most of the mammals including pig, sheep, rat, mouse, and human (Adelman et al., 1986, Seeburg et al., 1987). However, in non-mammalian vertebrates, especially in fishes the molecular variation found in the structure of the hormone (Sherwood et al., 1983). There are nine different molecular structure forms of GnRH has been identified in different vertebrates, and interestingly this structure is not conserved to the one particular group of vertebrates (Montero & Dufour, 1996). Among the different structure identified, the highest molecular diversity is exhibited by different fish group. Nine different forms of GnRH identified, and they are the decapeptides with conserved residues in positions 1, 2, 4, 9 and 10 and with variable amino acid in potion 8. The both C & N terminal of the decapeptides are conserved and indicates that these regions are involved in the interaction with receptors (Millar et al., 1994). The GnRH synthesized in the preoptic-hypothalamic area and acts on the pituitary gland where it stimulates the synthesis and release of gonadotrophic hormones in all vertebrates (Millar et al., 2004). GnRHs are members of the rhodopsin-like G-protein coupled receptor (GPCR) family and exert their actions through GnRH receptors (GnRHRs). The hypothalamic GnRH neurones innervate the adenohypophysis in teleost fishes, to control the release of gonadotrophins (Somoza et al., 2002), growth hormone, prolactin, somatolactin and thyroid stimulating hormone (Moncaut et al., 2005). A study on gold fish indicated that two forms of GnRH (salmon GnRH & Chicken GnRH II) could stimulate GH release in both *in-vivo* and *in-vitro*. Similarly, the stimulation of GH release by GnRH was also observed in another cyprinid fish, *Cyprinus carpio* (Lin *et al.* 1993) as well as in a cichlid fish, *Oreochromis* sp (Melamed et al. 1995). In contrast, GnRHs were not able to stimulate GH release in a salmonid fish, *Oncorhynchus mykiss* (Blaise et al. 1995). The evidences from the experiments showed the GnRH plays an important role in the control of GH release and it may vary among teleosts. However, in other vertebrate classes including mammals, direct effects of GnRH on GH secretion have not been seen, suggesting that this role may be limited to some teleost species.

Growth Hormone Releasing Factors

The biosynthesis and secretion of growth hormone (GH) in the anterior pituitary gland is under complex hormonal regulation of two neuro peptides such as growth hormone releasing hormone (GHRH) and pituitary adenylate cyclase activating polypeptide (PACAP). Growth hormone-releasing hormone (GHRH), also known growth hormone-releasing factor, stimulates both release and synthesis of GH, as well as transcription of the GH gene (Fukata et al., 1985). The existence of a human growth hormone-releasing hormone (GHRH) was demonstrated in 1964 by the ability of hypothalamic extracts to release growth hormone (GH) from pituitary tissue. In mammals, growth hormone-releasing hormone (GHRH) is the most important neuroendocrine factors that stimulate the release of growth hormone (GH) from the anterior pituitary. Initially, the growth hormone stimulating role of GHRF was confirmed by induction of GRF secretion in the human and transgenic mice resulted with high levels of plasma GH, pituitary enlargement, increased somatic growth, and acromegaly (Billestrup et al., 1986). GHRH was originally isolated from human pancreatic tumours and later it was isolated and characterized in human hypothalamic tissue (Ling et al., 1984). GHRH interacts with somatostatin (SRIF) to regulate GH release from somatotroph cells in the anterior pituitary through the dual antagonistic regulation of adenylate cyclase and Ca21 influx (Harvey, 1991). In mammals, GHRH is mainly expressed and released from the arcuate nucleus of the hypothalamus.

Miyata et al. (1989) reported for the first time that the identification of pituitary adenylate cyclase-activating polypeptide (PACAP) from the ovine hypothalamus due to its ability to stimulate cAMP formation in anterior pituitary cells in rats. PACAP was first isolated by its stimulatory action on adenylate cyclase activity in rat pituitary cells (Miyata et al., 1989). Two biologically active forms have been derived from the same precursor, PACAP-38 and PACAP-27. The PACAP exhibits number of structural and functional similarities with GHRH, which includes, presence of considerable numbers of similar sequences, appearance of both in very early during evolution prior to the emergence of vertebrates, both have widely distributed in peripheral organs, especially in the gut, pancreas, ovary and testis and finally the biological effects of both are mediated through 7-transmembrane domain receptors coupled to adenylyl cyclase. However, while the GHRH sequence is poorly conserved even in closely related species, the sequence of PACAP has been extremely well preserved during evolution, indicating a few major differences between these hormones. In mammals, GHRH and PACAP are encoded by separate genes, but in non-mammalian vertebrates and protochordates, the GHRH and PACAP were believed to be encoded by the same gene and hence, processed from the same transcript and pre-pro polypeptide (Lee et al., 2007). In fish, the expression of PACAP mRNA has been demonstrated in heart, liver, gastrointestinal tract, brain, pituitary, spinal cord, kidney, ovary, testes, gill (Wong et al., 2000), and eye (Wu et al., 2006). Consistent with the wide range of tissue distribution, PACAP is considered a pleiotropic neuropeptide with various functions (Wong et al., 2000). In several fish species, the growth stimulating effect of GRF & PACAP has been well studied and it is found similar to other mammals. The neuropeptides are having the stimulatory action on growth. In gold fish, the intraperitoneal injection of human GRF1-40 fragment caused an increase in serum GH levels. GRF-like immune-reactivity has been shown in the brain and pituitary of teleosts, such as cod fish, sea bass, rainbow trout, eel, carp, goldfish, and salmonids (Reverter & Canosa, 2009). The stimulatory effect of synthetic common carp GRF (cGRF) on GH release has also been observed in rainbow trout and tilapia, stimulation of GH release by human GRF was also confirmed in rainbow trout (Blaise et al., 1995).

Neuropeptide

The neuropeptide Y family is consisting of structurally related 36-amino acid peptides, which includes neuropeptide Y (NPY), gut endocrine peptide YY (PYY), pancreatic polypeptide (PP) and fish pancreatic peptide Y. Neuropeptide Y is expressed mainly in the central nervous system (CNS) of vertebrates (Tatemoto et al., 1982), but the presence of mRNA transcript of the peptide is also been reported from pancreas and testis cells from rat. NPY was first isolated and characterised from porcine brains (Tatemoto et al., 1982). The networks of NPY neurons have been characterized well in mammals (Hendry, 1993) and the hypothalamus appears to be the main synthesis site (Stanley, 1993). It shows remarkable sequence homology from fish to mammals (Larhammar, 1996), and several studies strongly suggest that NPY is also an orexigenic peptide in fish. So far, the identified structure of the NPY is conserved well across the vertebrate phylogeny. NPY is well known to play major role in physiological functions including maintaining blood pressure (Zukowska-Grojec and Wahlestedt, 1993), circadian rhythms, food intake and pituitary hormone secretion (McDonald and Koenig, 1993). In addition to its major role, the peptide is involved in the regulation of pituitary hormones related to somatic growth and reproduction in teleost fish, in different stages of the reproductive cycle (Yorio et al., 2015). Accordingly, it is reported that the neuropeptide Y (NPY) affects growth hormone (GH) secretion in several animal species. However, some studies have reported that the administration of NPY has decreased the GH concentration, but this hypothesis was disproved by other researchers. In several fishes, the growth and growth hormone stimulating effect has been well studied and the results suggested that administration of NPY could potentially increased the aquaculture production. NPY is a highly potent stimulant of growth hormone release in fish (Peng and Peter 1997), but growth enhancement can only be achieved when there is sufficient food available (Volkoff et al., 2005). Studies on intraperitoneal administration of recombinant full-length NPY at the dose of 1 μg/g body weight has expressed significantly higher growth, feed utilisation and growth hormone level in olive flounder (*Paralichthys olivaceus*). In cichlid fish the NPY administration, modulates the expression and release of GH and gonadotropins in both sexes of *Cichlasoma dimeru*, and finally concluded that NPY acts as a GH-, LH- and FSH-releasing factor in a dose- and sex-dependent way. Studies in sea bass have shown that central or peripheral NPY administration stimulates both growth hormone (GH) and luteinizing hormone (LH) secretion directly on the gonadotropic cells. Administration of neuropetide Y has stimulated the secretion of growth hormone and gonadotropin, gonadotropin II release from the goldfish pituitary in *in vitro* conditions (Peng & Peter, 1997)

Thyrotrophin Releasing Hormones (TRH)

Thyrotropin-releasing hormone (TRH) is also called as thyrotropin-releasing factor (TRF) or thyroliberin is a releasing hormone, produced in hypothalamus by hypothalamic neurons and travels along portal vessels to adenohypophysis and stimulates the release of thyrotropin and prolactin from anterior pituitary. Thyrotropin-releasing hormone (TRH) was the first hypothalamic hormone isolated and purified in 1969, and the purified hormone stimulated the TSH release from the mammalian pituitary both *in vivo* and *in vitro* (Groef et al., 2006). In different studies, it has been reported that the thyrotropin-releasing hormone act as potent stimulating agent for thyrotrophin, however in non-mammalian vertebrates, especially in fishes it is not clearly observed. In bighead carp (*Aristichthys nobilis*) and in Japanese eel (*Anguilla japonica*) the TRH increases *in vitro* hypophyseal TSH mRNA expression, but the TRH treatment did not stimulate TSH release in teleosts, lung fish or hag fish (Larsen et al., 1998). Apart from the stimulating role for secretion of thyrotropin (TSH) and prolactin secretion in mammals, it has also been implicated as a stimulator of growth hormone (GH) release. Other than mammals, the growth hormone stimulatory effect on secretion of GH has been reported in frogs and turtles (Denver and Licht 1989a). In fishes such as sailfin molly, common carp, goldfish, the *in-vitro* and *in-vivo* studies showed that the TRH stimulate the secretion of GH (Lin *et al.*, 1993). Following intraperitoneal injection in goldfish, TRH increases serum GH levels and stimulates GH release from perifused pituitary fragments (Trudeau et al. 1992).

Gherlin

Ghrelin is a non-amidated peptide hormone, existing in both acylated and non-acylated forms in mammals, and non-mammals. It is primarily produced from ghrelin cells present in the mucosal layer of the stomach, and also with minor levels in the brain and various other tissues. Ghrelin is a 28-amino acid peptide that was discovered first in rat stomach by Kojima et al. (1999) as the endogenous ligand to the growth hormone secretagogue-receptor (GHSR) (Tinoco et al., 2014). After isolation from rat, it has been isolated from several non-mammalian species such as chicken (*Gallus gallus*), blackhead seabream (*Acanthopagrus schlegelii*), rainbow trout (*Oncorhynchus mykiss*) and goldfish (*Carassius auratus*).

In mammals, ghrelin generally promotes food intake, body weight gain and adiposity through central and peripheral modes of action. It also has been suggested that it is having an important role in meal initiation because it increases just before feeding, and decreases immediately after food intake. In fishes, it has been studied that ghrelin is involved in the regulation of a number of physiological functions, including the regulation of pituitary hormone release and the stimulation of food intake in fish. In fishes, several studies have been carried out to identify gherlin gene and protein and ultimately studied the impact of gherlin on growth improvement in fishes. The ghrelin gene from several fish species has been cloned, and three receptor gene species belonging to the GHS-R family have been identified in the pufferfish (Palyha et al., 2000). Observation of growth improvement studies in tilapia demonstrated that plasma GH concentrations were increased by a growth hormone secretagogue (GHS) (Shepherd et al., 2000). Furthermore, release of GH from the tilapia pituitary *in vitro* is stimulated by rat ghrelin (Riley et al., 2002). In another study, it was demonstrated that the eel ghrelin stimulated the secretion of GH and prolactin (PRL) from organ-cultured tilapia pituitary at a dose of 0.1 nM. The results from several studies in fishes strongly suggest that a ghrelin–GHS-R system is exist and have a role in the release of GH in fish, and it is also suggested that it is present in fish stomach but has the ability to stimulate the secretion of GH from fish pituitary.

INHIBITORY FACTORS OF FISH GROWTH HORMONE

Somatostatin

Somatostatin (SS) are the structurally diverse family of peptide and it also known as somatotropin release-inhibitory factor (SRIF). The molecular heterogeneity of the SS family arises from differential tissue specific processing of the hormone precursor (preprosomatostatin, PPSS) as well as from the existence of multiple SS genes that express alternate PPSSs such as PPSS-I, PPSS-II & PPSS-III. Somatostatins are the cyclic 14-amino acid peptide, and mainly participate in various aspects of growth, development, and metabolism through five to seven trans-membrane G-protein-coupled receptors (GPCRs) (Sheridon & Hagemeister, 2010). SRIF is well established as the primary inhibitor of GH release in many species of teleosts and plays a major role in the regulation of growth in fish (Very and Sheridan, 2002). Somatostatin-14-1, which derived from PPSS-I is well established as the primary inhibitor of growth hormone (GH) in vertebrates (Canosa et al., 2007). Like mammals, in fish too, while SS-14-I may inhibit both basal and stimulated GH secretion, it has no effect on the expression of GH gene. Several studies have reported that all SSs do not exert similar effect on GH secretion and also suggest important structure-function relationships among the diverse elements of the SS signalling system. It is the existence of multiple SS receptor (SSTR) subtypes that display selective ligand binding characteristics and forms the basis of the structure-function relationships. SSTR are linked with G protein, and many of the actions of SS are mediated through adenylate cyclase. The somatostatin (SS) inhibits growth by decreasing GH release at the pituitary level and antagonizing the release and action of GHRH in the hypothalamus. Most of the earlier reports in different fishes showed that the effects of SRIF on the regulation of growth are confined at pituitary level by inhibiting the GH release from the pituitary gland (Hagemeister & Sheridan, 2008). However, increasing evidence suggests that SSs also regulates the growth outside of the pituitary by affecting complex of GH-insulin like growth factor I (IGF-I) system. In trout, it was studied that the SS has shown the inhibiting effect on the GH-IGF-I system and indicate that SS regulates organismal growth in both a central (pituitary) and peripheral (extrapituitary) manner (Hagemeister & Sheridan, 2008).

Noradrenaline

Norepinephrine (NE) is an organic chemical in the catecholamine family having with multiple roles as a hormone and a neurotransmitter, and it is otherwise also called as noradrenaline. It is synthesized from dopamine by dopamine β-hydroxylase and released from the adrenal medulla into the blood as a hormone. In central and sympathetic nervous system, it acts as a neurotransmitter, where it is released from noradrenergic neurons (Rang, 2003). In fish, the norepinephrine has shown to inhibit GH secretion in both in vivo and in vitro condition in goldfish and actions of the hormone is mainly carried out via the binding to adrenergic receptors. Chang et al. (1985) demonstrated that the serum GH levels are suppressed by intraperitoneal injection of the hormone. Thereafter, Wong (1993) found that treatment of dispersed goldfish pituitary cells in primary culture with NE significantly decreased GH release, indicating that NE can act directly on somatotrophs to inhibit GH secretion. However, it has been suggested that the growth hormone regulations by these regulator is status dependent such as development stage or nutritional status.

Serotonin

Serotonin, or otherwise called as 5-hydroxy tryptamine belongs to a group of signal substances called monoamines, and is a neurotransmitter. It is synthesized from the essential amino acid, L-tryptophan with help of the enzymes amino acid decarboxylase and tryptophan hydroxylase (TPH) (Fitzpatrick, 1999). In the vertebrates, the organization of serotonin in the central nervous system is well conserved evolutionarily. In teleost, the serotonin-positive cells are also present in the medulla oblongata, postrema, pineal gland, and the spinal cord (Lillesaar, 2011). In mammals, serotonin is involved in a wide range of reproductive functions such as secretion of GnRH, release of gonadotropin, gonadal maturation and socio-sexual behaviours. Similarly, several evidences have suggested that the serotonin also have the unique role in the variety of reproductive strategies in teleost (Um et al., 2010; Zohar et al., 2010), added to this, the role of serotonin in social behaviour has been well demonstrated in fish (Winberg and Nilsson, 1993) which is evidenced by its primary inhibitory role in aggressive behaviour of fishes (Perreault et al., 2003). Serotonin is also known to regulate the release of growth hormone (GH) in some fishes (Somoza and Peter, 1991; Wong et al., 1998). *In-vitro* studies using the dispersed goldfish pituitary cells in primary culture showed that the intraperitoneal injection of 5-HT decreases serum GH levels in a dose dependent fashion. The study suggests that in goldfish the serotonin exhibited the inhibitory action on growth hormone secretion (Wong, 1993) by directly acting on somatotrophs. However, further more information is not available on the effect of serotonin on the regulation of GH from other fishes.

OTHER NEUROPEPTIDES AND NEURO TRANSMITTERS IN GROWTH AND GROWTH HORMONE REGULATION

The regulation of GH is also influenced by the group of other neuropeptides and neurotransmitters mentioned earlier in this chapter. They include bombesin, activin, glutamate, N-methyl-D-aspartate (NMDA), gamma - aminobutyric acid (GABA) and other sex steroids. Like other neurotransmitters and neuropeptides, they also act on at the hypothalamic and supra-hypothalamic levels and then as a result GH secretion is influenced by hypophysiotropic, exogenous and endogenous factors and ultimately that lead to growth enhancement in fishes. Bombesin (BBS) is a tetra-decapeptide, in goldfish BBS like IR is present in the brain and pituitary and suggesting it plays role in neuroendocrine regulation of pituitary hormone secretion. In goldfish, intraperitoneal injection or brain intra-ventricular injection of BBS appears to significantly increase serum GH levels (Himick et al., 1993). Ge and Peter (1994) have reported activin production in goldfish somatotrophs, and stimulation of GH release through the administration of porcine activin, as well as release of gonadotropin-II from perifused pituitary fragments in goldfish. Glutamate act as an inhibitor of growth hormone and presence of the glutamate immunoreactive fibers are found in the fish pituitary and studies on intraventricular injection of glutamate suppressed serum GH level in goldfish and suggest that it is having negative regulation on GH. N-methyl-D-aspartate (NMDA) inhibits GH secretion in goldfish, but also shown stimulating effect of GH secretion in steroid-primed immature rainbow trout (Dai et al., 2015). Gamma-aminobutyric acid (GABA) inhibits GH secretion *in vivo* but not *in vitro*. However, administration of GABA through feed has shown to improve growth of pacific white shrimp (*L. vannamei)* (Xie et al., 2017). In addition, the expression of the core components of the somatotropic axis is also regulated by sex steroids. In rainbow trout, administration of estrogen hormone significantly reduces food conversion, body growth, and the expression of GHRs, IGFs, and IGFRs (Peng & Peter, 1997).

SUMMARY

It is well understood that in vertebrates including fish, the growth regulation is multifactorial and controlled by complex networks of hormones produced by both brain and peripheral tissues and also by the neurotransmitters. Studies in different groups of vertebrates showed that, the growth hormone regulation by some of the factors is similar in both vertebrates and mammals; however, some showed defences in both the systems. Thus, the impact of some the hormones and neurotransmitters need to be studied well in different fishes for better understanding. Understanding of the growth hormone regulation by different hormones may help for the increasing the growth of fish in a culture system through application of particular hormones or neurotransmitters.

REFERENCES

1. Adelman, J.P.; Mason, A.J.; Hayflick, J.S.; Seeburg, P.H. Isolation of the gene and hypothalamic cDNA for the common precursor of gonadotropin-releasing hormone and prolactin release-inhibiting factor in human and rat. *Proc. Natl. Acad. Sci. USA*. 1986, 83: 179-183.
2. Ágústsson, T.; Ebbeson, L.O.E.; Bjogrnsson, B.T. Dopaminergic innervation of the rainbow trout pituitary and stimulatory effect of dopamine on growth hormone secretion in vitro. *Comp. Biochem. Physiol A*. 2000, 127, 355–364.
3. Agustsson, T.; Sundell, K.; Sakamoto, T.; Johansson, V.; Ando, M.; Bjornsson. B.Th. Growth hormone endocrinology of Atlantic salmon (Salmo salar): pituitary gene expression, hormone storage, secretion and plasma levels during parr-smolt transformation. *J Endocrinol*. 2001, 170, 227-234
4. Alison L. Hagemeister and Mark A. Sheridan. Somatostatin inhibits hepatic growth hormone receptor and insulin-like growth factor I mRNA expression by activating the ERK and PI3K signalling pathways. *Am J Physiol Regul Integr Comp Physiol.* 295: R490–R497, 2008.
5. Baker, J.; Liu, J.P.; Robertson, E.J., Efstratiadis, A. Role of insulin-like growth factors in embryonic and postnatal growth. *Cell.* 1993, 75, 73-82.
6. Beckman, B. R.; Shimizu, M.; Gadberry, B. A.; Cooper, K. A. Response of the somatotropic axis of juvenile coho salmon to alterations in plane of nutrition with an analysis of the relationships among growth rate and circulating IGF-I and 41 kDa IGFBP. *Gen. Comp. Endocrinol.* 2004, 135, 334–344.
7. Billestrup, N.; Larry, W.; Swanson.; Vale, W. Growth hormone-releasing factor stimulates proliferation of somatotrophs in vitro.

Proc. Natl. Acad. Sci. USA. 1986, 83, 6854-6857,

8. Björnsson, B.T. The biology of salmon growth hormone: from daylight to dominance. *Fish Physiol. Biochem, 1997,* 17: 9-24.
9. Blaise, O.; Le Bail, P.Y.; Weil,C.; Lack of gonadotropin releasing hormone action on *in vivo* and *in vitro* growth hormone release, in rainbow trout *(Oncorhynchus mykiss). Compo Biochem. Physiol. Part C*. 1995, 110C, 133-141.
10. Canosa, L.F.; Chang, J.P.; Peter, R.E. Neuroendocrine control of growth hormone in fish. *General and Comparative Endocrinology.* 2007, 151, 1-26.
11. Chang, J.P.; Marchant, T.A.; Cook, A.F.; Nahorniak, C.S.; Peter, R.E. 1985. Influences of catecholamines on growth hormone release in female goldfish, *Carassius auratus. Neuroendocrinology*. 1985, 40, 463-470.
12. Chen, Y.; Wang, Y.; He, S.; Zhu, Z., Cloning and sequencing of the growth hormone gene of large yellow croacker and its phylogenetic significance. *Biochem. Genet*. 2004, 42 (9-10), 365-375.
13. Clements, M.D.; Bart Jr. H.L.; Hurley, D.L. Isolation and characterization of two distinct growth hormone cDNAs from the tetraploid smallmouth buffalow fish (*Ictiobus bubalus*). *Gen. Comp. Endocrinol.* 2004, 136, 411-418.
14. Dai, X.Y.; Zhang, W.; Zhuo, Z.J.; He, J.Y.; Yin, Z. Neuroendocrine regulation of somatic growth in fishes. *Sci China Life Sci.* 2015, 58, 137–147.
15. Denver, R.J.; Licht, P. Neuropeptides in Xuencing pituitary hormone secretion in hatchling turtles. *J. Exp. Zool.* 1989a, 251, 306–315.
16. Donaldson, E.M.; Fagerlund, U.M.H.; Higgs, D.A.; McBride, J.R. Hormonal enhancement of growth in fish. In: Fish physiology, Vol. III. Hoar. W.S, Randall D.J.; Brett J.R. Ed.; Academic press, New York; 1979; pp., 455-597.
17. Duan, C. 1997. The Insulin-like Growth Factor System and Its Biological Actions in Fish. Aner. Zool. 1997, 37, 491-503.
18. Duan, C. The Insulin-like Growth Factor System and Its Biological Actions in Fish. *Aner. Zoo.* 37, 491-503.

19. Duan, C.; Hirano, T. EVects of insulin-like growth factor-I and insulin on the in vitro uptake of sulphate by eel branchial cartilage: Evidence for the presence of independent hepatic and pancreatic sulphation factors. *J. Endocrinol.* 1992, 133, 211–219.

20. Farmanfarmaian, A.; Sun, L.Z. Growth hormone effects on essential amino acid absorption, muscle amino acid profile, N-retention and nutritional requirements of striped bass hybrids. *Gent. Anal. 1999,* 15: 107-113.

21. Fine, M.; Sakal, E.; Vashdi, D.; Daniel, V.; Levanon, A.; Lipshitz, O. Recombinant carp (*Cyprinus carpio*) growth hormone: Expression, purification, and determination of biological activity *in vitro* and *in vivo*. *Gen. Comp. Endocrinol.* 1993, 89, 51-61.

22. Fitzpatrick, P.F. Tetrahydropterin-dependent amino acid hydroxylases. *Annu. Rev. Biochem. 1999,* 68, 355–381.

23. Fukata, J.; Diamond, D.J.; Martin, J.B. Effects of rat growth hormone (rGH)-releasing factor and somatostatin on the release and synthesis of rGH in dispersed pituitary cells. Endocrinology. 1985, 117, 457–467

24. Ge, W.; Peter, R.E. Evidence for non-steroidal regulator(s) of gonadotropin release in the goldfish, *Carassius auratus. Zool. Sci.* 1994, 11, 717-724.

25. Ghigo, J.M.; Létoffé, S.; Wandersman, C. A new type of hemophore-dependent heme acquisition system of *Serratia marcescens* reconstituted in *Escherichia coli. J. Bacteriol.* 1997, 179, 3572–3579.

26. Groef, B.D.; Geyten, S.V.; Darras, V.M.; Kühn, E.R. Role of corticotropin-releasing hormone as a thyrotropin-releasing factor in non-mammalian vertebrates. *General and Comparative Endocrinology.* 2006, 146, 62–68.

27. Hagemeister, A.L.; Sheridan, M.A. Somatostatin inhibits hepatic growth hormone receptor and insulin-like growth factor I mRNA expression by activating the ERK and PI3K. *American Journal of Physiology-Regulatory, Integrative and Comparative Physiology*. 2008, 295(2):R490-497.

28. Harvey, S.; Fraser, R.A.; Lea, R.W. Growth hormone secretion in poultry. *Crit. Rev. Poult.Biol.* 1991, 3, 239–282.

29. Hendry, S.H.C. Organization of the neuropeptide Y neurons in the mammalian central nervous system. In The Biology of Neuropeptide Y and Related Peptides,; Colmers, W.F.; Wahlestedt, C. Ed. Humana Press, Totowa, NJ; 1993, pp. 65–156.

30. Himick, B. A.; Golosinski, A.A., Jonsson, A-C.; Peter. R.E. CCK/gastrin-like immunoreactivity in the goldfish pituitary: regulation of pituitary hormone secretion by CCK-Iike peptides *in vitro. Gen. Compo Endocrinol*. 1993, 92: 88-103.

31. Hutchings, J. A. Life histories of fish. In Handbook of Fish Biology and Fisheries; Hart, P.J.B.; Reynolds, J.D., Ed.; 1st edn., Blackwell Publishing, London; pp. 149–174.

32. Jeh, H.S.; Kim, C.H.; Lee, H.K.; Han, K. Recombinant flounder growth hormone from *Escherichia coli*: Overexpression, efficient recovery, and growth promoting effect on juvenile flounder by oral administration. *J. Biotechno.l. 1998,* 60, 183-193.

33. Jones, J.I.; Clemmons, D. Insulin like growth factors and their binding proteins: Biological actions. *Endocrine Rev. 1995,* 16, 3-34.

34. Kaplan, S.L. Hormonal regulation of growth and metabolic effects of growth hormone. In: Handbook of Physiology, Section 7: The endocrine system, Volume V: Hormonal control of growth.; Kostyo, J.L.; Goodman, H.M. Ed. 1999; Oxford University press, NY; pp. 129-143.

35. Kojima, M.; Hosoda, H.; Date, Y.; Nakazato, M.; Matsuo, H.; Kangawa, K. Ghrelin is a growth-hormone-releasing acylated peptide from stomach. *Nature.* 1999, 402, 656–660.

36. Krick, B.J.; Boyd, R.D.; Roneker, K.R.: Beermann, D.H.; Bauman, D.E.; Ross, D.A.; Meisinger, D.J. 1993. Porcine somatotropin affects the dietary lysine requirements and net lysine utilization for growing pigs. *J. Nut.* 1993, 132, 1913-1922.

37. Larhammar, D. Evolution of neuropeptide Y, peptide YY and pancreatic polypeptide. *Regul. Pe*p. 1996, 62, 1–11.

38. Larsen, D.A.; Swanson, P.; Dickey, J.T.; Rivier, J.; DickhoV, W.W. In vitro thyrotropin-releasing activity of corticotropin-releasing hormone- family peptides in coho salmon, *Oncorhynchus kisutch. Gen. Comp. Endocrinol*. 1998, 109, 276–285.

39. Lee, L.T.O.; Siu, F.K.Y., Tam, K.K.V.; Lau, I.T.Y.; Wong, A.O.L.;

Lin, M.C.M.; Vaudry, H.; Chow, B.K.C. Discovery of growth hormone-releasing hormones and receptors in nonmammalian vertebrates. PNAS. 2007, 104(7), 2133–2138.

40. Lillesaar, C. The serotonergic system in fish. *J Chem Neuroanat.* 2011, 41, 294–308.
41. Lin, C.; Lin, S.C.; Chang, C.P.; Rosenfeld, M.G. Pit-1-dependent expression of the receptor for growth hormone releasing factor mediates pituitary cell growth. *Nature. 1992,* 360:765–768.
42. Lin, W.W.; Lin, H.R.; Peter, R.E. Growth hormone and gonadotropin secretion in the common carp (*Cyprinus carpio* L.): in vitro interactions of gonadotropin-releasing hormone, somatostatin, and the dopamine agonist apomorphine. *Gen. Compo Endocrinol.* 1993, 89, 62-71.
43. Lin, X.W.; Lin, H.R.; Peter. R.E. The regulatory effects of thyrotropin-releasing hormone on growth hormone secretion from the pituitary of common carp *in vitro. Fish Physiology and Biochemistry.* 1993, 11, 71-76.
44. Ling, N.; Esch, F.; Bohlen, P.; Brazeau, P.; Wehrenberg, W.B.; Guillemin, R. Isolation, primary structure, and synthesis of human hypothalamic somatocrinin: growth hormone-releasing factor. *Proc Natl Acad Sci USA. 1984,* 81, 4302–4306
45. Market, J.R.; Higgs, D.A.; Dye, H.M.; MacQuarrie, D.W. Influence of bovine growth hormone on growth rate, appetite, and food conversion of yearlings coho salmon (*Oncorhynchus kisutch*) fed two diets of different composition. *Can J. Zoo.l.* 1977, 55, 74-83.
46. Matty, A.J. Nutrition, hormones and growth. *Fish physiol. Biochem.*1986, 2, 141-150.
47. McCormick. S.D. Endocrine control of osmoregulation in teleost fish. *Amer.Zool.* 2001, 41 (4), 781-794.
48. McLean, E.; Delvin, R.H.; Donaldson, E.M.;and Byatt, J.C. Bovine placental lactogen invokes rapid growth in chinook salmon. *Aquaculture Nutrition.* 1996, 2, 243-248.
49. McLean, E.; Donaldson, E.M. The growth of poiklotherms. In: The Endocrinilogy of growth, development and metabolism in vertebrates.; Schreibman M.P.; Scanes, C.G.; Pang, P.K.T. Ed. Academic Press, New York.; 1993; pp. 43-71.

50. McLean, E.; Donaldson, E.M.; Teskeredzic, E.; Souza, L.M. Growth enhancement following dietary delivery of recombinant porcine somatotropin to diploid and triploid of coho salmon (*Oncorhynchus kisutch*). *Fish Physiol. Biochem.* 1993, 11, 363-369.

51. Melamed, P.; Eliahu, N.; Levavi-Sivan,B,; fir, M.O,; Farchi-Pisanty, O,; Rentier-Delrue, F,; Smal, J,; Varon, Z,; Naor, Z. Hypothalamic and thyroidal regulation of growth hormone in tilapia. *Gen. Compo Endocrinol.* 1995, 97, 13-30.

52. Mendez, E.; Planas, J. V.; Castillo, J.; Navarro, I.; Gutierrez, J. Identification of a type II insulin-like growth factor receptor in fish embryos. *Endocrinology*. 2001a, 142, 1090–1097.

53. Millar, R.P.; Davidson, J.S.; Flanagan, C.; Iling, N.; Becker, I.; Jacobs, G.; Wakefield, I. Gonadotropin-releasing hormone receptor structure and function. In Perspectives in comparative endocrinology. Davey, K.G.; Peter, R.E.; Tobe, S.S.; Ed. Natl. Res. Council Canada, Publ.; 1994, pp. 264-268.

54. Millar, R.P.; Lu, Z.L.; Pawson, A.J.; Flanagan, C.A.; Morgan, K.; Maudsley, S.R. Gonadotropin-releasing hormone receptors. *Endocr Rev*. 2004, 25(2), 235-75.

55. Miyata, A.; Arimura, A.; Dahl, R.R., Minamino, N., Uehara, A., Jianh, L., Culler, M.D., Coy, D.H. **Isolation of a novel 38 residue-hypothalamic polypeptide which stimulates adenylate cyclase in pituitary cells.** *Biochem Biophys Res Commun.*1989, 164, 567-574

56. Mommsen, T.P. Paradigms of growth in fish. *Comp. Biochem. Physiol.* 2001, 129, 207-219.

57. Moncaut, N.; Somoza, G.; Power, D. M.; Canario, A. V. Five gonadotropin-releasing hormone receptors in a teleost fish: isolation, tissue distribution and phylogenetic relationships. *J. Mol. Endocrinol.* 2005, 34, 767–779.

58. Montero, M.; Dofour, S. Gonadotropin-releasing Hormones (GnRH) in Fishes: Evolutionary Data on Their Structure, Localization, Regulation, and Function. *Zoological Studies. 1996,* 35(3), 149-160.

59. Moriyama, S.; Ayson, F. G.; Kawauchi, H. Growth regulation by insulin-like growth factor-I in fish. *Biosci. Biotechnol. Biochem.* 2000, 64, 1553–1562.

60. Muller, F.; Lele, Z.; Varadi, L.; Menezel, L.; Orban, L. Efficient transient expression system based on square pulse electroporation and *in vivo* luciferase assay of fertilized fish eggs. *FEBS Letters*. 1993, 324, 27-32.

61. Mustapha, S.B.; Undiandeye, U.C.; Gwary, M.M. The role of extension in agricultural adaptation to climate change in the Sahelian Zone of Nigeria. *Journal of environment and earth science*. 2012, 2(6), 48-58

62. Nelson, J. S. 1984. *Fishes of the world.* 2d edition; John Wiley and Sons: New York; PP 523.

63. Palyha, O.C.; Feighner, S.D.; Tan, C.P.; McKee, K.K.; Hreniuk, D.L.; Gao, Y.D.; Schleim, K.D.; Yang, L.; Morriello, G.J.; Nargund, R.; Patchett, A.A.; Howard, A.D.; Smith, R.G. Ligand activation domain of human orphan growth hormone (GH) secretagogue receptor (GHS-R) conserved from pufferfish to human. *Molecular Endocrinology.*2000, 14,160–169

64. Peng, C.; Peter, R.E. Neuroendocrine Regulation of Growth Hormone Secretion and Growth in Fish. *Zoological Studie.* 1997, 36(2), 79·89.

65. Peng, C.; Peter, R.E. Neuroendocrine regulation of growth hormone secretion and growth in fish. *Zool. Stud.*, 1997, 36: 79-89.

66. Perreault, H.A.; Semsar, K.; Godwin, J. Fluoxetine treatment decreases territorial aggressionina coral reef fish. *Physiol.Behav. 2003,* 79,719–724.

67. Pozios, K. C., Ding, J., Degger, B., Upton, Z., and Duan, C. (2001). IGFs stimulate zebrafish cell proliferation by activating MAP kinase and PI3-kinase-signaling pathways. *Am. J. Physiol.* 280, R1230–R1239.

68. Promdonkoy, B., Warit, S. and Panyim, S., 2004. Production of a biologically active growth hormone from giant catfish (*Pangasianodon gigas*) in *Escherichia coli. Biotechnology Letters.* 1998, 26, 649-653.

69. Promdonkoy, B.; Warit, S.; Panyim, S. Production of a biologically active growth hormone from giant catfish (*Pangasianodon gigas*) in *Escherichia coli. Biotechnology Letters.* 2004, 26, 649-653.

70. Rang, H. P. (2003). *Pharmacology.* Edinburgh: Churchill

Livingstone. ISBN 0-443-07145-4. Page 167.

71. Reinecke, M.; Björnsson , B.T.; Dickhoff, W.W.; McCormick, S D.; Navarro, I.; Power, D. M.; Gutiérrez, J. Growth hormone and insulin-like growth factors in fish: where we are and where to go. *Gen Comp Endocrinol.* 2005, 15, 142(1-2), 20-4

72. Reverter, J.M.C.; Canosa, L.F. Neuroendocrine systems of the fish brain. In Fish Physiology: Fish Neuroendocrinology. 1-55 edited by Bernier,N.J.; Kraak, G.V.D.; Farrell, A.P.; Brauner, C.J. Ed; Academic press, Elsevier, USA; 2009, pp 527

73. Riley, L.G.; Hirano, T.; Grau, E.G. Rat ghrelin stimulates growth hormone and prolactin release in the tilapia, *Oreochromis mossambicus. Zoological Science. 2002,* 19, 797–800.

74. Rowlerson, A.; Veggetti, A. Cellular mechanisms of post-embryonic muscle growth in aquaculture species. In Muscle Development and Growth; Johnston, I.A., Ed.; Academic Press, London, 2001; pp. 103–140.

75. Sato, N.; Murata, K.; Watanabe, K.; Hayani, T.; Kariya, Y.; Sakaguchi, M.; Kimura, S.; Nonaka, M.; Kimura, A. Growth promoting activity of tuna growth hormone and expression of tuna growth hormone cDNA in *Escherichia coli. Biotechnol. Appl. Biochem.* 1988, 10, 385-393.

76. Sciara, A.A.; Rubiolo, J.A.; Somoza, G.M.; Arranz, S.E. Molecular cloning, expression and immunological characterization of pejerrey (*Odontesthes bonariensis*) growth hormone. *Comp. Biochem. Physiol C.* 2006, 142, 284-292.

77. Seeburg, P.H.; Mason, A.J.; Stewart, T.A; Nikolics, K. 1987. The mammalian GnRH gene and its pivotal role in reproduction. *Recent Prog. Horm. Res.* 1987, 43, 69-91.

78. Seeburg, P.H.; Sias, S.; Adelman, J.; De, B.H.; Hayflick, J.; Jhurani, P.; Goeddel, D.V.; Heyneker, H.L. Efficient bacterial expression of bovine and porcine growth hormones. *DNA.* 1983, 2, 37-45.

79. Shamblott, M. J.; Chen, T. T. Identification of a second insulin-like growth factor in a fish species. *Proc. Natl. Acad. Sci. USA. 1992,* 89, 8913–8917.

80. Shepher, B.S.; Eckert, S.M.; Parhar, I.S.; Vijayan, M.M.; Wakabayashi, I.; Hirano, T.; Grau, E.G.; Chen, T.T. The

hexapeptide KP-102 (D-Ala-D-beta-Nal-Ala-Trp-D-Phe-Lys-NH2) stimulates growth hormone release in a cichlid fish (*Oreochromis mossambicus*). *Journal of Endocrinology.* 2000, 167, R7–R10

81. Sheridan, M.A.; Hagemeister, A.L. Somatostatin and somatostatin receptors in fish growth. *Gen. Comp. Endocrinol.* 2010, 167, 360–365.

82. Sherwood, N.M.; Eiden, L.; Brownstein, M.; Spiess, J.; Rivier, J.; Vale, W. Characterization of a teleost gonadotropin-releasing hormone. *Proc. Natl. Acad. Sci. USA*. 1983, 80, 2794-2798.

83. Somoza, G.M.; Miranda, L.A.; Strobl-Mazzulla, P.; Guilgur, L.G. Gonadotropin-releasing hormone (GnRH): From fish to mammalian brains. *Cellular and Molecular Neurobiology.* 2002, 22, 589–609.

84. Somoza, G.M.; Peter, R.E. Effects of serotonin on gonadotropin and growth hormone release from *invitro* perifused gold fish pituitary fragments. *Gen.Comp.Endocrinol. 1991,* 82,103–110.

85. Stanley, B.G., 1993. Neuropeptide Y in multiple hypothalamic sites controls eating behaviour, endocrine and autonomic systems for body energy balance. In: Colmers, W.F., Wahlestedt, C. (Eds.), The Biology of Neuropeptide Y and Related Peptides. Humana Press, Totowa, NJ, pp. 457–509.

86. Sugimoto, S.; Yamaguchi, K.; Yokoo, Y. Isolation and characterization of recombinant eel growth hormone expressed in *Escherichia coli. Y. Chromatogr.* 1990, 515, 483-494.

87. Tatemoto, K.; Carlquist, M.; Mutt, V. Neuropeptide Y-A novel brain peptide with structural similarities to peptide YY and pancreatic polypeptide. *Nature*. 1982, 296, 659–660.

88. Tinoco, A.B.; Näslund, J.; Delgado, M.J.; De Pedro, N.; Johnsson, J.I.; Jönsson, E. Ghrelin increases food intake, swimming activity and growth in juvenile brown trout (*Salmo trutta*). *Physiology & Behavior.* 2014, 124, 15-22.

89. Trudeau, V. L.; Somoza, G. M.; Nahorniak, C. S.; Peter, R. E. Interactions of estradiol with gonadotropin-releasing hormone and thyrotropin-releasing hormone in the control of growth hormone secretion in the goldfish. *Neuroendocrinology. 1992,* 56, 483–490.

90. Tsai, H.J.; Tseng, F.S.; Liao, L.C. Electroporation of sperm to introduce foreign DNA in to genome of loach (*Misgurnus*

anguillicaudatus). *Can. J. Fish Aquat. Sci.*, 1995, 52, 776-787.

91. Um, H.N.; Han, J.M.; Hwang, J.I.; Hong, S.I.; Vaudry, H.; Seong, J. Y. Molecular coevolution of kisspeptins and their receptors from fish to mammals. *Ann.N.Y.Acad.Sci.* 2010, 1200, 67–74.
92. Upton, Z.; Yandell, C. A.; Degger, B. G.; Chan, S. J.; Moriyama, S.; Francis, G. L.; Ballard, F. J. Evolution of insulin-like growth factor-I (IGF-I) action: In vitro characterization of vertebrate IGF-I proteins. *Comp. Biochem. Physiol.* B, 1998, 121, 35–41.
93. Very, N.M.; Sheridan, M.A. The role of somatostatins in the regulation of growth in fish. *Comp Biochem Physiol B. 2002,* 138, 323–330.
94. Volkoff H.; Canosa, L. F.; Unniappan, S.; Cerda-Reverter, J.M.; Bernier, N.J.; Kelly, S. P.; Peter, R. E. Neuropeptides and the control of food intake in fish. *Gen. Comp. Endocrinol.* 2005, 142, 3–19.
95. Walker, W.H.; Fitzpatrick, S.L.; Barrera-Saldana, H.A.; Resendez-Pérez, D.; Saunders, G.F. The human placental lactogen genes: structure, function, evolution and transcriptional regulation. *Endocr. Rev. 1991,* 12, 316-328.
96. Weatherley, A.H.; Gill, H.S.; Lobo, A.F. Recruitment and maximal diameter of axial muscle fibres in teleosts and their relationship to somatic growth and ultimate size. *J. Fish Biol.* 1988, 33, 851–859.
97. Winberg, S.; Nilsson, G.E. Roles of brain monoamine neurotransmitters in agonistic behavior and stress reactions, with particular reference to fish. *Comp Biochem Physiol Part C: Pharmacol, Toxicol Endocrinol.* 1993, 106, 597–614.
98. Wong, A.O.L.; Chang, J.P.; Peter, R.F. Dopamine functions as a growth hormone-releasing factor in the goldfish, *Carassius auratus. Fish Physiol. Biochem.* 1993b, 11, 77-84.
99. Wong, A.O.; Li, W.S.; Lee, E. K.; Leung, M.Y.; Tse, L.Y.; Chow, B.K., Lin, H.R.; Chang, J.P. Pituitary Adenylate Cyclase Activating Polypeptide as a Novel Hypophysiotropic Factor in Fish. *Biochem Cell Biol.* 2000, 78 (3), 329-343.
100. Wong, A.O.L.; Chang, J.P.; Peter, R.E.; Characterization of D1 receptors mediating dopamine-stimulated growth hormone release from pituitary cells of the goldfish, *Carassius auratus. Endocrinology*. 1993a, 133, 577-584.

101. Wong, A.O.L.; Murphy, C.K.; Chang, J.P.; Neumann, C.M., Lo,A., and Peter,R.E. Direct actions of serotonin on gonadotropin-II and growth hormone release from gold fish pituitary cells: interactions with gonadotropin- releasing hormone and dopamine and further evaluation of serotonin receptor specificity. *Fish Physiol. Biochem. 1998,* 19, 23–34.

102. Wood, A.W.; Duan, C.; Bern, H.A. Insulin-Like Growth Factor Signaling in Fish. *In ternational Review of Cytology*. 2005, 243, 215-285.

103. Wu. S.; Adams, B. A.; Fradinger, E. A.; Sherwood, N. M. Role of two genes encoding PACAP in early brain development in zebrafish. *Ann. NY Acad. Sci.* 2006, 1070, 602–621.

104. Xie, S.W.; Li, Y.T.; Zhou, W.W.; Tian, L.X.; Zeng, S.L.; Liu, Y.J. Effect of γ-aminobutyric acid supplementation on growth performance, endocrine hormone and stress tolerance of juvenile Pacific white shrimp, *Litopenaeus vannamei,* fed low fishmeal diet. *Aquaculture Nutrition*. 2015, 23(1), 54-62.

105. Yorio, M.P.D.; Delgadin, T.H.; Pérez Sirkin, D.I.; Vissio, P. G. Growth hormone, luteinizing hormone, and follicle-stimulating hormone regulation by neuropeptide Y in both sexes of the cichlid fish, *Cichlasoma dimerus*. *Fish Physiology and Biochemistry.* 2015, 41(4), 843-852.

106. Zohar,Y.; Munoz-Cueto, J.A.; Elizur, A.; Kah, O. Neuro endocrinology of reproduction in teleost fish. *Gen.Comp.Endocrinol.* 2010, 165, 438-455.

107. Zukowska-Grojec, Z., Wahlestedt, C. Origins and actions of neuropeptide Y in the cardiovascular system. In: Colmers, W.F., Wahlestedt, C. (Eds.), The Biology of Neuropeptide Y and Related Peptides. Humana Press, Totowa, NJ; 1993, pp. 315-388.

CHAPTER 2

GROWTH HORMONE GENE REGULATED GROWTH IN FISH

Sekar Megarajan[1], Prerna Pandey[2]

[1]**Dr Sekar Megarajan** graduated in fisheries science, and then obtained master degree in Fish Genetics and Biotechnology and PhD in Fish Nutrition and Biochemistry with specialization in fish molecular endocrinology from Central Institute of Fisheries Education (CIFE), India. He worked for two years with Transasia Bio-Medicals Ltd, Mumbai in the area of immunology for development of diagnostic kits for human infectious disease and joined Indian Agriculture Research service in 2013 as a scientist. He currently works as scientist in mariculture division at Central Marine Fisheries Research Institute, Cochin, India. Dr Megarajan has been working on breeding, seed production and culture of high value marine finfishes. He has published more than 10 research papers in peer valued journals in the area of fish population genetics, molecular endocrinology and fish culture.

[2]**Dr Prerna Pandey**, a biotechnologist with several years of wet lab research experience, worked at International Center for Genetic Engineering and Biotechnology, New Delhi. Her field of research in PhD included isolation and molecular characterization of geminiviruses, genome sequencing, gene annotation, and gene silencing using the RNA interference technology. She has also worked at Transasia Biomedicals and Advance Enzyme Technologies as a scientist. Prerna has published papers in peer-reviewed journals, and has submitted a number of annotated Geminiviral genome sequences in the GenBank, including two novel ones. She has also completed her editing and proof-reading courses from Society for the promotion of Editors and Proof-readers, UK and now works as a freelance scientific editor and writer. When Prerna is not busy with her assignments, she loves to explore historical places.

INTRODUCTION

A candidate gene is one which is located in a chromosome region and presumed of being involved in the expression of a particular trait. Its protein product suggests that it could be the gene of interest (Tian et al., 2014). Many candidate genes are present and these are commonly targeted based on prior knowledge of their role in regulation of specific metabolic pathways influencing a particular quantitative trait. These genes might be directly influencing variation in the trait, or it may simply be linked with the major gene due to chromosomal proximity and is acting as a quantitative trait locus (QTL) markers. Using this approach, candidate genes have been identified for economic traits in live-stocks such as pigs (litter size) (Rothschild et al., 1996), beef cattle (hair length and heat tolerance) (Olson et al., 2003), and sheep (ovulation rate) (Davis et al., 1991), etc. However, unlike agricultural and animal production sectors, the genetic research in identifying the candidate genes for economic traits is in infancy stage in the aquaculture industry although, many programs on livestock genomics have been centred around identifying important genes that influence economy in aquaculture, including disease resistance, growth rate, and the quality of meat.

Growth rate is a primary trait of interest in most finfish selection programs and is intrinsically linked to productivity and profitability of finfish enterprises. In aquaculture, the search for candidate genes influencing growth has been a major focus of research and several such genes have been isolated from the genome of livestock and their effects are quantified (De-Santis and Jerry, 2007). Many of these genes have been shown to excessively contribute to livestock growth variation and importantly to be part of the same somatotropic governing pathways as they occur in fish.

GROWTH HORMONE GENE IN FISHES

Somatogenesis is a polygenic trait resulting from a multitude of physiological pathways regulating energy metabolism and muscle growth. Among the many possible growth-regulating pathways in vertebrates, genes within the somatotropic axis and transforming growth factor have been most targeted as possible candidate genes in livestock and finfish

(De-Santis and Jerry, 2007). The somatotropic axis, basically consisting of growth hormone (GH), growth hormone-releasing hormone (GHRH), growth hormone inhibiting hormone (GHIH or somatostatin), insulin-like growth factors (IGF-I and IGF-II), and associated carrier proteins and receptors, all these known to play a key role in the regulation of metabolism and physiological processes. Moreover, other hormones often included in this axis are insulin, leptin, glucocorticoids and thyroid hormones, which modulate GH and/or IGF-I synthesis and availability.

Growth hormone (GH) is a single chain polypeptide that plays a major role in the growth development of vertebrates. It is the chief postnatal somatic growth regulator and stimulator of anabolic processes such as cell division, protein synthesis and skeletal growth. Growth hormone is also involved in the regulation of oxidation of fatty acids (lipolytic activity), inhibition of glucose transport to peripheral tissues (diabetogenic activity) and the regulation of ribosome translational activity, which subsequently affects protein synthesis (Goodman, 1993). In livestock, a number of studies have repeatedly implicated the GH gene as a candidate gene for the enhancement of production traits such as growth (Tambasco et al., 2003), disease resistance, egg production (Kuhnlein et al., 1997), fat content (Knorr et al., 1997) and milk production (Hoj et al., 1993; Dybus, 2002). These results across many diverse livestock species clearly demonstrate that, at least a small amount of the total variance seen in growth rate variation can be confidently associated with the additive effect of alleles at the GH gene.

The genes encoding GH have been studied extensively in many species. Nucleotide sequences have been determined for GH genes from human, rat, porcine, bovine and ovine (Chiou et al., 1990). Because of its potential in enhancing the growth rate of fishes in aquaculture, the growth hormone encoding gene has been cloned from several species of teleosts (Venkatesh and Brenner, 1997). The genomic region in vertebrate is roughly 2 kb long, with four to five blocks (exons) of protein coding regions representing less than one third of the total genomic region. The coding region in GH tends to be highly conserved across vertebrates, presumably because of functional constraints imposed on the hormone structure (Bart et al., 2010). However, rates of GH sequence evolution vary for other groups of vertebrates and the variability in pituitary GH has been detected in mammals (Forsyth and Wallis, 2002).

GH GENE ORGANISATION

The GH gene of mammalian origin consists of five exons, whereas in the osteichthyes class the GH genes contain either five or six exons. Fishes are the oldest vertebrates and are thought to have evolved around 500 million years ago (Powers, 1989). Using a limited number of GH gene sequences, several studies have been reported on GH gene sequence information in comparative teleost (modern bony fish)/higher vertebrates. According to earlier studies, GH gene structure is not conserved in teleosts, and varies in different groups and it is slightly related towards living environment of fishes such as freshwater and marine environments. The cypriniformes (carps) fishes belongs to freshwater origin such as common carp, grass carp, silver carp and catfishes (siluriformes) consist of GH genes with five exons and four introns, which similar to avian and mammalian GH genes (Tang et al., 1993). However, the GH gene from the fishes of perciformes (Almuly et al., 2000), salmoniformes (Delvin, 1993), and tetradontiformes (Venkatesh and Brenner, 1997) consists six exons and five introns and most of the fishes from the groups having the seawater origin. Size of exons are fairly conserved with the exception of exon five, which has been suggested to have slipped in two parts during the evolution of teleosts (Almuly et al., 2000). It has been hypothesized that the insertion of the fifth intron occurred just after the separation event that led to the divergence between cypriniformes and other teleosts such as salmoniformes and perciformes (Ber and Daniel, 1993).

Presence of six exons and five introns are found to constitute the GH gene in many fishes. For example, the primary transcript of *Lates calcarifer* growth hormone gene was found to be 1615nt long and consists of six exons and five introns. The invariant GT and AG dinucleotodes are present at donor and acceptor sites of 5' and 3' ends of spliced intron (Green, 1986) which is present in all eukaryotes, the same was observed for all intron/exons junction of GH gene in *L. calcarifer*. The rainbow trout growth hormone gene spans a region of ~4 kilobases, and is comprised of six exons in contrast with 5 exons in mammals. It is suggested that the translated regions analogous to the last exon of its mammalian counterpart may be interrupted by the presence of an additional intron in the fish gene (Agellon et al., 1988). The GH gene of fugu consists of six exons and five introns, and the first intron of the fugu GH is 608 bp, which is larger than its counterparts from other teleosts. The large size of this intron is due to the presence of a satellite sequence. This satellite sequence is not found in the corresponding introns of other teleosts. Thus, it seems to have been newly inserted in to fugu intron. A tetranucleotide repeat, (CTGT)28, and a trinucletide repeat, (GAT)79, are found in the second and fifth introns of the fugu GH gene, respectively. Shorter stretches of the latter repeat element are also found in the corresponding introns of tilapia (Ber and Daniel, 1992), flounder (Tanaka et al., 1995) and barramundi (Yowe and Epping, 1995), suggesting that this minisatellite was present in the fifth intron of a common ancestor of tetrodontiformes (fugu), pleuronectiformes (flounder) and perciformes (tilapia and barramundi).

Observations from above given fishes clearly indicated that the GH structure is not conserved among teleosts fishes, and showing a mammalian-like organization (five exons and four introns) in fishes of cypriniformes and siluriformes, and presenting one supplementary exons in the most other fishes from perciformes, pleuronectiformes and tetrodontiformes (Ohkubo *et al.*, 1996). Size of the exons is fairly conserved with the exception of exon five, which has been suggested to have slipped in to two parts during the evolution of teleosts (Almuly et al., 2000). Based on the available information from earlier findings, it has been hypothesised that the insertion of the 5th intron took place only within the ray finned fish, after the evolutionary separation of cypriniformes but before salmoniformes, perciformes and tetradontiformes evolution (Ber and Daniel, 1993).

DUPLICATION AND STRUCTURE OF FISH GROWTH HORMONE GENE

Gene duplication is the process by which a region of DNA coding for a gene is copied during molecular evolution. An error in recombination or a retro-transposition event can result in gene duplication. Often, duplicate genes withstand to the selective pressure under which genes normally exist, and may accumulate mutation in the duplicated gene code. This in turn makes the gene either non-functional or in some cases confers some benefit to the organism. Common sources of gene duplications include retro-transposition event, ectopic recombination, aneuploidy, polyploidy, and replication slippage (Zhang, 2003). The fate of duplicated genes may be different from the original gene and presence of two genes may have different nature in an animal. If the existence of two or more genes is advantageous in an animal, then it can lead to the preservation of function in duplicated genes (conservation). However, in most frequently the duplicated gene lose their function as a result of accumulation of mutations in one of the paralogs, so one of the copies become non-functional gene (pseudogene) (pseudogenization). It is also assumed that the duplicated sequences may play an important role in the eukaryotic genome, which protects it from harmful mutations and contributes it to the genetic stability of the organism. The importance of gene duplication in supplying raw genetic material to biological evolution has been recognized since the 1930s. Substantial evidence for the abundance of duplicated genes in all surveyed organisms has been made possible by the available genomic sequence data. Presently, use of computational analysis of genomic sequences and population genetic modelling used for uncover the mechanisms behind the evolution by gene duplication (Pankova et al., 2017).

It is believed that the whole genome duplication occurred in the history of all organisms. Gene duplication has found in most of fish species, and most of the researchers agree that number of polyploidizations observed in the evolution of the group of fishes. Growth-hormone gene in the genome of vertebrates is usually represented by one copy, although there are exceptions (Hirt et al., 1987). In salmonid fishes, gene duplication has been identified and reported the presence of two paralogous growth hormone genes (GH): GH-I and GH-II in genome and both genes are apparently functional (Von Schalburg et al., 2008; Pankova et al., 2017). In the Pacific salmon, in addition to the two functional genes, a pseudogene localized on the Y chromosome is also found (Du et al., 1993). The gene analysis suggested that, both genes exist throughout the time of species divergence in this group for 25–100 million years, and are probably both functionally active (Allendorf & Thorgaard, 1984). Similar to salmonid fishes, analysis of some tetraploid fishes including fishes in Asian cyprinids, fishes in sucker family, and smallmouth buffalo fish, has shown the presence of two GH genes (Clements et al., 2004). In buffalo fish, while the predicted GHI protein contains 4 Cys residues, homologous to other vertebrate GH sequences, GH-II has 5 Cys residues, homologous to other ostariophysan sequences. GH-I and GH-II are most similar to other cypriniform fishes for both in nucleotide and protein sequences, phylogenic relationship of the sequences of smallmouth buffalo fish GH is consistently grouped with fishes of Asian cyprinids.

ORGANIZATION OF GROWTH HORMONE GENE

Growth hormone from a variety of species has been studied extensively at the levels of protein, mRNA, and nucleotide sequences. The nucleotide sequence of GH genes has been determined from birds (Tanaka *et al.*, 1995), mammals (Vize and Wells, 1987), and fishes (Chiou et al., 1990; Ber and Daniel, 1992; Zhu et al., 1992; Devlin, 1993; Du et al., 1993; Yowe and Epping, 1995; Megarajan et al., 2014). The size of most of the identified fish GH gene primary transcript ranges from 1500 to 4000 nucleotides in average length and with small changes in the lengths. However, the general structure of the transcript remains same with other vertebrates. The structures of eukaryotic genes consisting several nested sequence elements and these elements have a specific function in the multi-step process of gene expression. While DNA is a double-stranded molecule, only one of the strands is read by RNA polymerase to produce protein-coding mRNAs or non-coding RNAs. This sense or coding strand runs in the 5' to 3' direction and flanked by 5' untranslated region (UTR) and 3' UTR, respectively. GH and/or its cDNA have been identified in more than 110 species from all taxonomic groups of vertebrates (Venugopal et al., 2002). Among which, fishes from more than 5 orders have been identified and well characterised, some of them are common carp, grass carp, silver carp and mud loach, etc., from cypriniforms; sutchi catfish, pangas catfish and channel catfish, etc., from siluriformes; fishes from salmoniformes such as rainbow trout, Atlantic salmon, chinook salmon, and sockeye salmon: tilapia, flounder, barramundi yellowtail and gilthead seabream from perciformes; pufferfish from tetradontiformes are the few examples (Moriyama et al., 2006; Sekar et al., 2014). In a gene, the coding region of GH gene is divided in to three parts such as 3'UTR, Open reading frame (ORF) and 5' UTR and the total length of the complete region is varying around 1-1.5 kb among different groups. However, the genomic structure of the sea lamprey GH-encoding gene was determined and it consists of 13,604 bp, making it the largest among all known GH gene (Moriyama et al., 2006). In silurids fishes, 1193 bp long sequence was observed by Lemaire et al. (1994), which includes 59 bp in 5' UTR region, 600 bp in ORF and 515 bp in 3' UTR region and in

H. fossilis it was 1132 bp (58 bp in 5' UTR, 603 bp in ORF & 456 bp in 3' UTR) (Anathy et al., 2001). In cyprinids group of fishes, a 1156 bp long cDNA sequence, (14 bp in 5' UTR, 636 bp in ORF and 491 bp in 3' UTR) was observed in catla by Venugopal et al. (2002) and in rohu it was 1169 bp long (54 bp in 5' UTR, 621 bp in ORF & 494 bp in 3' UTR) (Rajesh & Majumdar, 2008). The cyprinid fishes have high number of nucleotides in ORF: 624 bp in rohu, 633 bp in mrigal and 636 bp in catla (Venugopal et al., 2002) and 624 bp in common carp (Chio et al., 1990). Similarly, ORF region with 630 bp was observed in *Chanos chanos* (Jesus et al., 2002) and 612 bp in *Lates calcarifer* (Yowe and Epping, 1995). However, in some of the fishes, the nucleotide sequences in the ORF region is less than 600 nucleotides: 570 bp in flounder (Tanaka et al., 1995), and 588 bp in puffer fish (Venkatesh & Brenner, 1997). The general schematic diagram for the gene structure of the growth hormone is given below.

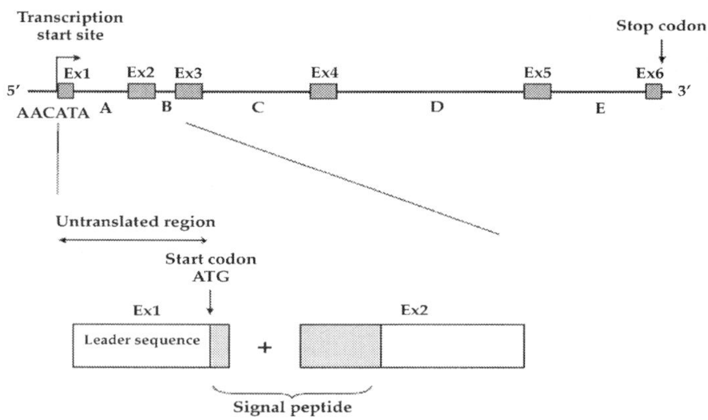

Schematic representation of the hormone gene structure (Pankova et al., 2017)

REGULATORY SEQUENCES IN GROWTH HORMONE GENE

In vertebrates, the somatotroph cells constitute the major site in the anterior pituitary gland for GH expression, and this phenomenon suggesting that some developmental mechanisms determining tissue-specific expression of the genes and it has been conserved in the course of evolution. In mammals and other vertebrates, the specific expressions of the genes are regulated by the different transcription factor. Transcription factors are the protein that binds to specific DNA sequences, thereby regulates the gene expression. *In-silico* analysis of the GH gene from different taxa revealed the presence of many number of transcription factors in the gene and they regulate the GH gene expression (Andersen and Rosenfeld, 2001). In mammals, for the transcription of the GH gene, the regulatory factor Pit-1, also called as growth hormone factor 1 (GHF-1), needs to interact, and is present with two *cis*-acting elements within the proximal promoter (Andersen and Rosenfeld, 2001). It is well known that in similar to mammals, several transcription factors present in fish growth hormone gene and that helps to regulate the gene for the tissue specific expression. Though GH genes have been cloned from numerous fish species, research on the mechanisms involved in transcriptional regulation of the GH gene in fish has been limited to a few species, mainly salmonids and tilapia. However, studies in rainbow trout, chinook salmon, and tilapia revealed that the Pit-1 proximal binding site is conserved among all vertebrates and is crucial for the tissue-specific activation of GH gene transcription (Almuly et al., 2005). The specific study on tilapia and salmonids GH gene also revealed that the gene is regulated by some more transcription factors other than Pit-1, which is in similar with human. Moreover, the extensive study on GHcDNA sequences of fishes from different orders has showed the presence of many transcription factors including Pit-1a, CRE, CREB, CREBP, Ap-1 SP1 and TBP ((Almuly et al., 2005).

It is known that the specific expression of the GH is controlled by the pituitary specific transcription factor Pit-1 which binds to AT-rich cis elements in the GH promoter (Castrillo et al., 1989). There is only one Pit-1 binding site with motifs ATTCATAAA located in the 5' region (-113 to -121), thereby making the GH expression tissue specific as in mammals (Yamada et al., 1989; Argenton et al., 1996). Presence of Pit-1 transcription factor in the promoter region has been observed in many fishes such as rainbow trout (Agellon et al., 1988), barramundi (Yowe and Epping, 1995), common carp (Chiou et al., 1990), flounder (Tanaka et al., 1995), rohu (Rajesh and Majmudar, 2008), and puffer fish (Venkatesh and Brenner, 1997) sutchi catfish (Megarajan et al., 2014). The impact of transcription factor on the expression of GH gene was studied and observed that the pituitary gland specific expression of rat (Elsholtz et al., 1990) and human (Lefevre et al., 1987) GH genes was due to the interaction of the transcription factors GHF-1/Pit-1 with two *cis*-acting elements in their promoter regions. Accordingly, three core binding sites for mammalian GHF-1/Pit-1 were identified within the barramundi fish GH gene promoter regions and the role for such elements in the control of tissue-specific expression was suggested by the RT-PCR detection of GH mRNA from pituitary, but not from non-pituitary gland tissues (Yowe and Epping, 1995). Transcriptions of GH genes are also affected by changes in cAMP levels in response to a variety of external signals and are mediated via a cAMP response element (CRE). This DNA sequence is recognized by a diverse family of DNA binding proteins, of which the CRE-binding protein (CREB) has been characterized in fishes (Brindle and Montminy, 1992).

Activation of the protein kinase A (PKA) pathway leads to phosphorylation of CREB at serine, which is required for CREB to initiate transcription of the target gene. GH cDNA analysis in sutchi catfish showed the presence of cAMP related transcription factors in different location of the GH gene: CRE (GTCATG: -182 to -187, TGTGAA: 1674 TO 1679), CREB (ATGACATCAC: 1626 to 1636), and CREBP (AAACGTCACA: -177 to -168, CACGTCAGTG: 1261 to 1270, ATGACATCAC: 1927 to 1936) (Megarajan et al., 2014). Presence of the transcription factors suggest that the possible involvement of cAMP in the regulation of GH expression in this fish and motif sequence of this CRE factors is identical to those found in several eukaryotic gene. A similar result is also observed in barramundi (Yowe and Epping, 1995), *Nibea coibor* (Zhang et al., 2009) and rainbow trout (Argenton et al., 1993) promoter gene, which can mediate the cAMP induction of transcription. The presence of Ap-1 transcription factor motif was identified in the promoter region (GAAGTCATG: -182 to -190) of the fish. The TATA box binding protein which specifically binds to the TATA box is also observed in GH gene of the fish. The TBP identified in three different locations of the gene (GTGTTTATAA: -346 to -337, ATTACAAAAA: -31 to -328, CGTATAAAAA: -23 to -31) in striped catfish. Interestingly, it is noted that the third TBP overlaps the TATA box, which may increase the binding capacity. The other *cis* regulators TATAAA and CCAAT boxes recognized by TBP (Patikoglou et al., 2002) and Sp1 (Morgan et al., 1988) respectively, were identified in upstream to the transcriptional start point in a gene. The TATA box is located 23 nt upstream to the start codon, and CAAT located 129 nt and 100 nt upstream from the start codon and TATA box, respectively. The TATA box is the highly conserved sequence that is identified between the GH gene and promoters in sutchi catfish. Presence of all the regulating factors similar in different fishes and ultimately they show similarity with their higher eukaryotic gene promoter and upstream element consensus sequences (Bucher, 1990). Studied sequences of different GHcDNA showed that the higher number of conserved regions presents within the mammalian class, as compared to osteichthyes class GH gene promoters (Yowe and Yepping, 1995).

POLYMORPHISM IN GROWTH HORMONE GENES

A gene is said to be polymorphic when more than one allele occupies in that gene's locus within a population. Polymorphism is caused by events like duplications, deletions, as well as by changes in the introns or in regions of one or multiple DNA bases that are between genes. Changes in the gene's coding sequence may result in different phenotypes as a result of protein variation caused by sequence changes. Presence of polymorphism in the coding region of gene liked to the important economic traits of an animal, which may keep the animal or fish in the advantages position and this need to be correctly identified and utilised in the breeding programme to exploits particular character in the animal. Significant association of polymorphic candidate gene with economic traits will help the breeders to search out some genetic marker for economic traits (Tian et al., 2014). Thus, a very powerful method to investigate associations of gene polymorphisms is the candidate gene approach leading to economically important traits in aquaculture and other farming sectors. For the implementation of an any genetic improvement program in fish that focus on the selection of important economic traits, determination of the genetic variability is an essential step including faster growing fish, lower rates of feed conversion and disease resistance (Lupchinski Jr. et al., 2011). Improvements in the some of these economic important traits are essential in commercial aquaculture practices because these desirable economic traits have declined with the rapid development of the culture industry for some of the commercially cultured fishes. Thus, in a strategy known as gene-assisted selection (GAS) (De-Santis and Jerry, 2007), most of the recent aquaculture studies aim to search for variations in genomes associated with productive phenotypes, use them as tools for assisting the offspring selection at any early stage and predict their productivity (Na-Nakorn and Moeikum, 2009). This Application of GAS requires molecular markers. Currently, productive traits in many aquatic species are being improved by using single nucleotide polymorphisms (SNP) as molecular markers from GAS (De-Santis and Jerry, 2007).

Generally, SNPs cause most sequence variations, and with the increase in speed and efficiency of SNPs identification, studies promote exhaustive, broad range of tests for the hypothesis that molecular nature of quantitative traits are influenced significantly by the common variants (Tian et al., 2014). In aquaculture species, the most important growth traits include growth rate, body weight and length. Growth hormone (GH) is a very important factor, which regulates the growth of vertebrates. Several studies have reported that the GH gene as a candidate for enhancing several such a production traits in live stocks and aquaculture (Sawayama and Takagi, 2015). Therefore, polymorphisms in GH gene are thought to be useful for marker-assisted selection (MAS) to improve growth performance in different cultured fishes. Polymorphisms in the GH gene have been reported and some SNPs have been shown to be associated with economic traits in several fishes, such as Atlantic salmon, brown trout, chinook salmon, large yellow croaker, bleak, and common bream (Tian et al., 2014).

Apart from the coding region of a gene, polymorphisms also occur in the intron region of the GH gene and these helps for the tracing out the evolutionary history of the different fish species. In nature, occurrences of polymorphisms associated with coding region in the GH gene are very rare because majority of mutations in the GH gene occurs within introns (Almuly et al., 2005). Most common polymorphisms found in introns of GH gene are either microsatellite repeat indels, or SNPs. However, the presence of this microsatellite has also been described in the 5′ flanking region of several other fish species, including barramundi (Yowe and Epping, 1995), olive flounder (Tanaka et al., 1995) and yellowtail kingfish (Ohkubo et al., 1996). Therefore, sequence analysis of GH gene from different groups of teleosts should help in tracing the origin of this intron and the evolution of teleost GH genes. The intron pattern of GH genes might serve as a natural marker in understanding the evolutionary relationships of various teleost groups.

GROWTH HORMONE GENE IN FISH GROWTH IMPROVEMENT

Aquaculture is one of the fastest growing food-producing sectors globally, having the greatest potential to meet the ever-growing demand of aquatic food around the world. Worldwide, decline of fish catches from the marine sector in recent years has led to rapid growth of fish and shellfish farming. Fish production through aquaculture has achieved phenomenal growth in last few decades, and the contribution of aquaculture to global supplies of fish, crustaceans, molluscs and other aquatic animals for human consumption continues to grow every year, and has reached 73.8 million tonnes in 2014 (FAO, 2016). Increase in the domestication of fishes for culture has reduced the growth potential of fishes in many number of fish species and thus the production is constrained mainly by the growth efficiency of cultured fish species (Sekar et al., 2015). Growth rates of many fish species used in aquaculture are naturally slow, but are being enhanced by methods of domestication, nutritional approach, genetic selection, biotechnological approach of transgenic fish production and endocrinological approaches (Delvin et al., 2001). Use of molecular biology and genetic engineering technologies adopted in transgenic fish production and this technology can be employed to enhance the targeted beneficial characters in transgenic fish (Hafsa et al., 2016).

The production and characteristics of conventional fish strains can be improved by genetically modifying (GM) fish (transgenic fish), a technique currently being exploited in aquaculture (Fu et al., 2005). Genetic modification can refer to changes in the genetic makeup of organisms not found in nature, which utilises recombinant DNA techniques to insert genetic material from one organism into the genome of a fish or other aquatic organism. One of the important approaches for development of transgenic fish has been through enhancing the growth rate; the growth efficiency can be increased in finfish by creating transgenic fish that incorporate a gene construct that encodes growth hormone. The research and development of transgenic fish focused initially on obtaining huge increases in growth rates from 2 to 11 times faster than normal fish using growth hormone genes. In the growth-enhanced transgenic fish production, the inserted genetic constructs typically contain both a growth hormone gene and DNA sequences that control the gene. The DNA control sequences known as promoters, which direct the activation and expression of growth hormone gene at particular location in a particular time. The growth hormone gene may come from either the same fish species as the one being engineered or from a different fish species. However, the key to stimulating dramatically faster growth rates is not changing the growth hormone gene, but rather the promote DNA sequence that controls it (MacLean and Laight 2000). For transgenic fish development, several important steps are routinely taken to produce a desired transgenic fish:

- Selection of an appropriate fish species;
- Preparation of a suitable gene construct;
- Introduction of transgene into fish embryos; and
- Selection and characterization of the resulting transgenic fish (Chen, 1998).

With aim of producing transgenic fish, at first Zhu *et al.* (1985) produced the first batch of fast-growth transgenic fish and thereafter many laboratories throughout the world have been successful in generating transgenic fish in a variety of species. In the initial study, growth hormone gene of non-piscine source was introduced in fish for growth improvement, where the common carp (*Cyprinus carpio*) was the host fish and a recombinant human growth hormone (hGH) gene was the transgene. Thereafter, Fu et al (2005) reported successful transfer of recombinant human growth gene under the control of mouse MThGH into the fertilised goldfish eggs, which subsequently developed into loach, and to the birth of fast-growth transgenic fish. After the initial success in non piscian GH gene transfer, the all fish gene or auto gene concept had developed and it was considered to produce transgenic fish in order to eliminate misgiving on food safety and to benefit expression of the transferred gene (Ye et al., 2011). Different numbers transgenic fish with all fish gene has been developed for aquaculture for growth improvement and research purposes. Among the food fishes, the major food fish species used for the transgenic fish development are carp (*Cyprinus sp.*), tilapia (*Oreochromis sp.*), salmon (*Salmo sp., Oncorhynchus sp.*) and channel catfish (*Ictalurus punctatus*) (Dunham & Liu, 2006). For most of the transgenic fish, the growth performance evaluation has showed that the growth rate of transgenic fish can be increased by 400% to 600% by improving food conversion ratios while simultaneously reducing feed input by up to 25% per unit of out.

Generally, the growth performance of different transgenic fish has showed four different levels of responses. In some of the transgenic fishes there was no difference in the growth after transfer of the gene construct. For examples in salmonid fishes the mammalian gene constructs (mMT/rGH) failed to affect growth of the fish (Guyomard et al., 1989ab; Penman *et al.,* 1990), even though the fish were very responsive to growth stimulation by exogenously administered mammalian GH protein (McLean and Donaldson, 1993). In the second level of response, the moderate growth enhancement was reported for some fish species (Zhu et al.,1986; Zhu, 1992; Wu et al., 1994). The gene constructs containing fish GH sequences with non-piscine promoters elicited growth enhancement in transgenic carp, catfish, zebrafish, and tilapia (Martinez et al., 1996). Also in several other fish species including loach, common carp, crucian carp, Atlantic salmon, channel catfish, tilapia, medaka, and northern pike the faster growth of about 10-80% was observe after transferring the gene construct containing either human, bovine, or salmonid fish growth hormone. However, the growth was not observed in the non-transgenic sibling fish and moreover the observed growth improvement was the first response after development of transgenic fish. The next level of enhancement was a 2-6-fold increase in growth from GH transfer. All fish GH gene construct used in transgenic Atlantic salmon by Du et al. (1992) and showed a 2-6-fold increase of the transgenic fish growth rate. Similarly, Nile tilapia possessing one copy of an eel (ocean) pout promoter-chinook salmon growth hormone fusion grew 2.5-4-fold faster and could convert feed upto 20% better than non-transgenic siblings (Rahman et al., 2001; Rahman and Maclean, 1999). Rohu, *Labeo rohita,* containing CMV- or *B* actin-rohu GH gene had a 4-5-fold increase in growth rate (Venugopal et al., 2004). Transgenic Atlantic salmon containing the ocean pout antifreeze promoter-chinook salmon growth hormone (GHcDNA1) gene construct had 3-6-fold accelerated growth compared to non-transgenic salmon (Du et al., 1992; Cook et al., 2000a). The last level of observation was the hyper levels of growth enhancement. When transgene introduced into coho salmon, cutthroat trout, rainbow trout, and chinook salmon, the GH gene constructs using either an ocean pout antifreeze promoter driving a Chinook salmon GH cDNA or a sockeye salmon metallothionein promoter driving the full-length sockeye GHI gene elevated circulating GH levels by as much as 40-fold (Devlin, 1997), resulting in a 5-30-fold increase in weight after one year of growth (Devlin et al., 2001) and allowing precocious development of physiological capabilities necessary for marine survival (smoltification). In some of the studies, the largest transgenic fish were mated and produced offspring with extraordinary growth and similar results were also obtained with mud loach (Nam et al., 2001). In this way, different research trials have explained that development of transgenic fish for improving the growth potential of the animal using growth hormone gene could be possible and thus it may lead to the improving the aquaculture fish production.

REFERENCES

1. Agellon, L.B.; Davies, S.L.; Lin, C.M.; Chen, T.T; Powers, D.A. Rainbow trout has two genes for growth hormone. *Mol. Reprod. Dev.* 1988b, 1: 11-17.
2. Allendorf, F.W.; Thorgaard, G.H. Tetraploidy and the evolution of salmonid fishes, in *Evolutionary Biology of Fishes,* Turner B.J., Ed., New York: Plenum; 1984, pp. 1-53.
3. Almuly, R.; Cavari, B.; Ferstman, H.; Kolodny, O.; Funkenstein, B. Genomic structure and sequence of the gilthead seabream (*Sparus aurata*) growth hormone-encoding gene identification of minisatellite polymorphism in intron I. *Genome.* 2000, 43, 836-845.
4. Almuly, R.; Poleg-Danin, Y.; Gorshkov, S.; Gorshkova, G.; Rapoport, B.; Soller, M.; Kashi, Y.; Funkenstein, B. Characterization of the 5′ flanking region of the growth hormone gene of the marine teleost, gilthead sea bream *Sparus aurata*: analysis of a polymorphic microsatellite in the proximal promoter. *Fisheries Science.* 2005, 71, 479-490.
5. Anathy, V.; Venugopal, T.; Koteeswaran, R.; Pandian, T.J.; Mathavan, S. Cloning, sequencing and expression of cDNA encoding growth hormone from Indian catfish (*Heteropneustes fossilis*). *J. Biosci.* 2001, 26 (3): 315-324.
6. Andersen, B.; Rosenfeld, M.G. POU domain factors in the

neuroendocrine system: lessons from developmental biology provide insights into human disease. *Endocr. Rev.* 2001, 22, 2-35.

7. Argenton, F.; Ramoz, N.; Charlet, N.; Bernardini, S.; Colombo, L.; Bortolussi, M. Mechanisms of transcriptional activation of the promoter of the rainbow trout prolactin gene by GHF1/Pit1 and glucocorticoid. *Biochem. Biophys. Res. Commun.* 1996, 224, 57-66.
8. Argenton, F.; Vianello, S.; Bernardini, S.; Jacquemin, P.; Martial, J.; Belayew, A.; Colombo, L; Bortolussi, M. The transcriptional regulation of the growth hormone gene is conserved in vertebrate evolution. *Biochem. Biophys. Res. Commun.* 1993, 192, 1360-1366.
9. Bart, H.L.; Reneau, P.C.M.; Doosey, M.H.; Bell, C.D. Evalutionary divergence of duplicate copies of the Growth Hormone gene in suckers (Actinopterygii: Catostomidae). *Int. J. Mol. Sci.* 2010, 11, 1090-1102.
10. Ber, R.; Daniel, V. Sequence analysis suggests a recent duplication of growth hormone encoding gene in *Tilapia nilotica*. *Gene.* 1993, 125, 143-150.
11. Ber, R.; Daniel, V. Structure and sequence of the growth hormone encoding gene from *Tilapia nilotica*. *Gene.* 1992, 113, 245-250.
12. Brindle, P.K.; Montminy, M.R. The CREB family of transcription activators. *Curr. Opin. Genet. Dev.* 1992, 2, 199-204.
13. Castrillo, J.L.; Bodner, M.; Karin, M. Purification of growth hormone-specific transcription factor GHF-1 containing homeobox. *Science.* 1989, 243, 814-817.
14. Chiou, C.S.; Chen, H.T.; Chang, W.C. The complete nucleotide sequence of the growth-hormone gene from the common carp (*Cyprinus carpio*). *Biochim. Biophys. Acta.* 1990, 1087, 91-94.
15. Clements, M.D.; Bart Jr, H.L.; Hurley, D.L. Isolation and characterization of two distinct growth hormone cDNAs from the tetraploid smallmouth buffalowfish (*Ictiobus bubalus*). *Gen. Comp. Endocrinol.* 2004, 136, 411-418.
16. Cook, J.T.; McNiven, M.A.; Richardson, G.F.; Sutterlin, A.M. Growth rate, body composition and feed digestibility/conversion

of growth-enhanced transgenic Atlantic salmon (*Salmo salar*), Aquaculture. 2000a, 188, 15-32.

17. Cuevas-Rodríguez, B.L.; Sifuentes-Rincón, A.M.; Ambriz-Morales, P.; García-Ulloa.M,; Valdez-González.F.J.; Rodríguez-González.H. Novel single nucleotide polymorphisms in candidate genes for growth in tilapia (*Oreochromis niloticus*). *R. Bras. Zootec,* 2016, 45(6):345-348.
18. Davis, G.H.; McEwan, J.C.; Fennessy, P.F.; Dodds, K.G.; Farquhar, P.A. Evidence for the presence of a major gene influencing ovulation rate on the X chromosome of sheep. *Biology of Reproduction.* 1991, 44, 620-624.
19. De-Santis, C.; Jerry, D.R. Candidate growth genes in finfish-where should we be looking. *Aquaculture.* 2007, 272, 22-38.
20. Devlin R.H., 1997. Transgenic salmonids. In: L.M. Houdebine (ed.) *Transgenic Animals: Generation and Use.* Harwood Academic Publ., Amsterdam, Netherlands; 1997, Pp 105-117
21. Devlin, R.H. Sequence of and the relationship of rainbow trout with atlantic salmon and pacific salmon. *Can. J. Fish. Aquat. Sci.* 1993, 50, 1738-1748.
22. Devlin, R.H.; Biagi, C.A.; Yesaki, T.Y.; Smailus, D.E.; Byatt, J.C. Growth of domesticated transgenic fish. *Nature*. 2001, 409:781-782.
23. Du, S.J.; Devlin, R.H.; Hew, C.L. Genomic structure of growth hormone genes in Chinook salmon (*Oncorhynchus tshawytscha*): presence of two functional genes, GH-I and GH-II and a male specific pseudogene, *DNA Cell Biol.* 1993, 12, 739–751.
24. *Du, S.J.; Gong, Z.; Fletcher, G.L.; Shears, M.A.; King, M.J.; Idler, D.R.; Hew, C.L. Growth Enhancement in Transgenic Atlantic Salmon by the Use of an "All Fish" Chimeric Growth Hormone Gene Construct. Bio/Technology. 1992, 10, 176–81.*
25. Dunham, R.A.; Liu, Z. Transgenic Fish - Where We Are and Where Do We Go?. *The Israeli Journal of Aquaculture – Bamidgeh.* 2006, 58(4), 297-319
26. Dybus, A. Associations between Leu/Val polymorphism of growth hormone gene and milk production traits in black and white cattle. *Archiv Fur Tierzucht-Archives of Animal Breeding.* 2002, 45, 421-428.

27. Elsholtz, H.P.; Albert, V.R.; Treacy, M.N.; Rosenfeld, M.G. A two-base change in a POU factor-binding site switches pituitary-specific to lymphoid-specific gene expression. *Genes Dev.* 1990, 4, 43-51.

28. Forsyth, I.A.; Wallis, M. Growth hormone and prolactin molecular and functional evolution. *J. Molecular Gland Biol. Neopal.* 2002. 7, 21-312.

29. Fu, C.; Hu, W.; Wang, Y.; Zhu. Z. Developments in transgenic fish in the People's Republic of China. *Rev. sci. tech. Off. int. Epiz.* 2005, 24 (1), 299-307

30. Goodman, H.M., 1993. Growth hormones and metabolism. *In*: The Endocrinology of growth, development and metabolism in vertebrates: In Schreibman, M.P.; Scanes, C.C.; Pang, P.K.T., ed.). 1993, Academic Press, San Diego; pp. 93-115.

31. Guyomard, R.; Chourrout, D.; Lerox, C.; Houdebine, L.; Pourrain, F. Integration and germ line transmission of foreign genes microinjected into fertilized trout eggs. *Biochimie.* 1989, **71**: 857–63

32. Hafsa, A.B.; Nabi, N.; Zellama,M.S,; Said, K.; Chaouachi, M. A new specific reference gene based on growth hormone gene (GH1) used for detection and relative quantification of Aquadvantage GM salmon (*Salmo salar L.*) in food products. *Food Chemistry.* 2016, 190, 1040-1045.

33. Hirt, H.; Kimelman, J.; Birnbaum, M.J.; Chen, E.Y.; Seeburg, P.H.; Eberhardt, N.L.; Barta A. The human growth hormone locus: Structure, evolution and allelic variation. *DNA*. 1987, 6, 59-70.

34. Hoj, S.; Fredholm, M.; Larsen, N.J.; Nielsen, V.H. Growth hormone gene polymorphism associated with selection for milk fat production in lines of cattle. *Animal Genetics.* 1993, 24, 91-96.

35. Knorr, C.; Moser, G.; Muller, E.; Geldermann, H. Associations of GH gene variants with performance traits in F2 generations of European wild boar, Pietrain and Meishan pigs. *Animal Genetics.* 1997, 28, 124-128.

36. Kuhnlein, U.; Ni, L.; Weigend, S.; Gavora, J.S.; Fairfull, W.; Zadworny, D. DNA polymorphisms in the chicken growth hormone gene: Response to selection for disease resistance and association

with egg production. *Animal Genetics.* 1997, 28, 116-123.

37. Lefevre, C.; Imagawa, M.; Dana, S.; Grindlay, J.; Bodner, M.; Karin, M. Tissue-specific expression of the human growth hormone gene is conferred in part by the binding of a specific *trans-acting* factor. *EMBO J.* 1987, 6, 971-981.
38. Lemaire, C.; Writ, S.; Panyin, S. Giant catfish (*Pangasianodon gigas*) growth hormone-encoding cDNA cloning and sequencing by one sided polymerase chain reaction. *Gene.* 1994, 49, 271-276.
39. Lupchinski Jr., E.; Vargas, L.; Lopera-Barrero, N. M.; Ribeiro, R. P.; Povh, J. A.; Gasparino, E.; Braccini, G. L. Genetic characterization of three Nile tilapia (*Oreochromis niloticus*) strains. *Archivos de Zootecnia.* 2011, 60, 985-995.
40. Martinez, R.; Estrada, M.P.; Berlanga, J.; Guillin, I.; Hernandez, O.; Cabrera, E.; Pimentel, R.; Morales, R.; Herrera, F.; Morales, A.; Pina, J.; Abad, Z.; Sanchez, V.; Melamed, P.; Lleonart, R.; de la Fuente, J. Growth enhancement of transgenic tilapia by ectopic expression of tilapia growth hormone. *Molecular Marine Biology and Biotechnology.* 1996, 5, 62-70
41. McLean, E. and Donaldson, E.M. The role of somatotropin in growth in poikilotherms. The Endocrinology of Growth, Development and Metabolism in Vertebrates. In Schreibman, M.P., Scanes, C.G., and Pang, P.K.T. (eds.), Academic Press, New York, NY, USA; pp. 43-71.
42. Megarajan, S.; Singh, S.D.; Gupta, S. Cloning and characterization of Pangasianodon hypophthalmus growth hormone gene and its heterocogous expression. *Applied Biochemistry and Biotechnology*, 2014, 173, 1446- 1468.
43. Morgan, J.G.; Courtois, G.; Fourel, G.; Chodosh, L.A.; Campbell, L.; Evans, E.; Crabtree, G.R. Sp1, a CAAT-binding factor, and the adenovirus major late promoter transcription factor interact with functional regions of the gamma-fibronogen promoter. *Mol. Cell. Biol.* 1988, 8, 2628-2637.
44. Moriyama, O.M.; Takahash, A.; Sower, S.A.; Kawauchi, H. Genomic structure of the sea lamprey growth hormone encoding gene. *General and comparatuive endocrinology.* 2006, 148, 33-40.

45. Nam, Y.K.; Noh, J.K.; Cho Y.S.; Cho, H.J.; Cho, K.-N.; Kim, C.G.; Kim, D.S. Dramatically accelerated growth and extraordinary gigantism of transgenic mud loach *Misgurnus mizolepis*. *Transgenic Res*. 2001, 10, 353-362.

46. Na-Nakorn, U.; Moeikum, T. Genetic diversity of domesticated stocks of striped catfish, *Pangasianodon hypophthalmus* (Sauvage, 1878), in Thailand: Relevance to broodstock management regimes. *Aquaculture*. 2009, 297, 70-77.

47. Ohkubo, T.; Araki, M.; Tanaka, M.; Sudo, S.; Nakashima, K., 1996. Molecular cloning and characterization of the yellowtail GH gene and its promoter: A consensus sequence for teleost and avian Pit-1/GHF-1 binding sites. *Journal of Molecular Endocrinology*. 1996, 16, 63-72.

48. Olson, T.A.; Lucena, C.; Chase, C.C.Jr.; Hammond, A.C. Evidence of a major gene influencing hair length and heat tolerance in *Bos taurus* cattle. *Journal of Animal Science. 2003,* 81, 80-90.

49. Pankova, *M.V.;* Kukhlevsky*ab*, A.D.; Brykov V. A. Fish Growth Hormone Genes: Divergence of Coding Sequences in Salmonid Fishes. *Genetika*. 2017, 53, 201-213

50. Patikoglou, G.A.; Kim, C.; Kim, W.B.; Kim, Y.K.; Lee, S.Y; Yang, J.M. Efficiency of promoter and cell line in high-level expression of erythropoietin. *Biotechnol. Appl. Biochemi.* 2002, 32, 167-172.

51. Penman, D.; Varadi, L.; Maclean, N. Growth performance of transgenic tilapia containing an exogenous piscine growth hormone gene. *Journal of Fish Biology*. 2001, *59*, 62-78.

52. Penman, D.J.; Beeching, A.J.; Penn, S.; Maclean, N. Factors affecting survival and integration following microinjection of novel DNA into rainbow trout eggs. *Aquaculture,* 1990, **85**, 35-50.

53. Powers, D. A Fish as model systems. *Science*. 1989, 246, 352-358.

54. Rahman, M.A.; Maclean, N. Growth performance of transgenic tilapia containing an exogenous piscine growth hormone gene. *Aquaculture,* 1999, 173, 333-346.

55. Rahman, M.A.; Ronyai, A.; Engidaw, B.Z.; Jauncey, K.; Hwang, G.; Smith, A.; Roderick, E.; Penman, D.; Varadi, L.; Maclean, N. Growth performance of transgenic tilapia containing an exogenous piscine growth hormone gene. *Journal of Fish Biology*. 2001, 59,

62-78.

56. Rajesh, R.; Majumdar, K.C. The growth hormone encoding gene isolated and characterized from *Labeo rohita* (Hamilton) is expressed in CHO cells under the control of constitutive promoters in autotransgene constructs. *Fish Physiol Biochem.* 2008, 34 (4), 13-36.
57. Robinson-Rechavi, M.; Laudet, V. Evolutionary rates of duplicate genes in fish and mammals, *Mol. Biol. Evol.* 2001, 18 (4), 681–683.
58. Rothschild, M..; Jacobson, C.; Vaske, D.; Tuggle, C.; Wang, L.; Short, T.;Eckardt, G.; Sasaki, S.; Vincent, A.; McLaren, D.; Southwood, O.; Van der Steen, H.; Mileham, A; Plastow, G., 1996. The estrogen receptor locus is associated with a major gene influencing litter size in pigs. *Proceedings of the National Academy of Sciences of the United States of America.* 1996, 93, 201-205.
59. Sawayama, E.; Takagi, M. Parental contribution and growth hormone gene polymorphism associated with growth phenotypes of red sea bream Pagrus major in mass production: A case study. *Aquaculture Reports.* 2015, 2, 144–151.
60. Von Schalburg, K.R.; Yazawa, R.; de Boer, J. Isolation, characterization and comparison of Atlantic and Chinook salmon growth hormone 1 and 2. *BMC Genomics.* 2008, 9. p223.
61. Tambasco, D.D.; Paz, C.C.P.; Tambasco-Studart, M.; Pereira, A.P.; Alencar, M.M.; Freitas, A.R.; Coutinho, L.L.; Packer, I.U.; Regitano, L.C.A. Candidate genes for growth traits in beef cattle crosses *Bos taurus*×*Bos indicus*. *Journal of Animal Breeding and Genetics.* 2003, 120, 51-56.
62. Tanaka, M.; Toma, Y.; Ohkubo, T.; Sudo, S.; Nakashima, K. Sequence of the flounder (*Paralichthys olivaceus*) growth hormone encoding gene and its promoter region. *Gene.* 1995, 165, 321-322.
63. Tang, Y.; Lim, C.M.; Chen, T.T.; Kawauchi, H.; Dunham, R.A.; Powers, D.A. Structure of the channel catfish (*Ictalurus punctatus*) growth hormone gene and its evolutionary implications. *Mol. Mar. Biol. Biotechnol.* 1993, 4, 198-206.
64. Tian, C.; Yang, M.; Lv, L.; Yuan, Y.; Liang, X.; Guo, W.; Song, Y.; Zhao, C. Single Nucleotide Polymorphisms in *Growth Hormone* Gene and Their Association with Growth Traits in *Siniperca chuatsi*

(Basilewsky). *Int. J. Mol. Sci.* 2014, 15, 7029-7036.

65. Venktesh, B.; Brenner, S. Genomic structure and sequence of the pufferfish (*Fugu rubripes*) growth hormone-encoding gene: A comparative analysis of teleost growth hormone gene. *Gene.* 1997, 187, 211-215.
66. Venugopal, T.; Mathavan, S.; Pandian, T.J. Molecular cloning of growth hormone encoding cDNA of Indian major carps by a modified rapid amplification of cDNA ends strategy. *J. Biosci.* 2002, 27, 261-272.
67. Von Schalburg, K.R.; Yazawa, R.; de Boer, J. Isolation, characterization and comparison of Atlantic and Chinook salmon growth hormone 1 and 2. *BMC Genomics.* 2008, 9. p223.
68. Wu, T.; Yang, H.; Dong, Z.; Xia, D.; Shi, Y.; Ji, X.; Shen, Y.; Sun, W. The integration and expression of human growth gene in blunt snout bream and common carp. *Journal Fisheries China Shuichan Xuebao.* 1994, *18*, 284-289.
69. Yamada, S.; Hata, J.; Yamashita, S. Molecular cloning of fish Pit-1 cDNA and its functional binding to promoter of gene expressed in the pituitary. *J. Biol. Chem.* 1989, 268, 24361-24366.
70. Ye, X.; Tian, Y.Y.; Gao, F.Y. Progress in transgenic fish techniques and application. 2011. 33(5):494-503
71. Yowe, D.L.; Epping, R.J. Cloning of the barramundi growth hormone-encoding gene: A comparative analysis of higher and lower vertebrate GH genes. *Gene.* 1995, 162, 255-259.
72. Zhang J. Evolution by gene duplication: an update. *Trends in Ecology & Evolution*. 2003, 18 (6), 292–8.
73. Zhang, D.; Shao, Y.; Jiang, S.; Li, J.; Xu, X. *Nibea coibor* growth hormone gene: Its phylogenetic significance, microsatellite variation and expression analysis. *General and Comparative Endocrinology.* 2009, 163, 233-241.
74. Zhu, Z. Generation of fast growing transgenic fish: methods and mechanisms, *Transgenic Fish*, World Scientific Publishing, Singapore. In Hew, C.L. and Fletcher, G.L., (eds)., 1992 . pp. 92-119.

75. Zhu, Z.; Xu, K.; Li, G.; Xie, Y.; He, L. Biological effects of human growth hormone gene microinjected into the fertilized eggs of loach, *Misgurnus anguillicaudatus*. *Kexue Tongbao Academia Sinica.* 1986, *31*, 988-990.

CHAPTER
3

NUTRITION AND FISH GROWTH

D. Linga Prabu[1], Sanal Ebeneezar[2], S. Chandrasekar[3];

[1]D. Linga Prabu With Masters in Fish Nutrition and Biochemistry, he joined as Agricultural Research Service Scientist in Marine Biotechnology Division of ICAR-Central Marine Fisheries Research Institute, Kochi, India in 2014. Before joining into ARS, Mr Prabhu worked with Growel Feeds Pvt Ltd., Andhra Pradesh, India, a leading fish and shrimp feed producer, as Assistant Manager- Fish Nutrition and Formulation for a period of two years since 2012 and gained rich experience in commercial aspects of aquafeed formulation and manufacture. Currently he is working in marine food fish nutrition with nutrigenomic approach, and also involving in the development of cost effective best performance feed for marine food fishes. Mr Prabhu has great interest in aquatic disease management in mariculture species through nutraceuticals and its effects on immune development.

[2]Mr Sanal Ebeneezar joined as Scientist in the division of Marine Biotechnology at ICAR-Central Marine Fisheries Research Institute, Kochi, India. He has completing Masters in Fisheries Science (MFSc) from Central Institute of Fisheries Education, Mumbai, India and qualified the Agricultural Research Service Examination (ARS). Mr Ebeneezar has been working in the area of Fish Nutrition and Biotechnology for the last two years. Currently he is working on the aspects of enhancement of colour and growth in ornamental fish and also on the nutrient requirement studies of marine fin fish.

[3]Mr S Chandrasekar is Masters in Fish Physiology and Biochemistry from Central Institute of Fisheries Education, Mumbai, India. He joined as Agricultural Research Service Scientist in Marine Biotechnology Division of ICAR-Central Marine Fisheries Research Institute, Kochi, India in 2014. Currently he works in marine food fish nutrition mainly focusing on fishmeal replacement and immunomodulation. He is also working on the nutrition of marine ornamental fish with the aim of colour enhancement.

INTRODUCTION

Growth is a parameter of obvious importance in any culture activity. Fish growth involves accretion of muscle, through deposition of protein, fat and other bio-molecules. Better understanding of the growth processes results in significant benefits in terms of productivity, sustainability and profitability for aquaculture operations. The growth in fish follows S-shaped or sigmoid curve (Quinn and Deriso, 1999). During the initial phase, fish growth is slower due to the smaller size and less consumption of feed. Hence, this phase is known as preparatory phase (Lovell, 1998). With the progression of time, fish starts to consume more feed and grow bigger and faster. After attaining certain growth, the rate of increment of growth begins to slow down as the total weight of fish reaches a maximum point and thus the growth occurs at a diminishing rate. Although fish growth will get slow down at the later stage of life, but it never stops growing. The growth rate is slower in older fish and proportional increase in size is greatest in younger fish (Bureau et al., 2000). Fish feed containing different nutrients is one of the important factors that helps for growth improvement in fishes. However, some fish can remain without feed for a long time even months together and lose considerably amount of weight and when conditions become favorable it can able to resume their growth (Weatherley and Gill, 1987).

Growth requires dietary input of required nutrients and certain nutrients especially minerals and some vitamins are directly extracted from water and used as building blocks for accretion of body components (Bureau et al., 2000; Brett, 1979). Fish needs a balanced diet with all the essential nutrients in adequate quantity in appropriate form for augmenting the growth. However, the requirement of these nutrients varies with species to species, at different phase of life in the same species due to variation of metabolic rate and physiological status of fish.

The availability of essential nutrients varies significantly in the natural feeds of fish due to their surrounding environment. Some nutrients are available from water, which needs to be considered critically while formulating feed for fish. Finfish require around 40 nutrients for optimal growth and well-being. The dietary proteins, lipids and carbohydrates are considered as macro nutrients and vitamins and minerals are considered as micronutrients which are required for the normal growth of fish under favorable conditions. Therefore, it is necessary to develop a best cost effective diet with balanced nutrients to maximize the growth and minimize environmental pollution based on the knowledge of nutritional requirements of species being cultured and also appropriate feeding practices (NRC, 2011). The present chapter deals with nutrients utilization and fish growth for understanding the aquaculture activities efficiently.

THE FISH FEED

The wild fish fully depends on the natural environment where it lives for its feed. As the fish is cultured in ponds or cages it becomes less dependent on natural feed available in the pond or cage but more and more dependent on the prepared feed offered to them and hence the nutritionally complete feed is most warranted (Lovell, 1998). The compounded diets predominantly consumed by herbivores and omnivores, fishes particularly carps which accounts more than 42% of total volume produced but the diet with high fish meal content is mainly consumed by salmonids and shrimps (NRC, 2011). The nutrient requirement of fish for optimum growth depends on the stages of life and the environment where it lives. While compared with warm blooded animals the energy requirement of fish is low and therefore it needs high protein to energy ratio in the diet. Fish can absorb most of the minerals and some of the vitamins from the water and hence, these nutrients may be minimized while formulating the diet. Since fish feed is offered in water; if the feed is not consumed within a reasonable time the fish may not get the offered amount of nutrients from the feed due to leaching of feed. For a particular species, the nutrient requirement should be known precisely to design a formulated feed which should offer optimum growth in a cost-effective manner (NRC, 2011). Therefore, nutrient requirement study will be an unending process whenever a new species introduced into aquaculture until completely knowing its nutrient requirements.

ENERGY

Dietary lipid, protein and carbohydrate are source of energy in fishes (Popma, 1982). Energy is released during metabolic oxidation of the nutrients such as carbohydrates, fats and protein. The energy requirement varies with different stages of life. The energy requirements of fish can be quantified by measuring heat production or oxygen consumption (NRC, 1993). Young fish needs less energy and high protein whereas adult fish requires high energy low protein in the diet (Ali, 2002). One gram of carbohydrate, protein and lipid are equal to 17.2, 23.6, and 39.5 kJ/g of gross energy, respectively (Blaxter, 1989). Dietary excess or deficiency of energy may reduce the growth of the fish. In case of excess energy in feed, the fish may not consume once its energy requirement is satisfied which may curb the fish from consuming sufficient feed to satisfy its other nutrient requirements (Chou and Shiau, 1996) and hence affect the growth. High energy content in the diet may also adversely affect the body composition of fish due to excess deposit of fat which may adversely affect the slaughter yield and keeping quality of fish (Macgoogan and Reigh, 1996). Similarly, lower energy content in the diet affect the protein accretion in fish because protein will be used for energy requirement of fish (NRC, 1993).

Generally, commercial fish feeds that contain optimum protein and fat content do not show deficiency of energy. It may supply about 3-4 kcal digestible energy (DE) per gram of feed which is adequate for energy requirement of most of the fishes. According to Cowey and Cho (1993) the energy requirement increases with increase in fish size and found 12 KJ g^{-1} weight gain in small rainbow trout where as it is increased to 20 KJ g^{-1} weight gain for larger trout. In case of Atlantic salmon, *Salmo salar*, DP/DE ratio was about 19 g MJ^{-1} for 1 to 2.5 kg size group and 16–17 g MJ^{-1} for 2.5 to 5 kg size groups (Einen and Roem, 1997). The fish of 300-400 g size require about 270-320 KJ/kg/day of DE for maximum growth (Ali, 2002). The reduction in DP/DE ratios proves to be enormously efficient in improving protein utilization and decreasing nitrogenous loses in fishes (Kaushik and Cowey, 1991). Alike DP/DE, the DE/DP value increases with increase in size of the fish. The DE/DP ratio required for maximum weight gain in several fishes was reported in the range of 9 to 11 kcal/g (Lovell et al., 1994). It was in the range of 8.7-11.6 kcal/g for different stages of channel catfish with the least value during earlier stage of life and higher value during latter phase of life (Prather and Lovell, 1973; Garling and Wilson, 1977; Mangalik, 1986) and 8.3 kcal/g for common carp (Takeuchi et al., 1979).

Fishes evolve from an aqueous environment where the carbohydrate sources are in paucity, the digestive and metabolic systems of most of the fishes are well adopted to utilize dietary lipid and protein as energy source rather than carbohydrates (Lovell, 1998). Tropical herbivorous and omnivorous fishes may better utilize dietary carbohydrates than other fishes. Therefore, in fishes, lipid is considered as a major source of energy. Fish utilize protein efficiently as energy source because fish can metabolize higher percentage of energy from protein consumed and the heat increment for protein ingested is lower and also lower energy requirement for nitrogen excretion in fishes than land animals. These freak characters make the fishes to readily utilize protein than carbohydrates. It was also reported that if the fishes were fed with excess of dietary protein than the non-protein energy, it may repress their growth (Hossain et al., 1998). In addition, supplying excessive level of protein is more expensive. Therefore, balancing the non-protein energy sources such as carbohydrates and lipid may have protein sparing action if the protein is available as optimum or slightly suboptimum level but should be balanced with amino acid composition to ensure adequate growth.

Fish possesses superior feed conversion efficiency than the land animals due to their efficiency to assimilate feed with higher percentage of protein because of their lower energy requirement (Bowen, 1987). It also has lower energy cost of protein gain than any terrestrial animals and hence have more protein gain per calories of energy consumed i.e. less amount of energy is required for protein synthesis (Lovell, 1998). Fish also have 10-20% less energy requirement than other terrestrial animals because it doesn't require maintain body temperature (poikilotherms), need very less energy to maintain its position in water column and being ammonotelic need less energy for nitrogenous waste excretion as ammonia through gills (Brett and Groves, 1979; Trotter et al., 2001). Basal metabolic energy of fish is just 0.01-0.07 $MJ/kg^{0.75}$ which is very high in other animals like chicks 0.36 $MJ/kg^{0.75}$, cow 0.32 $MJ/kg^{0.75}$ and human 0.29 $MJ/kg^{0.75}$ (Ali, 2002). In fishes, during the somatic growth phase of life around 40% of energy may be diverted for growth but after maturation hardly 10% of energy will be allotted for growth (Vinberg,

1960). Therefore, fishes cultured for consumption should be harvested before attaining sexual maturation. When the energy supplied from the feed is not sufficient, the fishes mobilize their body reserved fat for maintenance.

ROLE OF PROTEIN AND AMINO ACIDS ON FISH GROWTH

Generally, the nutrient requirement of fish depends on their natural feeding habits such as herbivorous, omnivorous or carnivorous. Dietary protein is the key factor that affects the growth performance of fish and feed cost as well (Lovell, 1989; Lee et al., 2000; Luo et al., 2004). Among the three major nutrients, protein comprises 65–75% of the dry matter in fish tissues, and is therefore protein more indispensable nutrient in fish nutrition (Wilson, 2002). As discussed earlier, fish tends to utilize more protein as an energy source over lipid or carbohydrate in spite of their feeding habits (Walton and Cowey, 1982). Protein requirement of any fish is higher than the terrestrial animals. But fish can utilize the protein for growth more efficiently than the other animals (Lovell, 1998). In most of the fin fishes, protein requirement for maintenance is 1.5-2 g/kg/day and for maximum growth is 7-11 g/kg/day (Kaushik et al., 1981; Ali, 2002). Carnivorous fish require relatively higher protein diet than herbivorous and omnivorous fish species (NRC, 1993). Due to efficient production of protease and lipase enzymes, carnivorous fish species are efficient in using dietary protein and lipid for energy rather dietary carbohydrates (Gatlin, 2010). Dietary protein in excess quantity is wasteful and stresses the fish (Hossain et al., 1998).

Protein being the most expensive nutrient of fish feeds (Wilson, 2002), umpteen numbers of studies have been conducted to elucidate the optimum dietary protein requirement for several fish species to develop more cost-effective, nutritionally balanced practical diets. The dietary protein requirements of fish are influenced by various factors such as fish species and size, the quality and bioavailability of the protein source, and the amount of non-protein energy present in the diet (Bowyer et al., 2013). The protein requirement of fin fish species varies between 25-55% as per its nature of feeding (herbivorous, omnivorous and carnivorous), stage of life such as fry, fingerling, grow-out and adult fish and environmental parameters (NRC, 2011). In case of salmonids the optimal dietary protein requirement is 45-50%, 40% and 35% respectively in fry, juveniles and yearlings (Hilton and Slinger, 1981; NRC, 1981). The protein requirement may vary according to the changes in water temperature in most of the fishes. Millikin (1982), reported that striped bass require 47% protein while reared at 20 °C and 55% protein at while the temperature is elevated to 24 °C. According to Prabu et al. (2016a), the crude protein requirement of juveniles of snubnose pompano, *Trachinotus blochii* was assessed based on weight gain percentage, specific growth rate, feed conversion ratio and protein conversion efficiency and was 40%. Chou et al. (2001) conducted an experiment in *Rachycentron canadum* with seven diets of different level of protein ranging from 36-60% crude protein and found that 44.5% seems to be optimal for maximum growth. Therefore, the diet should be prepared with sufficient quantity of non-protein energy nutrients, especially dietary lipids that may have protein sparing action and thus reduce the unwanted protein utilization as energy that enhance economic returns and reduce water pollution (Mohanta et al., 2013). Generally, insufficient level of dietary protein cause reduction or retardation of growth and loss of body weight and emaciation due to elimination of protein from less vital organs particularly fish muscle in order to maintain the functions of more vital organs. Excess dietary protein leads to extra energy costs and increased nitrogenous excretions (Monentcham et al., 2009).

Protein is composed of several amino acids. Protein as such may not be required for fish growth but a mixture of indispensable and dispensable amino acids in a balanced manner are required to synthesize new proteins or maintain existing proteins in tissues (Wilson and Halver, 1986). Like other animals, fish can synthesize their body proteins from amino acids in the diet or from amino acid pool in their body. Among the 20 amino acids, at least 10 amino acids such as arginine, histidine, isoleucine, leucine, lysine, methionine, phenylalanine, threonine, tryptophan and valine may not be synthesized or inadequately synthesized de novo by fishes and must be supplemented through the diet; they are known as indispensable amino acids (Li et al., 2008). The other 10 amino acids are dispensable, can be synthesized from the body and are also available plenty in the diet. Both the essential and non-essential amino acids are vital for muscle formation and enzymatic function and also offer energy for body maintenance (Yang et al., 2002). Several studies in aquatic animals show that certain amino acids which can regulate key metabolic pathways that are essential for growth, reproduction, maintenance and immune responses are known as 'functional amino acids'. If the amino acid composition of diet matches the amino acid composition of fish, it augments the growth (Li et al., 2008). The requirement of amino acid varies in different species at different stages. Further, the amino acid requirement estimation by several authors in different species showed apparently high degree of variation (Kaushik and Seiliez, 2010). The underlying reasons for such variations are depends on the methodology used in the study (Cowey, 1995), the composition and type of feed used (purified or practical diet), criterion used for the estimation of requirement, the statistical method used to analyze the data obtained and the mode of expression of data either relative to dietary dry matter or dietary DP or DE level, or in absolute terms per unit metabolic mass per day (Kaushik and Seiliez, 2010). The main objective of supplementation of dietary amino acid is for increasing whole body protein accretion and therefore, Ogino (1980) used the calculation of daily amino acid increment for estimating the essential amino acid requirements of fish. However, the whole-body protein bound amino acid profile of different species and the essential amino acids deposited during the growth phase is apparently similar. Therefore, it is clear that there is possibly more apparent variability in amino acid requirement profile than in the whole body amino acid profile of fish (Akiyama et al., 1997). The rainbow trout fed with the diets containing different ratio of essential/non-essential amino acids showed a ratio of 57:43 was found to be the most appropriate based on several criteria on protein utilization (Green et al., 2002). In gilthead sea bream, a ratio of 1.1 was found to be better than a ratio of 0.8 essential to non-essential amino acids in the diet (Gomez-Requeni et al., 2003).

In fish, the muscle protein accounts nearly 50% of the whole-body protein but muscle protein synthesis rates represent only 20% of the whole-body protein synthesis (Kaushik and Seiliez, 2010). Besides, their roles as building blocks in protein synthesis, amino acids also have several other roles in metabolism of fish. Amino acids are part of coenzymes, precursor for synthesis of structural compounds (chitin, purine etc.), neurotransmitters, hormones and metabolic intermediates which are essential for a variety of functions in an organism (NRC, 2011). According to Cowey and Walton (1989) amino acid accretion in the body is about 25-55% of the total amino acid consumed. Amino acid imbalances in the diet of fish during larval stage significantly cause higher mortalities and skeletal deformities (Pozios et al., 2001; Felip et al., 2012) which also have a negative impact on fish growth. Dietary excess of amino acids leads to inevitable and referential catabolism of these amino acids (NRC, 2011). Alanine and aspartate are serving as a major glucogenic precursors and important energy substrates in fish. Aspartate is essential for synthesis of purine nucleotides in all cells. Alanine is a favored carrier of nitrogen for inter-organ amino acid metabolism in fish (Mommsenet al., 1980) and stimulates feeding response in some fishes (Shamushaki et al., 2007). According to Wu and Morris (1998), arginine is a multipurpose amino acid and serves as a precursor for the synthesis of protein, nitric oxide, urea, polyamines, proline, glutamate and creatine in land animals. Arginine is an indicator of acute stress and its level in plasma was found to decrease during stress in Senegalese sole (Aragao et al., 2008).

Leucine is considered as a functional amino acid in mammals which stimulates muscle protein synthesis and inhibits proteolysis (Nakashima et al., 2007). In rainbow trout, excessive dietary leucine increased feed efficiency but depressed feed intake, growth, and protein deposition (Choo et al.,1991). Glutamine is the most profuse free amino acid in fish plasma and muscle which is essential for the synthesis of purine and pyrimidine nucleotides and plays crucial role in regulating the acid–base balance in the body (Li et al., 2008). It also stimulates muscle protein synthesis in mammals (Wu et al., 2007) and regulates cytokine and Nitrous Oxide (NO) production there by augmenting the immune response in fish (Buentello and Gatlin, 1999). Lin and Zhou (2006) reveal that glutamine inclusion in the diets of Jian carp elicits weight gain, feed intake, intestinal development and digestive enzyme activities. Glutamate and its decarboxylation product gamma amino butyrate (GABA) are neurotransmitters present in brain and GABA is an inducer of metamorphosis in gastropods (Morse et al., 1979).

Glycine and *serine* are involved in gluconeogenesis, sulfur containing amino acid metabolism, fat digestion and stimulate feed intake in many fishes (Shamushaki et al., 2007). In rainbow trout, dietary inclusion of glycine has been shown to increase hepatic thyroxine 5'monodeiodinase activity and enhance the efficiency of nutrient absorption (Riley et al.,

1996). According to Takeuchi (2007), glycine has a key role in the osmoregulatory responses and enhances the survival of oysters after transfer to freshwater from seawater. Histidine involves in one-carbon unit metabolism and thus affecting DNA and protein synthesis and also serves as an energy fuel during starvation (Li et al., 2008). Dietary supplementation of histidine enhances flavor of aqua cultured seafood (Ogata, 2002).

Lysine is available in higher quantity in marine source ingredients and its availability is poor in plant based ingredients. Therefore, lysine will be the first limiting amino acid in the fish meal replaced with plant protein based diets (NRC, 2011). Inclusion of lysine in plant protein-based diets significantly reduces the dietary crude protein requirement without affecting fish growth (Mai et al., 2006). Lysine deficiency not only reduces protein deposition but also cause dorsal and caudal fin erosion in some fishes (Guillaume et al., 1999). Phenylalanine and tyrosine jointly enhance pigmentation, feed intake, growth performance, immunity, survival of fish and metamorphosis (Chang et al., 2007; Pinto et al., 2008). Hydroxyproline level was positively correlated to the growth rate in salmon (Sunde et al., 2001) and dietary supplementation of hydroxyproline increased growth rate and modified bone composition (Aksnes et al., 2008). Hydroyproline incorporation in the plant protein based fish feeds maximize survival and growth. Plant protein based diets are generally deficient in methionine level (Mai et al., 2006). Taurine plays an important role in fat digestion, antioxidative defense, cellular osmoregulation and development of neural and muscular systems (Fang et al., 2002). Taurine is abundantly available in fish meal and animal source protein but completely deficient in plant source protein. Supplementation of taurine in all-plant protein diets, promote growth of carnivorous fish through efficient feed utilization (Gaylord et al., 2007) and promote intestinal development of cobia larvae (Salze et al., 2008). Dietary supplementation of L-tryptophan curbs the aggressive behavior of rainbow trout juveniles and thus reduces cannibalism (Hseu et al., 2003). It also averts the surge of stress-induced cortisol which otherwise negatively affects growth, feed intake, protein accretion, immunity and disease challenge (Lepage et al., 2003; Tejpal et al., 2009).

LIPIDS AND FATTY ACIDS

Dietary lipids play a key role as a source of energy and essential fatty acids necessary for growth of fish. Dietary lipid is a source of essential fatty acids, phospholipids, sterols, fat-soluble vitamins, steroid hormones and prostaglandins that is required for normal physiological processes and maintenance of cell membranes and its function as well (Ghanawi et al., 2011). Dietary lipid is a principal source of metabolic energy in fish because of high digestibility (Ellis and Reigh, 1991). Now-a-days, lipids are incorporated in the formulated feeds to take advantage of their protein sparing effect to reduce the feed cost (De Silva and Anderson, 1995). The protein sparing action of lipids suggested that it may limit the ammonia production in several fish species (McGoogan and Gatlin, 1999). Inclusion of lipid as a non-protein energy source rather than carbohydrate is most preferred for increasing energy level in the feed because lipid is an energy-dense nutrient and is readily metabolized by most of the fishes, especially carnivorous fish (Shimeno et al., 1996).

Lipid is generally digested by several lipases that have been recognized in the digestive tract of juvenile and adult fishes. In several species, pancreatic lipases are the major enzymes for lipid digestion and are activated by colipase in the presence of bile salts. This enzyme specifically digests triglycerides (Léger et al., 1977; Divakaran and Ostrowsky, 1990). Non-specific bile salt activated lipase (BAL) is another lipase that plays an important role in digestion of neutral lipids in several fish species (Izquierdo et al., 2000). This enzyme catalyzes the hydrolysis of carboxyl ester bonds of acyl glycerols, cholesterol esters and vitamin esters (Wang and Hartsuck, 1993). Phospholipase A2 found in several fish species, involves in the the hydrolysis of the fatty acid ester bond at the *sn*-2 position of phosphoacylglycerides from dietary and biliary origin and yields free fatty acids and lysophospholipids (Izquierdo and Henderson, 1998). This enzyme is Ca^{2+} dependent and secreted as a zymogen in the intestinal tract and activated by tryptic proteolysis. After digestion, the fat is taken to the epithelial cells of the intestine by diffusion of a micellar form of monoglycerides and fatty acids. After which reacilation is takes place in two ways either monoglyceride pathway or L-α-glycerophosphate pathway (Sire et al., 1981). The former one occurs in the smooth endoplasmic reticulum that synthesizes triglycerides and the later pathway takes place in both rough and smooth endoplasmic reticulum which produces triglycerides and phospholipids. Finally the absorbed lipids will be discharged as very low density liopoproteins or chylomicron-like particles into the submucosa (Léger et al., 1988).

Generally, plant source oils with mostly saturated or low level of unsaturated fatty acid profile, are added as a source of lipid which does not apparently influence the digestibility or utilization of these lipids as an energy sources for cold water or warm water fish (Steffens, 1989). Carnivores like trout have natural diets rich in triglycerides and can adapt to high fat diets even up to 35%. The lipid requirement for most of the freshwater fishes is well below 15 % for optimum growth without producing an excessively fatty carcass (Cowey and Sargent, 1979) but the dietary lipid levels of up to 35 % has been incorporated in the diets of carnivorous fishes particularly some salmon and trout primarily with the aim of protein sparing (New, 1996). Generally, temperate fishes need more dietary lipid than the tropical or warm water fishes (Bowyer et al., 2013). In cobia, Chou et al. (2001) conducted an experiment with graded level of dietary lipid ranging from 3 to 18% and found that lipid requirement is 5.76% beyond this point there was no significant increment of growth was observed. Prabu et al. (2016b) revealed the dietary lipid requirement of snubnose pompano is in the range of 6-6.38% based on the polynomial regression of weight gain and specific growth rate. Besides this, other beneficial effects of the high-energy diets over low-energy diets such as better growth, feed utilization, nitrogen retention and reduced nutrient discharges (Hillestad and Johnsen, 1994). According to Lee and Sinnhuber (1972) if the diet is enriched with ample amounts of choline, methionine and tocopherol, the fish can utilize 20-30% of dietary lipid and supports protein sparing action.

As discussed formerly, excess supply of dietary lipid than the requirement may limit feed consumption, thereby curbs the intake of the adequate quantity of protein and other essential nutrients required for growth and also enhance body fat deposition (Ghanawi et al., 2011). Increased dietary lipid level enhances the weight gain in carnivorous fishes such as salmon and trout (Arzel et al., 1994; Hemre and Sandnes, 1999). Though increased lipid level in the diets of some of the fishes enhances its utilization, generally excess dietary lipid may cause extreme metabolic burden to the liver and thereby impede the growth (Gonzalez-Felix et al., 2015). Excessive deposition of lipid in the flesh of fish due to dietary excess lipid content has impact on carcass quality particularly oily texture which leads to low consumer preference (Bell et al., 1997) and also earlier sexual maturation at least in males (Shearer and Swanson, 2000). Therefore, it is crucial to determine the optimal dietary balance between protein and lipid for accomplishing maximum growth with least cost (Sankian et al., 2017). But some fish species such as white seabream, *Diplodus sargus* (Ozorio et al., 2006) and tiger puffer, *Takifugu rubripes* (Kikuchi et al., 2009) did not show obvious protein-sparing effect.

Lipid level is a deciding factor of palatability in the diet of a fish (Boonyaratpalin, 1991). Therefore, it is necessary to offer adequate quantity of lipid in diet for better utilization of feed but excess quantity causes problems connected with feed manufacturing besides producing fatty fish (Cowey and Sargent, 1977). During the early phase of life (fry and fingerlings stage), fish requires more dietary lipid because of its faster growth than the later stage i.e. adults. After digestion, the dietary lipid is breakdown into fatty acids. The fatty acids can be classified as essential fatty acids and non-essential fatty acids. The essential fatty acids such as α-linolenic acid (LNA, 18:3ω3), linoleic acid (LA, 18:2 ω6), eicosapentaenoic acid (EPA, 20:5ω3) and docosahexaenoic acid (DHA, 22:6ω3) cannot be synthesized by the body, therefore need to be supplemented in the diet.

The specific role of *DHA* is maintaining the structural and functional integrity of cell membranes, especially in the neural cells. The presence of DHA in the fish brain and eye as a major component suggests its importance during the early developmental stages of fish larvae (Bell et al., 1991). The arachidonic acid (ARA, 20:4 ω6) is considered as a precursor of several eicosanoids which have key role during the process of ovulation (Sargent et al., 2002). The ARA is preferentially incorporated into phosphatidylinositol that plays a crucial role in signal transduction in the cell membrane of fish (Geurdenet al., 1997). The physiological role of EPA is to modulate the formation of eicosanoids from ARA. Thus, sufficient quantities of EPA and ARA are required for better production of eicosanoids, which in turn produce prostaglandins and leukotrienes (Watanabe, 1993).

Fish cannot synthesize either 18:2 (ω6) or 18:3 (ω3) *de novo*. Therefore, these fatty acids have to be provided through the diet, depending upon the level of requirement. Freshwater fish requires either linoleic acid or linolenic acid or both, whereas marine fish requires EPA and DHA (NRC, 1993). The reason is freshwater fish are capable of elongating and desaturating 18:2 (ω6) to 20:4 (ω6) and 18:3(ω3) to 20:5 (ω3) and 22:6 (ω3). Thus, the essential fatty acid requirements of freshwater fishes can be met by a combination of dietary 18:2 (n-6) and 18:3 (n-3) fatty acids at least 1% of dry weight of the diet (NRC, 2011). Many marine fishes lack or produce inadequate level of elongation (Ghioni et al., 1999) desaturase (Tocher and Ghioni, 1999) enzymes and hence require EPA and DHA in their diets (Sargent et al., 2002). In case of marine fishes, LC-PUFA such as EPA and DHA should be supplied at least 1% of dry weight of its diet to satisfy their essential fatty acid requirements in most of the marine fishes but in silver bream, striped jack and yellow tail flounder it goes above 1%. Generally, the EPA requirement is higher than DHA requirement and it is almost double the requirement of latter (Brinkmeyer and Holt, 1998, Wu et al., 2007). The quantitative requirement of ARA is not well studied in most marine fishes. The ARA requirement is about 0.3% of dry diet in turbot (Castell et al., 1994).The essential fatty acid (EPA, DHA and ARA) requirement is almost double during the early phase of life than the later stage of fish (Tocher, 2010).

Fish oil is a predominant source of lipid with ω3 poly unsaturated fatty acid (PUFA) such as α-linolenicacid, linoleic acid, eicosapentaenoic acid anddocosahexaenoic acid. Though plant or vegetable oils are proposed as alternative to fish oil, they are poor sources of ω3 PUFA in comparison to fish oil, and totally deficient in long chain (LC) ω3 PUFA (Hixson, 2014). In contrast, plant oils are rich sources of ω6 and ω9 fatty acids, especially linoleic acid and oleic acid (18:1ω9) respectively, except few oilseeds. It results in lower levels of DHA and EPA in the flesh of fish fed with any of the plant oils such as vegetable oil (Bell et al., 2010), echium oil (Alhazzaa et al., 2011), camelina oil (Hixson et al., 2013), which compromises the health benefits of humans who consumes these fish. The high levels of ω-6 fatty acids in the human food is considered as undesirable due to their negative health impacts such as cardiovascular and autoimmune diseases (Simopoulos, 2006).

CARBOHYDRATES

Carbohydrates are the inexpensive source of dietary energy and are not utilized by several fish species. Generally, it is classified according to the type of sugars, structure, composition, degree of polymerization and glycosidic linkage into monomer sugars (glucose, galactose, fructose), oligosaccharides (lactose, maltose, sucrose, raffinose) and polysaccharides (starch, chitin, cellulose) (Englyst and Hudson, 1996). The sources of carbohydrates are ranging from seaweed, algae and plankton to refined grains and agricultural by-products (Krogdahl et al., 2005). In nutrition point of view, it is classified into two categories such as energy reserve polysaccharides such as starch and structural polysaccharides or non-starch polysaccharides (NSP) such as cellulose, chitin, pectin etc. Carbohydrate of starch origin often referred as digestible carbohydrate, whereas NSPs are often referred as non-digestible carbohydrate (De Silva and Anderson 1995; Stone, 2003). In general, NSPs are complex group, composed of monomers linked by β- glycosidic bonds (eg. galactose, mannose, arabinose etc.). Whereas starch composed entirely of glucose monomers linked exclusively by α- glycosidic bonds. Dietary starches are useful in the manufacturing of floating feeds through extrusion technology that makes starch more biologically available to fish. Although fish do not have a specific dietary requirement for carbohydrates; however, using the appropriate level of carbohydrates in aquafeeds is of great importance, mainly to reduce the feed cost and for their protein sparing activity.

The carbohydrates utilization in fish varies between fish species, carbohydrate sources (Krogdahl et al., 2005), and appears to be associated with complexity of carbohydrate and presence of carbohydrate-metabolizing enzymes (NRC, 1983; Wilson, 1994). Fish are known to have limited ability to digest and utilize the dietary carbohydrates; hence, excess intake resulted in poor feed utilization and reduced growth (Hemre et al., 2002). In the diets of herbivorous / omnivorous fishes it offers a protein sparing effect (New, 1987). Warm-water herbivorous and omnivorous species may digest and utilize more than 70 percent of the gross energy in non-cooked starch while carnivorous fish may digest less than 50 percent of non-cooked starch (Cho and Slinger, 1979). According to Furuichi and Yone (1980), dextrin as a dietary carbohydrate source showed growth retardation and low feed efficiency when fed with the diets containing more than 40, 30 and 20% dextrin in common carp, red sea bream and yellowtail, respectively. Cooking enhances the digestion and utilization of carbohydrates through gelatinization process (Wilson and Poe, 1985). The dietary carbohydrate requirement of Indian major carp, *Labeo rohita* has been reported to be 26% (Sen et al., 1978). Mohapatra et al. (2002) revealed that rohu can utilize up to 40% gelatinised carbohydrate without any adverse effect on growth and health. The paucity of amylolytic enzyme activities in carnivorous fish species curbs their ability to digest complex carbohydrates. Therefore, an increase in dietary carbohydrate level in the diets of several carnivorous species such as salmon, trout, grouper etc. significantly reduce the digestion of complex carbohydrates and feed as a whole (Steffens, 1989; van Barneveld et al., 1997). Proper cooking during extrusion of feed can improve the digestibility and utilization of carbohydrates by any fish species.

Energy provided by dietary carbohydrate spares amino acids from catabolic energy providing process and conserve them in for body growth and maintenance is referred as protein sparing effect and it is very much reported in fish (Shiau and Peng, 1993; Wilson, 1994; Mohanta et al., 2007; Zeng et al., 2015). The decrease in dietary protein level from 28% to 24% and increase in starch or dextrin in the diet from 37% to 41% did not affect weight gain in tilapia, indicating that starch or dextrin can spare some protein when the dietary protein is low (Shiau and Peng, 1993). This protein sparing effect of carbohydrates may be related to the fact that glucose is the preferred oxidative substrate for nervous tissue and blood cells and carbohydrate present in the fish diets can depress gluconeogenic activity, thereby diverting amino acids away from oxidative pathways (Cowey et al., 1977; Sanchez-Muros et al., 1996). The balance between dietary lipid and carbohydrate also influence the protein sparing, yielding best growth rate when dietary lipid and carbohydrates appear in equal calorific quantity (Halver and Hardy, 2002). The protein sparing effect of dietary carbohydrates may also influenced by various environmental factors, including temperature, light and season, or particular strain or the type and amount of carbohydrate (sugar/starch) supplied in the diet (Thibault et al., 1997; Medale et al., 1999). For example, Atlantic salmon shows significantly better protein sparing activity and energy utilization from dietary carbohydrate at 12.5 °C than at 20 °C (Hemre et al., 1995).

MINERALS

Minerals are inorganic elements present in ash when a body tissue or food is burnt. All the minerals present in an animal's body may not be essential in its diet. Based on the requirement, minerals are classified as macro/major, which are required in large quantities and micro/ trace minerals, which are required only in trace amounts. The macro minerals are calcium (Ca), phosphorus (P), magnesium (Mg), sodium (Na), chlorine (Cl), potassium (K) and sulphur (S). Trace minerals are iron (Fe), copper (Cu), zinc (Zn), manganese (Mn), selenium (Se), iodine (I), cobalt (Co), fluorine (F), molybdenum (Mo), aluminium (Al), iodine (I), nickel (Ni), vanadium (V), silicon (Si), tin (Sn), chromium (Cr) etc. Minerals are generally involved in functions such as (i) structural components of skeletal system (Ca, P, Mg, Na, K), (ii) enzyme activators (Zn, Cu), (iii) components of organic compounds and (iv) maintaining osmotic and acid-base balance (Na, K, Cl). Fish can absorb dissolved minerals from the water across the gill membrane or through the digestive tract. Even though, supplementation of minerals in fish diet is necessary because of its vital roles in skeleton formation, the regulation of the acid-base balance, osmotic and ionic regulation and functioning of several enzymes.

The major function of *calcium* is to provide strength to the skeleton. About 99% of the calcium found in fish body is in bones and scales. Calcium is essential for bone formation, muscle contraction, nerve impulse transmission, blood clotting and enzyme activation (Lall, 2002). Most of the Ca requirement of fish is satisfied from the ambient water in which it lives. Around 90% of phosphorus in fish is in the bones, complexed with calcium to form apatite or tri calcium phosphate. It is also a major component of adenosine triphosphate (ATP). The phosphate buffer system maintains normal pH of body fluids. *Phosphorus* plays important roles in fish growth, bone mineralization and lipid and carbohydrate metabolism (Lall, 2002). Phosphorous is a limiting nutrient in freshwater bodies that dictates its supplementation in the diet of fish (Lall, 2002). Diet deficient in P content or with low bioavailability leads to poor absorption of P from the diet and in turn influences poor growth, delayed bone mineralization and skeletal deformities such as deformed head and spinal column, abnormal calcification of ribs and soft rays (Sugiura et al., 2004). According to Hansen et al. (2010) high fat diets offer faster growth rates than the normal diet, in such diets if the phosphorous level is fixed or on par with low fat diet that can affect growth allometry resulting in vertebral deformities especially under dietary phosphorus insufficiency.

About 70 % of magnesium in a fish body is present in hard parts. It is also involved in carbohydrate metabolism, osmoregulation and protein synthesis reactions. Deficiency symptoms include reduced growth, vertebral curvature, convulsions, cataracts, degeneration of epithelial tissue and muscle fibres (NRC, 2011). Sodium, potassium and chloride are involved in osmoregulation, electrolyte balance and pH balance in body, nerve impulse transmission etc. Dietary deficiencies of these minerals are not readily observed in fishes. Iron is an important element in the formation of hemoglobin and as a component of heme enzymes such as cytochromes, involved in electron transport chain. Deficiency of iron results in microcytic anaemia. Dietary excess doses result in phosphorus deficiency and iron toxaemia. Manganese is involved in blood clotting, bone formation, enzyme systems including super oxide dismutase, amino acid, fatty acid and glucose oxidation. Deficiency of manganese causes depressed growth, abnormal tail growth and shortening of body.

Selenium is mainly involved in superoxide dismutase antioxidant system along with vitamin E and protects the cells from free radicals and thus prevents the peroxidation of polyunsaturated fatty acids of phospholipids and cholesterol in cellular and sub-cellular membranes (Kolb et al., 1997). Se deficiency in fish causes reduced growth, muscular dystrophy, and suppressed glutathione peroxidase activity. Copper plays a role in hematopoiesis and is a component of enzyme systems such as super oxide dismutase, tyrosinase and cytochrome C oxidase. Deficiency of copper causes growth depression and anemia. Iodine is a component of the thyroid hormones, thyroxine and triiodothyronine, which regulate the rate of metabolism. Deficiency of iodine results in hyperplasia of thyroid gland or goiter in fish. Chromium functions as a cofactor of enzymes in carbohydrate, lipid, protein, sterol and nucleic acid metabolism. It also potentiates insulin for glucose metabolism. Deficiency results in reduced glucose tolerance and impaired lipid and glycogen synthesis from glucose. Generally, fish meal is a good source of minerals and especially of phosphorus to fish (Lall, 2002). Replacement of fish meal by plant sources ingredient reduces the availability of several minerals in the diet of fish and it is inevitable to ensure an adequate supply of minerals to meet the physiological demands. Diets with high levels of plant protein and low levels of fish meal, supplemented with inorganic phosphorous resulted in adequacy of available phosphorous to fish (Kaushik et al., 2004).

VITAMINS

Vitamins are organic compounds required in trace amounts from an exogenous source (diet) for normal growth, reproduction, health and function in organisms. Although the requirements for vitamins are small, their deficiency can cause symptoms such as poor appetite, growth and tissue deformities. Fish lack the physiological mechanisms to synthesize vitamins in sufficient quantities required for normal functioning hence they must be supplied through the diet. In some fish species, microbial synthesis of some vitamins in the gut supplements the requirement. The quantitative requirement of vitamins in fish varies with several aspects such as species, life stage, feeding habits, health condition, environmental parameters and nutrient interrelationships. Vitamins can broadly be classified as water soluble and fat soluble (Table 1).

Table 1. Classification of vitamins

Water soluble vitamins	Fat soluble vitamins
B- vitamins:	
Vitamin B1 (Thiamine)	Vitamin A (Retinol)
Vitamin B2 (Riboflavin)	Vitamin D (Cholecalciferol)
Vitamin B3 (Nicotinic acid (niacin)	Vitamin E (Tocopherol)
Vitamin B5 (Pantothenic acid)	Vitamin K (Phylloquinone)
Vitamin B6 (Pyridoxine)	
Vitamin B7 (Biotin)	
Vitamin B9 (Folic acid)	
Vitamin B12 (Cyanocobalamin)	
Vitamin C (Ascorbic acid)	
Inositol	
Choline	

The water-soluble vitamins are easily absorbed from the water via the intestine and the excess quantity is excreted out. Since the water-soluble vitamins are not stored in the body tissues, a constant supply in adequate level is required in order to prevent the deficiency symptoms. The fat-soluble vitamins are absorbed in the intestine along with dietary fats and are stored in body tissues if the dietary intake exceeds metabolic requirements. The accumulation of fat soluble vitamins in the tissues results in a toxic condition known as hypervitaminosis (NRC, 1993). Hence the inclusion of excessive vitamins in the diets of fish is not only uneconomical but also compromises the well-being of fish.

Thiamine is present in many grains and seeds, and is mostly concentrated in the seed coat. It is synthesized only by higher plants and not by animals. The deficiency of thiamine results in neurological disorders in fish, as it affects the central nervous system. The tissues of most fish contain thiaminase, an enzyme capable of destroying thiamine. Hence the use of raw or uncooked fish tissues for feeding results in the development of deficiency symptoms. Several studies have reported neurological disorders due to thiamine deficiency such as hyper irritability (Comacho, 1978; Ikeda et al., 1988), subcutaneous hemorrhages and congested fins (Arai et al., 1972) in several fish species. Riboflavin is widely found in nature and the sources include green plants, yeast and seed coat of grains. The co-enzymes of riboflavin include flavin mononucleotide (FMN) and Flavin adenine di nucleotide (FAD), which are components of prosthetic groups of oxidation-reduction reactions involved in the metabolism of keto acids, fatty acids and amino acids. Hughes et al. (1981) reported photophobia, cataract, corneal vascularization and hemorrhages in salmonids and Huang et al. (2010) reported high oxidative stress and low anti-oxidative activity in riboflavin deficient grouper. Niacin is found in plant and animal tissues as a component of two co-enzymes of hydrogen transport system, nicotinamide adenine di nucleotide (NAD) and nicotiamide adenine dinucleotide phosphate (NADP), which are involved in oxidation-reduction reactions. The deficiency of niacin exhibited anorexia, poor growth, poor feed conversion, photosensitivity, intestinal lesions, abdominal edema, muscular weakness, spasms and increased mortality in salmonids (Halver, 1957; Andrews and Murai, 1978), while a high dietary intake of niacin increased liver fat, decreased body fat and reduced growth in brook trout (Poston, 1969).

Pantothenic acid functions as a component of co enzyme A (CoA), which is involved in the transfer of two carbon atoms. Deficiency of this vitamin impairs the metabolism of mitochondria rich cells that undergo rapid mitosis and high energy expenditure. Gill lamellar hyperplasia or clubbed gills is a characteristic sign of pantothenic acid deficiency in most fish (Rucker et al., 1952; Hasokawa, 1989).

Pyridoxine is functional in a number of enzymes in which amino acids are metabolized such as decarboxylases, transaminases, sulfhydrases and hydroxylases. Increased protein content in the diet causes an increased requirement of pantothenic acid. Deficiency symptoms in fish include nervous disorders, convulsions, erratic swimming (Coates and Halver, 1958), poor growth and anorexia (Shiau and Hsieh, 1997).

Biotin functions as a co-enzyme in carboxylation reactions. The common deficiency symptoms in fish include, anorexia, reduced weight gain, higher feed conversion, changes in skin pigmentation, weight loss, convulsions and high mortality rates (Poston and McCartney, 1974; Poston and Page, 1982).

Folic acid functions as a pivoting intermediate carrier of one carbon groups in a number of complex enzymatic reactions. Poor growth and dark skin colouration in Japanese eels (Arai et al., 1972) and megaloblastosis in Rainbow trout (Cowey and Woodward, 1993) fed a folate deficient diet.

Cyanocobalamin is required for normal maturation and development of erythrocytes, metabolism of fatty acids, in the methylation of homocysteine to methionine and for the normal recycling of tetra hydrofolic acid (Halver, 1969). Reduced growth and lower haematocrit in rohu fed with vitamin B12 deficient diet was observed by John and Mahajan (1979).

Fish cannot synthesize *ascorbic acid* because it lacks L- gulonolactone oxidase, which is the terminal enzyme in ascorbic acid synthesis pathway (Ciereszko et al., 1999). Therefore, ascorbic acid should be supplemented in the diets of fish. Several roles are attributed to this vitamin which includes hydroxylation of proline and lysine to hydroxyl amino acids for collagen synthesis, bone calcification, iron synthesis, tyrosine metabolism, blood clotting, immune response and detoxification of xenobiotics (Gould, 1960; Johnson et al., 1971). Ascorbic acid requirements of fish showed variation with respect to their physiology and developmental stages of life cycle (Dabrowski, 2001). Deficiency symptoms include deformation of spinal column, fin erosion, hemorrhages, reduced rate of wound healing, deformed gill operculum, scoliosis and lordosis (Mahajan and Agrowl, 1979; Tucker and Halver, 1986).

Inositol is widely distributed in plant and animal tissues. In animals, it occurs as myoinositol. The major functions include lipotropic action, signal transduction and other cellular processes such as smooth muscle contraction, homeostasis of normal lipid metabolism, prevents the accumulation of cholesterol, platelet aggregation and histamine secretion (West et al., 1966). The deficiency symptoms include anaemia, fin erosion and slow rate of gut emptying (Aoe and Masuda, 1967).

The major metabolic functions of *choline* include synthesis of phospholipids, phosphatidylcholine, as a precursor of the neurotransmitter, lipotropic and anti-hemorrhagic factor, acetyl choline and as a precursor of betaine which serves as methyl donor for methylation reactions, in fat transport (West et al., 1966).

Vitamin A is crucial in a number of physiological processes necessary for optimal function of organisms. The major functions include cell differentiation and embryo development, bone development, retina formation, differentiation of immune cells and reproductive performance in fish (Blomhoff et al., 1992). Deficiency symptoms in fish include poor growth, pale skin colour, ascites, anaemia, exophthalmia, edema, eye lens displacement, skin hemorrhages, deformed opercula, thinning of cornea and degradation of retina (Poston and Levington, 1969; Hosokawa, 1989).

Vitamin D functions to maintain the calcium homeostasis. It is important for the development, growth and maintenance of a healthy skeletal system, alkaline phosphatase activity, intestinal absorption of calcium and influences the action of parathyroid hormone on bone (Bronner, 2009; Lock et al., 2010). Deficiency symptoms in trout include poor growth, elevated liver lipid, impaired calcium homeostasis and ultra-structural changes in white muscle fibres of epaxial musculature (Leatherland et al., 1980; George et al., 1981).

Vitamin E mainly functions as a component of anti-oxidant defense system, which protects the cells against free radicals. It has also roles in reproduction. Deficiency symptoms include, exudative diathesis, reduced percentage of fertilized eggs, decrease in the number, viability and motility of spermatoids, erythrocyte fragility, muscular dystrophy and necrosis of white muscle fibres (Hashimoto et al., 1966; Poston et al., 1976). Vitamin E requirement is depend on the dietary content of highly unsaturated fatty acids (HUFA), an increase in n-3 HUFA in the diet requires similar increase in vitamin E requirement also (Watanabe et al., 1991).

Vitamin K is mainly involved as a necessary factor for normal blood clotting. The synthesis of proteins such as prothrombin, proconvertin, plasma thromboplastin antecedent and Stewart Prower factor are dependent upon vitamin K (Krossoy et al., 2011). The supplementation of vitamin K deficient feed resulted in prolonged coagulation time (Olson, 1999), post translational modification and activation of vitamin K dependent protein (Lee et al., 2007).

SUMMARY

The role of nutrients in the diets of organisms for the maintenance of health and growth is well recognized. Adequate nutrition is an inevitable component for optimum productivity in aquaculture by the way of improved growth and maintaining health of fishes. The formulated feeds should be complete in nutritional aspects in order to accomplish the anticipated benefits. Another aspect in fish nutrition is that the nutrient requirements depends on various factors such as species, stage of maturity, age, feeding habits and environmental influences as well. Since fish lives in an aquatic environment, the formulation and manufacture of aquafeed involves additional level of skill so as to deliver ample levels of nutrients to the cultured organisms, without significant loss of nutrients and also considering the nutrients available from the environment. The future of fish nutrition aims in the development of well balanced, cost effective, low polluting feeds using ingredients which are sustainable for the growth of aquaculture industry.

REFERENCES

1. Akiyama, T.; Oohara, I.; Yamamoto, T. Comparison of essential amino acid requirements with A/E ratio among fish species. *Fish. Sci.* 1997, 63, 963-970.
2. Aksnes, A.; Mundheim, H.; Toppe, J.; Albrektsen, S. The effect of dietary hydroxyproline supplementation on salmon (*Salmo salar* L.) fed high plant protein diets. *Aquaculture*. 2008, 275, 242-249.
3. Alhazzaa, R.; Bridle, A.; Nichols, P.; Carter, C. Replacing dietary fish oil with echium oil enriched barramundi with C18 PUFA rather than long-chain PUFA. *Aquaculture.* 2011, 312, 162-171.
4. Ali, S.A. *Protein, lipid and energy requirements of shrimp and fish.* In *Training manual on shrimp and fish nutrition and feed management;* Ali, S.A., Ed.; CIBA special publication No. 15; CIBA: India, 2002; pp. 9-18.
5. Andrews, J.W.; Murai, T. Dietary niacin requirements of channel catfish. *J. Nutr*. 1978, 108, 1508-1511.
6. Aoe, H.; Masuda, I. Water-soluble vitamin requirements of carp. 2. Requirements for *p*-aminobenzoic acid and inositol. *Bull. Jpn. Soc. Sci. Fish.*1967, 33, 674-680.
7. Aragao, C.; Corte-Real, J.; Costas, B.; Dinis, M.T.; Conceicao, L.E.C. Stress response and changes in amino acid requirements in Senegalese sole (*Solea senegalensis* Kaup 1858). *Amino Acids*. 2008, 34, 143-158.
8. Arai, S.; Nose, T.; Hashimoto, Y. Qualitative requirements of young eels, *Anguilla japonica*, for water-soluble vitamins and their deficiency symptoms. *Bull. Freshwater Res. Lab. Tokyo*. 1972, 22, 69-83.
9. Arzel, J.F.X.; Lopez, M.; Métailler, R.; Stéphan, G.; Viau, M.; Gandemer, G.; Guillaume, J. Effect of dietary lipid on growth performance and body composition of brown trout (*Salmo trutta*) reared in seawater. *Aquaculture*. 1994, 123, 361-375.
10. van Barneveld, R.J.; Smart, A.; Clarke, S.M.; Carter, C.G.; Davis, B.J.; Tivey, D.R. et al. *Nutritional management of sea-caged southern bluefin tuna (Thunnus maccoyii)*. In *Recent Advances in Animal Nutrition in Australia*; Corbett, .JL., Choct, M., Nolan, J.V.,

Rowe, J.B., Eds.; University of New England: Armidale, Australia, 1997; pp. 88–97.

11. Bell, G.; Pratoomyot, J.; Strachan, F.; Henderson, R.; Fontanillas, R.; Growth, flesh adiposity and fatty acid composition of Atlantic salmon (*Salmo salar*) families with contrasting flesh adiposity: Effects of replacement of dietary fish oil with vegetable oils. *Aquaculture*. 2010, 306, 225-232.
12. Bell, J.G.; Tocher, D.R.; Farndale, B.M.; Cox, D.I.; McKinney, R.W.; Sargent, J.R. The effect of dietary lipid on polyunsaturated fatty acid metabolism in Atlantic salmon (*Salmo salar*) undergoing Parr-Smolt transformation. *Lipids*. 1997, 32, 515-525.
13. Bell, M.V.; Dick, J.R. Molecular species composition of the major glycerophospholipids from the muscle, liver, retina and brain of cod (*Gadus morhua*). *Lipids*. 1991, 26, 565–573.
14. Blaxter, K..; *Energy Metabolism in Animals and Man*. Cambridge University Press, Cambridge, 1989.
15. Blomhoff, R.; Green, M.H.; Norum, K.R. Vitamin A: Physiological and biochemical processing. *Annu. Rev. Nutr*. 1992, 12, 37-57.
16. Boonyaratpalin, M. *Nutritional studies on sea bass (Lates calcarifer)*. In *Fish Nutrition Research in Asia;* De Silva, S.S., Ed.; Proceedings of the Fourth Asian Fish Nutrition Workshop, Asian Fisheries Society Special Publication No: 5. Asian Fisheries Society: Philippines, 1991; pp. 33-44.
17. Bowen, H. Dietary protein requirements of fishes-A reassessment. *Can. J. Fish. Aquat. Sci.* 1987, 44, 1995-2001.
18. Bowyer, J.N.; Qin, J.G.; Stone, D.A.J. Protein, lipid and energy requirements of cultured marine fish in cold, temperate and warm water. *Rev. Aquac*. 2013, 5, 10–32.
19. Brinkmeyer, R.; Holt, G.J. Highly unsaturated fatty acids in diets forred drum (*Scianops ocellatus*) larvae. *Aquaculture*. 1998, 161, 253–268.
20. Brett, J.R.; Groves, T.D.D. *Physiological energetics*. In *Fish Physiology*; Hoar, W.S., Randall, D.J., Brett, J.R., Eds.; Academic Press: New York, 1979; pp. 279–352.
21. Bronner, F. Recent developments in intestinal calcium development. *Nutr. Rev.* 2009, 67, 109-113.

22. Buentello, J.A.; Gatlin, D.M. Nitric oxide production in activatedmacrophages from channel catfish (*Ictalurus punctatus*): influence of dietary arginine and culture media. *Aquaculture.* 1999, 179, 513-521.

23. Castell, J.D.; Bell, J.G.; Tocher, D.R.; Sargent, J.R. Effects of purified diets containing different combinations ofarachidonic and docosahexaenoic acid on survival, growth and fatty acid composition of juvenile turbot, *Scophthalmus maximus. Aquaculture.* 1994, 128, 315-333.

24. Cho, C.Y.; Slinger, S.J. *Effect of water temperature on energy utilization in rainbow trout*. In *Proceedings of the Eighth Symposium on Energy Metabolism in Farm Animals*; Mont, S.E., Ed.; Cambridge University Press: Cambridge, U.K., 1979; pp. 287-291.

25. Choo, P.; Smith, T.K; Cho, C.Y.; Ferguson, H.W. Dietary excesses of leucine influence growth and body composition of rainbow trout. *J. Nutr.* 1991, 121, 1932–1939.

26. Chou, B.S.; Shiau, S.Y. Optimal dietary lipid level for growth of juvenile hybrid tilapia, *Oreochromis niloticus* x *Oreochromis aureus*. *Aquaculture*. 1996, 143, 185-195.

27. Chou, R.L.; Su, M-S.; Chen, H-Y. Optimal dietary protein and lipid levels for juvenile (*Rachycentron canadum*). *Aquaculture.* 2001, 193, 81–89.

28. Ciereszko, A.; Dabrowski, K.; Lin, F.; Liu, L. Protective role of ascorbic acid against damage to male germ cells in rainbow trout (*Oncorhynchus mykiss*). *Can. J. Fish. Aquat. Sci.* 1999, 56, 178-183.

29. Coates, J.A.; Halver, J.E. Water-soluble vitamin requirements of silver salmon.*Spec. Sci. Rep. U.S. Fish Wildl. Serv. (Fish.).*1958, 281, 9.

30. Comacho, A.S. Characterization of thamine deficiency in channel catfish fed heated and non-heated catfish processing waste. Ph.D. dissertation, Auburn University, Auburn, Alabama, 1978.

31. Cowey, C.B.; Cho, C.Y. Nutritional requirements of fish. *Proc. Nutr. Soc.* 1993, 52, 417–426.

32. Cowey, C.B. Protein and amino acid requirements: a critique of methods. *J. Appl. Ichthyol.* 1995, 11, 199-204.

33. Cowey, C.B.; Sargent, J.R. Lipid nutrition in Fish. *Comp. Biochem. Physiol.* 1977, 457B, 269-273.
34. Cowey, C.B.; Woodward, B. The dietary requirement of young rainbow trout (*Oncorhynchus mykiss*) for folic acid. *J. Nutr.* 1993, 123, 1594-1600.
35. Cowey, C.B.; Sargent, J.R. *Nutrition.* In *Fish physiology*; Hoar, W.S., Randall, D.J. Brett, J.R. Ed.; Academic Press: New York, 1979; Vol. 8; pp. 1-69.
36. Cowey, C.B.; de la Higuera, M.; Adron, J.W.The effect of dietary composition and of insulin on gluconeogenesis in rainbow trout (*Salmo gairdneri*). *Br. J. Nutr.* 1977, 38, 385–395.
37. De Silva, S.S.; Anderson, T.A. *Fish Nutrition in Aquaculture.* Chapman and Hall: London, 1995.
38. Divakaran, S.; Ostrowsky, A.C. Enzymes present in pancreatic extracts of the dophin *Coryphaena hippuras*. *J. World Aquac. Soc.* 1990, 21, 35–40.
39. Einen, O.; Roem, A.J. Dietary protein/energy ratios for Atlantic salmon in relation to fish size: growth, feed utilization and slaughter quality. *Aquac. Nutr.* 1997, 3, 115–126.
40. Englyst, H.N.; Hudson, G.J. The classification and measurement of dietary carbohydrates. *Food Chem.* 1996, 57, 15-21.
41. Fang, Y.Z.; Yang, S.; Wu, G. Free radicals, antioxidants, and nutrition. *Nutrition.* 2002, 8, 872–879.
42. Felip, O.; Ibarz, A.; Fernandez-Borras, J.; Beltran, M.; Martin-Perez, M.; Planas, J.V.; Blasco, J. Tracing metabolic routes of dietary carbohydrate and protein in rainbow trout *(Oncorhynchus mykiss)* using stable isotopes ([13C] starch and [15N] protein): effects of gelatinisation of starches and sustained swimming. *Brit. J. Nutr.* 2012, 107, 834-844.
43. Furuichi, M.; Yone, Y. Effect of dietary dextrin levels on the growth and feed efficiency, the chemical composition of liver and dorsal muscle, and the absorption of dietary protein and dextrin in fishes. *Bull. Jpn. Soc. Sci. Fish.* 1980, 46, 225-229.
44. Gatlin, D.M. *Principles of Fish Nutrition.* SRAC Publication No. 5003, Southern Regional Aquaculture Center: Mississippi, USA,

2010.

45. Gaylord, T.G.; Barrows, F.T.; Teague, A.M.; Johansen, K.A.; Overturf, K.E.; Shephed, B. Supplementation of taurine and methionine to all-plant protein diets for rainbow trout (*Oncorhynchus mykiss*). *Aquaculture.* 2007, 269, 514–524.
46. George, J.C.; Barnett, B.J.; Cho, C.Y.; Slinger, S.J. Vitamin D3 and muscle function in the rainbow trout. *Cytobios.* 1981, 31, 7-18.
47. Geurden, I.; Coutteau, P.; Sorgeloos, P. Effect of a dietary phospholipid supplementation on growth and fatty acid composition of European sea bass (*Dicentrarchus labrax* L.) and turbot (*Scophthalmus maximus* L.) juveniles from weaning onwards. *Fish Physiol. Biochem.* 1997, 16, 259-272.
48. Ghanawi, J.; Roy, L.; Davis, D.A.; Saoud, I.P. Effects of dietary lipid levels on growth performance of marbled spine foot rabbit fish *Siganus rivulatus. Aquaculture.* 2011, 310, 395–400.
49. Ghioni, C.; Tocher, D.R.; Bell, M.V.; Dick, J.R.; Sargent, J.R. Low C18 to C20 fatty acid elongase activity and limited conversion of stearidonic acid, 18:4n-3, to eicosapentaenoic acid, 20:5n-3, in a cell line from turbot, *Scophthalmus maximus. Biochim. Biophys. Acta.* 1999, 1437, 170-181.
50. Gomez-Requeni, P.; Mingarro, M.; Kirchner, S.; Calduch-Giner, J.A.; Medale, F.; Corraze, G.; Panserat, S.; Martin, S.A.M.; Houlihan, D.F.; Kaushik, S.J.; Perez-Sanchez, J. Effects of dietary amino acid profile on growth performance, key metabolic enzymes and somatotropic axis responsiveness of gilthead sea bream (*Sparus aurata*). *Aquaculture.* 2003, 220, 749-767.
51. Gonzalez-Felix, M.L.; Minjarez-Osorio, C.; Perez-Velazquez, M.; Urquidez-Bejarano, P. Influence of dietary lipid on growth performance and body composition of the Gulf corvine, *Cynoscion othonopterus. Aquaculture.* 2015, 448, 401-409.
52. Green, J.A.; Hardy, R.W.; Brannon, E.L. The optimum dietary essential: nonessential amino acid ratio for rainbow trout (*Oncorhynchus mykiss*), which maximizes nitrogen retention and minimizes nitrogen excretion. *Fish Physiol. Biochem.* 2002, 27, 109-115.

53. Halver, J.E. Nutrition of salmonid fishes: Water-soluble vitamin requirements of chinook salmon. *J. Nutr.* 1957, 62, 225-243.

54. Halver, J.E. *Vitamin requirements.* In *Fish in Research*; Neuhaus, O.W., Halver, J.E., Eds.; Academic Press: New York, 1969; pp. 209-244.

55. Halver, J.E.; Hardy, R.W. *Nutrient flow and retention.* In *Fish Nutrition*; Halver, J.E., Hardy, R.W., Eds.; Academic Press: San Diego, CA, 2002; pp. 755–770.

56. Hansen, T.; Fjelldal, P.G.; Yurtseva, A.; Berg, A. A possible relation between growth and number of deformed vertebrae in Atlantic salmon (*Salmo salar* L.). *J. Appl. Ichthyol.* 2010, 26, 355-359.

57. Hasan, M.R. In Aquaculture in the Third Millennium. Technical Proceedings of the Conference on Aquaculture in the Third Millennium, Bangkok, Thailand, February 20-25, 2000; Subasinghe, R.P., Bueno, P., Phillips, M.J., Hough, C., McGladdery S.E., Arthur, J.R., Eds; NACA, Bangkok and FAO: Rome, 2001.

58. Hashimoto, Y.; Okaichi, T.; Watanabe, T.; Furukawa, A.; Umezu, T. Muscle dystrophy of carp due to oxidized oil and the preventive effective of vitamin E. *Bull. Jpn. Soc. Sci. Fish.* 1966, 32, 64-69.

59. Hemre, G.I.; Sandnes, K. Effect of dietary lipid level on muscle composition in Atlantic salmon *Salmo salar. Aquac. Nutr.* 1999, 5, 9-16.

60. Hemre, G.I.; Lie, O.; Lied, E.; Lambertsen, G. Starch as an energy source in feed for cod (*Gadus morhua*): Digestibility and Retention. *Aquaculture.* 1989, 80, 261-270.

61. Hemre, G.I.; Sandnes, K.; Lie, O.; Waagbø, R. Blood chemistry and organ nutrient composition in Atlantic salmon, *Salmo salar* L., fed graded amounts of wheat starch. *Aquac. Nutr.* 1995, 1, 37–42.

62. Hillestad, M.; Johnsen, F. High-energy/low-protein diets for Atlantic salmon: effects on growth, nutrient retention and slaughter quality. *Aquaculture.* 1994, 124, 109-116.

63. Hilton, J.W.; Slinger, S.J. Nutrition and feeding of rainbow trout. *Can. Spec. Publ. Fish. Aquat. Sci.* 1981, 55, 15.

64. Hixson, S.; Parrish, C.; Anderson, D. Effect of replacement of fish oil with camelina (*Camelina sativa*) oil on growth, lipid class and

fatty acid composition of farmed juvenile Atlantic cod (*Gadus morhua*). *Fish Physiol. Biochem.* 2013, 39, 1441-1456.

65. Hixson, S.M. Fish nutrition and current issues in aquaculture: The balance in providing safe and nutritious seafood, in an environmentally sustainable manner. *J. Aquac. Res. Dev.* 2014, 5, 234.
66. Hosokawa, H. The vitamin requirements of fingerling yellow tail, *Seriola quinqueradiata.* Ph.D. Dissertation, Kochi University, Kochi, Japan, 1989.
67. Hossain, M.A.; Begum, S.; Islam, M.N.; Shah, A.K.M.A. Studies on the optimum protein to energy ratio of African catfish (*Clarias gariepinus* Burchell) *Bangladesh J. Fish. Res.* 1998, 2, 47-54.
68. Hseu, J.R.; Lu, F.I.; Su, H.M.; Wang, L.S.; Tsai, C.L.; Hwang, P.P. Effect of exogenous tryptophan on cannibalism, survival and growth in juvenile grouper, *Epinephelus coioides*. *Aquaculture*. 2003, 218, 251–263.
69. Huang, W.Y.; Cai, Y.Z.; Corke, H.; Sun, M. Survey of antioxidant capacity and nutritional quality of selected edible and medicinal fruit plants in Hong Kong. *J. Food Compos. Anal.* 2010, 23, 510-517.
70. Hughes, S.G.; Riis, R.C.; Nichum, J.G.; Rumsey, G.L. Biomicroscopic and histologic pathology of the eye in riboflavin deficient rainbow trout (*Salmo gairdneri*). *Cornell Vet.* 1981, 71, 269-279.
71. Ikeda, S.; Ishibashi, Y.; Murata, O.; Nasu, T.; Harada, T. Qualitative requirements of the Japanese parrotfish for water-soluble vitamins. *Bull. Jpn. Soc. Sci. Fish.* 1988, 54, 2029-2035.
72. Izquierdo, M.S.; Henderson, R.J. The determination of lipase and phospholipase activities in gut contents of turbot (*Scophthalmus maximus*) by fluorescence-based assays. *Fish Physiol. Biochem.* 1998, 19, 153–162.
73. Izquierdo, M.S.; Socorro, J.; Arantzamendi, L.; Hernandez-Cruz, C.M. Recent advances in lipid nutrition in fish larvae. *Fish Physiol. Biochem.* 2000, 22, 97–107.
74. John, M.J.; Mahajan, C.L. The physiological response of fishes to deficiency of cyanocobalamin and folic acid. *J. Fish Biol.* 1979, 14,

127-133.

75. Kaushik S.J.; Cowey, C.B. *Ammoniogenesis and dietary factors affecting nitrogen excretion*. In *Nutritional Strategies & Aquaculture Waste*; Cowey, C.B., Cho, C.Y., Eds.; University of Guelph: Guelph, Canada, 1991; pp. 3-19.
76. Kaushik, S.J.; Seiliez, I. Protein and amino acid nutrition and metabolism in fish: current knowledge and future needs. *Aquac. Res.* 2010, 41, 322-332.
77. Kaushik, S.J.; Luquet, P.; Blanc, D. Usefulness of feeding protein and non-protein calories apart in studies on energy-protein interrelationships in rainbow trout. *Annu. Zootech.* 1981, 30, 3-11.
78. Kaushik, S.; Coves, D.; Dutto, G.; Blanc, D. Almost total replacement of fish meal by plant protein sources in the diet of a marine teleost, the European seabass, *Dicentrarchus labrax*. *Aquaculture.* 2004, 230, 391-404.
79. Khan, M.A.; Jafri, A.K. On the dietary protein requirement of *Clarias batrachus* Linnaeus. *J Aqua Trop.* 1990, 5, 191-198.
80. Kikuchi, K.; Furuta, T.; Iwata, N.; Onuki, K.; Noguchi, T. Effect of dietary lipid levels on the growth, feed utilization, body composition and blood characteristics of tiger puffer *Takifugu rubripes*. *Aquaculture*. 2009, 298, 111–117.
81. Kolb, E.; Kaskous, S.; Seehawer, J. Nutritional aspects of the importance, utilization, metabolism and the use of vitamin E and selenium in sheep. *Berl Munch Tierarztl Wochenschr.* 1997, 110, 178-184.
82. Krogdahl, A.; Hemre, G.I.; Mommsen, T.P. Carbohydrates in fish nutrition: digestion and absorption in postlarval stages. *Aquac. Nutr.* 2005, 11, 103-122.
83. Krossoy, C.; Waagbo, R.; Ornsrud, R. Vitamin K in fish Nutrition. *Aquac. Nutr.* 2011, 17, 585-594.
84. Lall, S.P. *The minerals*. In: *Fish Nutrition, 3rd Edn*; Halver, J.E., Hardy, R.W., Eds.; Elsevier Science: New York, 2002; pp. 259-308.
85. Leatherland, J.F.; Barnett, B.J.; Cho, C.Y.; Slinger, S.J. Effect of dietary cholecalciferol deficiency on plasma thyroid hormone

concentrations in rainbow trout, *Salmo gairdneri* (Pisces, Salmonidae). *Environ. Biol. Fish.* 1980, 5, 167-173.

86. Lee, D.J.; Sinnhuber, R.O. *Lipid requirements*. In *Fish Nutrition*; Halver, J.E., Ed.; Academic Press Inc.: New York, 1972; pp.145-180.
87. Lee, N.K; Sowa, H.; Hinoi, E.; Ferron, M.; Ahn, J.D.; Confavreux, C.; Dacquin, R.; Mee, P.J.; McKee, M.D.; Jung, D.Y.; Zhang, Z.; Kim, J.K.; Mauvais-Jarvis, F.; Ducy, P.; Karsenty, G. Endocrine regulation of energy metabolism by skeleton. *Cell.* 2007, 130, 456-469.
88. Lee, S.M.; Cho, S.H.; Kim, K.D. Effects of dietary protein and energy levels on growth and body composition of juvenile flounder *Paralichtys olivaceus*. *J. World Aquac. Soc.* 2000, 30, 306-315.
89. Léger, C.; Bauchart, D.; Flanzy, J. Some properties of pancreatic lipase in *Salmo gairdneri* Rich. Effects of bile salts and Ca^{2+}, gel filtration. *Comp. Biochem. Physiol.* 1977, 57, 359–363.
90. Léger, C. Digestion, absorption and transport of lipids. In *Nutrition and feeding in fish*; Cowey, C.B., Mackie, A.M., Bell, J.G. Eds.; Academic Press: London, 1988; pp. 299–331.
91. Lepage, O.; Tottmar, O.; Winberg, S. Elevated dietary intake of L-tryptophan counteracts the stress-induced elevation of plasma cortisol in rainbow trout (*Oncorhynchus mykiss*). *J Exp. Biol.* 2003, 205, 3679-3687.
92. Lin, Y.; Zhou, X. Dietary glutamine supplementation improves structure and function of intestine of juvenile Jian carp (*Cyprinus carpio* var. Jian). *Aquaculture.* 2006, 256, 389-394.
93. Lock, E-J.; Waagbo, R.; Bonga, S.W.; Flick, G. The significance of vitamin D for fish: a review. *Aquac. Nutr.* 2010, 16, 100-116.
94. Lorentzen, M.; Maage, A.; Julshamn, K. Effects of dietary selenite or selenomethionine on tissue selenium levels of Atlantic salmon (*Salmo salar*). *Aquaculture.* 1994, 121, 359-367.
95. Lovell, R. T. Nutrition and feeding of fish. Van Nostrand Reinhold, New York, p. 260. 1989.
96. Lovell, R.T.; Bush, L.J.; Getz, W.R. Energy Requirements of Fish. *The Professional Animal Scientist*. 1994, 10, 145-149.

97. Luo, Z.; Liu, Y.J.; Mai, K.S.; Tian, L.X.; Liu, D.H.; Tan, X.Y. Optimal dietary protein requirement of grouper, *Epinephelus coioides* juveniles fed isoenergetic diets in floating net cages. *Aquacult. Nutr.* 2004, 10, 247-252.

98. Mahajan, C.L.; Agrawal, N.K. Vitamin C deficiency in *Channa punctatus (*Bloch). *J. Fish Biol.* 1979, 15, 613-622.

99. Mai, K.; Zhang, L.; Ai, Q.; Duan, Q.; Zhang, C.; Li, H.; Wan, J.; Liufu, Z. Dietary lysine requirement of juvenile seabass *(Lateolabrax japonicus)*. *Aquaculture* 2006, 258, 535-542.

100. McGoogan, B.B.; Gatlin, D.M. Dietary manipulations affecting growth and nitrogenous waste production of red drum, *Sciaenops ocellatus*: I. Effects of dietary protein and energy levels. *Aquaculture.* 1999, 178, 333-348.

101. McGoogan, B.B.; Reigh, R.C. Apparent digestibility of selected ingredients in red drum (*Sciaenops ocellatus*) diets. *Aquaculture.* 1996, 141, 233-244.

102. Medale, F.; Poli, J.M.; Vallee, F.; Blanc, D. Utilisation of a carbohydrate-rich diet by common carp reared at 18 and 25 °C. *Cybium* 1999, 23, 139–152.

103. Millikin, M.R. Effects of dietary protein concentration on growth, feed efficiency and body composition of age-0 striped bass. *Trans. Am. Fish. Soc.* 198, 111, 373-378.

104. Mohanta, K.N.; Subramanian, S.; Korikanthimath, V.S. Effect of dietary protein and lipid levels on growth, nutrient utilization and whole-body composition of blue gourami, *Trichogaster trichopterus* fingerlings. *J. Anim. Physiol. Anim. Nutr.* 2013, 97, 126-36.

105. Mohanta, K.N.; Mohanty, S.N.; Jena, J.K. Protein-sparing effect of carbohydrate in silver barb, *Puntius gonionotus* fry. *Aquac. Nutr.* 2007, 13, 311–317.

106. Mohapatra, M.; Sahu, N.P.; Chaudhari, A. Utilization of gelatinized carbohydrate in diets in *Labeo rohita* fry. *Aquac. Nutr.* 2002, 8, 1-8.

107. Mommsen, T.P.; French, C.J.; Hochachka, P.W. Sites and patterns of protein and amino acid utilization during the spawning migration of salmon. *Can. J. Zool.* 1980, 58, 1785-1799.

108. Monentcham, S.E.; Pouomigne, V.; Kestemont, P. Influence of

dietary protein levels on growth performance and body composition of African bony tongue fingerlings *Heteriostis niloticus* (Cuvier, 1829). *Aquac. Nutr.* 2009, 16, 144-152.

109. Morse, D.E.; Hooker, N.; Duncan, H.; Jensen, L. c-Aminobutyricacid, a neurotransmitter, induces planktonic abalone larvae to settle and begin metamorphosis. *Science*. 1979, 204, 407-410.

110. Nakashima, K.; Yakabe, Y.; Ishida, A.; Yamazaki, M.; Abe, H. Suppression of myofibrillar proteolysis in chick skeletal muscles by a-ketoisocaproate. *Amino Acids*. 2007, 33, 499–503

111. NRC (National Research Council). Nutrient Requirements of Coldwater Fishes. National Academy of Science: Washington, DC, USA, 1981.

112. NRC (National Research Council). Nutrient requirements of fish. National Academy Press: Washington, DC, USA, 1993.

113. NRC (National Research Council). Nutrient requirements of fish and shrimp. National Academy Press: Washington, DC, USA, 2011.

114. New, M. Responsible use of aquaculture feeds. *Aquac. Asia*. 1996, 1, 3-12.

115. New, M.B. Feed and feeding of fish and shrimp; ADCP/Rep/87/26; UNDP/FAO: Rome, 1987.

116. Ogata, H.Y. Muscle buffering capacity of yellowtail fed diets supplemented with crystalline histidine. *J. Fish Biol.* 2002, 61, 1504-1512.

117. Olson, R. *Vitamin K*. In *Modern nutrition in health and disease, 9thEdn*; Shils, M., Olson, J., Shike, M., Ross, A., Eds.; Lippincott Williams and Wilkins: Maryland, USA, 1999; pp 363-380.

118. Ozorio, R.O.A.; Valente, L.M.P.; Pousao-Ferreira, P.; Oliva-Teles, A. Growth performance and body composition of white seabream (*Diplodus sargus*) juveniles fed diets with different protein and lipid levels. *Aquac. Res.* 2006, 37, 255-63.

119. Pinto, W.; Figueira, L.; Dinis, M.T.; Aragao, C. How does fish metamorphosis affect aromatic amino acid metabolism? *Amino Acids*. 2008, 36, 177-183.

120. Popma, T.J. Digestibility of selected feedstuffs and naturally occurring algae by tilapia. Ph.D Dissertation, Auburn University,

Auburn, Alabama, 1982.

121. Poston, H.A., and McCarthey, T.A. Effect of dietary biotin and lipid on growth, stamina, lipid metabolism and biotin-containing enzymes in brook trout (*Salvelinus fontinalis*). *J. Nutr.* 1974, 104, 315-322.

122. Poston, H.A. The effect of excess levels of niacin on the lipid metabolism of fingerling brook trout. *Fish. Res. Bull.* 1969, 32, 9-12.

123. Poston, H.A.; Page, J.W. Gross and histological signs of dietary deficiency of biotin and pantothenic acid in lake trout, *Salvelinus namaycush. Cornell Vet.* 1982, 72, 242-261.

124. Poston, H.A.; Combs, G.F. Jr.; Leibovitz, L. Vitamin E and selenium interrelations in the diets of Atlantic salmon (*Salmo salar*): gross, histological and biochemical deficiency signs. *J. Nutr.* 1976, 106, 892-904.

125. Poston, H.A.; Livingston. D.L. *Effects of massive doses of dietary vitamin E on fingerling brook trout. Fish. Res. Bull* 1969, 33, 6-12.

126. Pozios, K.C.; Ding, J.; Degger, B.; Upton, Z.; Duan, C. IGFs stimulate zebra fish cell proliferation by activating MAP kinase and PI3-kinase-signaling pathways. *Am. J. Physiol. Regul. Integr. Comp. Physiol.* 2001, 280, 1230-1239.

127. Prabu, D.L.; Ebeneezar, S.; Chandrasekar, S.; Sayooj, P.; Vijayagopal, P. Book of Abstracts, National Seminar on Aquaculture Diversification: The Way Forward for Blue Revolution, Bhubaneswar, India, December 01-03, 2016; ICAR-CIFA: Bhubaneswar, India, 2016b.

128. Prabu, D.L.; Ebeneezar, S.; Chandrasekar, S.; Vijayagopal, P.; Sayooj, P.; Rao, S.N.; Iyyapparajanarasimapallavan, G. Books of Abstracts, International conference on Science and Technology for National development, Kerala, India, October 25-26, 2016; Kerala University of Fisheries and Ocean Studies: Kerala, India, 2016a, AQFS-P 7.

129. Riley, W.W.; Higgs, D.A.; Dosanjh, B.S.; Eales, J.G. Influence of dietary arginine and glycine content on thyroid function and growth of juvenile rainbow trout, *Oncorhynchus mykiss* (Walbaum). *Aquac. Nutr.* 1996, 2, 235-242.

130. Rucker, R.R.; Johnson, H.E.; Kaydas, G.M. An interim report on

gill diseases. *Prog. Fish Cult.* 1952, 14, 10-14.

131. Sanchez-Muros, M.J.; Garcia-Rejon, L.; Lupianez, J.A.; De la Higuera, M. Long-term nutritional effects on the primary liver and kidney metabolism in rainbow trout (*Oncorhynchus mykiss)*. 2. Adaptive response of glucose-6-phosphate dehydrogenase activity to high carbohydrate/low-protein and high-fat/ non-carbohydrate diets. *Aquac. Nutr*. 1996, 2, 193-200.
132. Sankian, Z.; Khosravi, S.; Kim, Y-O.; Lee, S-M. Effect of dietary protein and lipid level on growth, feed utilization, and muscle composition in golden mandarin fish *Siniperca scherzeri. Fish. Aquat. Sci.* 2017, 20, 7.
133. Sargent, J.R.; Tocher, D.R.; Bell, J.G. *The lipids*. In *Fish Nutrition*; Halver, J.E., Hardy, R.W., Eds.; Academic Press: San Diego, CA, USA, 2002; pp. 181-257.
134. Sen, P.R.; Rao, N.G.S.; Ghosh, S.R.; Rout, M. Observation on the protein and carbohydrate requirement of carps. *Aquaculture*. 1978, 13, 245-255.
135. Shamushaki, V.A.J.; Kasumyan, A.O.; Abedian, A.; Abtahi, B. Behavioural responses of the Persian sturgeon (*Acipenser persicus*) juveniles to free amino acid solutions. *Mar. Freshw. Behav. Physiol.* 2007, 40, 219–224.
136. Shearer, K.D.; Swanson, P. The effect of whole body lipid on early sexual maturation of 1+ age male chinook salmon (*Oncorhynchus tshawytscha*). *Aquaculture.* 2000, 190, 343-367.
137. Shiau, S.Y.; Peng, C.Y. Protein-sparing effect by carbohydrate in diets for tilapia *Oreochromis niloticus* x *O. aureus*. *Aquaculture*. 1993, 117, 327-334.
138. Shiau, S.-Y.; Hsieh, H. L. Vitamin B6 requirements of tilapia *Oreochromis niloticus x O. aureus* fed two dietary protein concentrations. *Fish. Sci.* 1997, 63, 1002-1007.
139. Shimeno, S.; Hosokawa, H.; Takeda, M. Metabolic response of juvenile yellowtail to dietary carbohydrate to lipid ratios. *Fish. Sci.* 1996, 62, 945-949.
140. Simopoulos, A.P. Evolutionary aspects of diet, the omega-6 / omega-3 ratio and genetic variation: nutritional implications for chronic diseases. *Biomed. Pharmacother.* 2006, 60, 502-507.

141. Sire, M.F.; Lutton, C.; Vernier, J.M. New views on intestinal absorption of lipids in teleostean fishes: an ultra structural and biochemical study in the rainbow trout. *J. Lipid Res.* 1981, 22, 81-94.

142. Steffens, W. *Principles of fish nutrition*; Ellis Horwood Ltd.: West Sussex, UK, 1989.

143. Stone, D.A.J. Dietary carbohydrate utilization by fish. *Rev. Fish. Sci.* 2003, 11, 337-369.

144. Sugiura, S.H.; Hardy, R.W.; Roberts, R.J. The pathology of phosphorus deficiency in fish - a review. *J. Fish Dis.* 2004, 27, 255-265.

145. Sunde, J.; Taranger, G.L.; Rungruangsak-Torrissen, K. Digestive protease activities and free amino acids in white muscle as indicators for feed conversion efficiency and growth rate in Atlantic salmon (*Salmo salar* L.). *Fish Physiol. Biochem.* 2001, 25, 335-345.

146. Terrance, J.Q.; Richard B. D. *Quantitative Fish Dynamics*; Oxford University Press: Oxford, UK, 1999.

147. Thibault, M.; Blier, U.; Guderley, H. Seasonal variation of muscle metabolic organization in rainbow trout (*Oncorhynchus mykiss*). *Fish Physiol. Biochem.* 1997, 16, 139-155.

148. Tocher, D.R. Fatty acid requirements in ontogeny of marine and freshwater fish. *Aquac. Res.* 2010, 41, 717-732.

149. Tocher, D.R.; Ghioni, C. Fatty acid metabolism in marine fish: low activity of fatty acyl D5 desaturation in gilthead seabream (*Sparus aurata*) cells. *Lipids* 1999, 34, 433-440.

150. Trotter, A.J.; Pankhurst, P.M.; Hart, P.R. Swim bladder malformation in hatchery-reared striped trumpeter *Latris lineate* (Latridae). *Aquaculture* 2001, 198, 41-54.

151. Trushenski, J.T.; Kasper, C.S.; Kohler, C.C. Challenges and opportunities in finfish nutrition. *North Am. J. Aquac.* 2006, 68, 122-140.

152. Tucker, B.W.; Halver, J.E. Utilization of ascorbate-2-sulfate in fish. *Fish Physiol. Biochem.* 1986, 2, 151-160.

153. Vinberg, G.G. *Rate of metabolism and food requirements of fishes*;

Fisheries Research Board of Canada, Biological Station: Nanaimo, B.C, Canada, 1960.

154. Walton, M.J.; Cowey, C.B. Aspects of intermediary metabolism in fish. *Comp. Biochem. Physiol.* 1982, 73B, 59-79.

155. Wang, C.S.; Hartsuck, J.A. Bile salt-activated lipase. A multiple function lipolytic enzyme. *Biochim. Biophys. Acta* 1993, 1166, 1-19.

156. Watanabe, T.; Lee, M.; Mizutani, J.; Yamada. T.; Satoh, S.; Takeuchi, T.; Yoshida, N.; Kitada, T.; Arakawa, T. Effective components in cuttlefish meal and raw krill for improvement of quality of red sea bream *Pagrus major* eggs. *Nippon Suisan Gakkaishi.* 1991, 57, 681-694.

157. Watanabe, T. Importance of docosahexaenoic acid in marine larval fish. *J. World Aquac. Soc.* 1993, 24, 152-161.

158. West, E.S.; Todd, W.R.; Mason, H.S.; Van Bruggen, J.T. *Textbook of Biochemistry*. Macmillan Co: New York, 1966.

159. Wilson, R.P. Utilization of dietary carbohydrate by fish. *Aquaculture.* 1994, 124, 67–80.

160. Wilson, R.P. *Amino acids and proteins*. In *Fish Nutrition*; Halver, J.E., Hardy, R.W., Eds.; Academic Press: San Diego, CA, 2002; pp. 144-179.

161. Wilson, R.P.; Poe, W.E. Apparent digestibility protein and energy coefficients of feed ingredients for channel catfish. *Prog. Fish Cult.* 1985, 47, 154-158.

162. Wilson, R.P.; Halver, J.E. Protein and amino acid requirements of fishes. *Ann. Rev. Nutr.* 1986, 6, 225-244.

163. Wu, A.G.; Ying, Z.; Gomez-Pinilla, F. Omega-3 fatty acids supplementation restores mechanisms that maintain brain homeostasis in traumatic brain injury. *J. Neurotrauma.* 2007, 24, 1587-1595.

164. Wu, G.; Bazer, F.W.; Davis, T.A.; Jaeger, L.A.; Johnson, G.A.; Kim, S.W.; Knabe, D.A.; Meininger, C.J.; Spencer, T.E.; Yin, Y.L. Important roles of the arginine family amino acids in swine nutrition and production. *Livest. Sci*. 2007, 112, 8-22.

165. Wu, G.; Morris, S.M., Jr. Arginine metabolism: nitric oxide and beyond. *Biochem. J.* 1998, 336, 1-17.

166. Yang, S.D.; Liou, C.; Liu, F. Effects of dietary protein level on growth performance, carcass composition and ammonia excretion in juvenile silver perch, *Bidyanus bidyanus*. *Aquaculture.* 2002, 213, 363-372.

167. Zeng, L.; Lei, J.; Ai, C.; Hong, W.; Liu, B. Protein-sparing effect of carbohydrate in diets for juvenile turbot *Scophthalmus maximus* reared at different salinities. *Chin. J. Oceanol. Limnol.* 2015, 33, 57-69.

CHAPTER 4

GROWTH REGULATION IN FISH BY ENVIRONMENTAL PARAMETERS

Shubhadeep Ghosh[1], Ritesh Ranjan[2] and Biji Xavier[3]

[1]Author has completed his Master of Fishery Science (M F Sc) in Fishery Pathology and Microbiology and his Ph.D. on the role of probiotics in enhancing growth and survival in live-bearing ornamental fishes. He was inducted in Agriculture Research Service (ARS) in 2005 and since then working in Pelagic Fisheries Division of CMFRI. He is currently acting as Scientist-in-charge of Visakhapatnam Regional Centre of CMFRI. He has studied in detail of the biology, population parameters and stock structure of most commercially important pelagic fishery resources from the west and east coast of India. He has been actively involved in pop-up satellite tagging of yellowfin tunas from the Indian seas and contributed to the expansion of cage mariculture in the country. Recently, under his guidance, Visakhapatnam RC has bred and produced seed of Indian pompano (Trachinotus mookalee) for the first time in the world.

[2]Author has graduated from College of Fisheries, Mangalore in 2003 and completed his post-graduation in 2005 and Ph. D. in 2012 in Fish Pathology and Microbiology from Central Institute of Fisheries Education, Mumbai. Joined as scientist in Central Marine Fisheries Research Institute (CMFRI) in 2008 and since then he has been working in the field of mariculture, particularly on marine finfish breeding. His research work mainly focused on the sex reversal of grouper and it's broodstock development. Presently, he is working on breeding and seed production of grouper and Indian Pompano, and live feed culture for marine finfish larval rearing.

[3]Dr Biji Xavier obtained graduation in fisheries science followed by master degree and PhD in Fish Nutrition and Biochemistry from Central Institute of Fisheries Education (CIFE), India. She joined in Agriculture Research Service in 2009 as Scientist and at present working in mariculture. She has been working on breeding, seed production and culture of high value marine finfishes. Dr. Biji has published research papers in peer reviewed

journals in the area of fish nutrition and Biochemistry as well as fish culture especially various aspects related to live feed culture.

INTRODUCTION

Information about fish age, development and growth are pivotal for fishery research and management. Growth is a characteristic feature of living beings. Every organism grows and attains its normal size. Growth is actually an increase in any dimension of any organism in relation to time. While a pattern of determinate growth is displayed by fish that have a shorter life-span and inhabit warmer locations, an indeterminate pattern is exhibited by those that occupy colder areas and a longer life span. Length and weight are used to measure growth of which specific growth rate is the best representation. A condition factor "K" is used that is an expression between length and weight of a fish. An essential regulator of growth is the amount and the quality of nutrients (Sankian et al., 2017). The growth rate is also controlled by several factors of the environment that include the level of oxygen, exposure to light, temperature and level of salt. At an endogenous level, the genotype along with the conditions at a physiological level including hormones plays vital roles as growth regulators. While an increase of size of the entire body structure or a part is referred to as growth, the change in size with reference to time is referred to as growth rate. The food quality and level dictates the growth rate as being dependant on the nutrition as less amounts of nutrition cause hampered growth along with impaired development such as maturation at a sexual level (Hutchings, 2002). The levels of nutrition along with the growth in size, reserve of energy, sexual maturation as well as reproduction are all connected in an intimate fashion. This is manifested as a population with the growth of fish as an indicator of health proportional to the water quality of the habitat. The growth of fish is also dependent on physical factors such as the temperature and the levels of solids in the water or chemical factors such as the pH, the metals dissolved and the water hardness (Enayatmehr and Jamili, 2013).

The quality of water is critical for the production of fish that is also challenging to comprehend and co-ordinate. The water quality can be a factor (chemical or physical or biological) that influences the production of fish. Apart from the habitation of fish, water pays a role in the health, growth, efficiency of feeding as well as survival of the fish. In pond management, the management of water quality is the main aim which in turn fulfils the requirements of optimal health and production of fish with all the parameters within expected ranges.

WATER QUALITY PARAMETER AND FISH GROWTH

Temperature

The temperature of water is the most integral factor that influences the growth and proliferation of fish. Fish exhibits temperature optimum for growth and survival. The body temperature of fish is equivalent to the environment by virtue of their cold-bloodedness. This affects metabolic activity, and is a major factor controlling muscle growth and muscle phenotype (Johnston et al., 1975). The development and growth of muscles is influenced by the temperature of the environment as the cells of muscles are affected by fluctuations especially for fish in initial stages of development (Sumpter, 1992). An optimum temperature dictates a better growth profile when compared to extremes of temperature towards either end. Fluctuations beyond the optimum range can cause a reduction of the immune system and a drop in the tolerance of shifts in certain compounds like ammonia in the waters (Viadero, 2005). This is reflected in a lowered growth or productivity or even in extreme cases death. For instance, *Oncorhynchus mykiss* or rainbow trout has an optimum temperature range of between 1 to 26 °C for survival. However, depending on the reference source, the optimum temperature range for the growth of the fish is between 13 to 21°C (Jobling, 1994; Viadero, 2005). Water temperature is directly influencing the growth rate at all the stages from egg to till death of the fish. The temperature of water plays a role in shaping the meristic characters of finfishes. This refers to influencing the formation of segments by exerting control on the lengthening and differentiation of the body. The differentiation of muscles and timing of formation of organs is also dictated by temperature. The meristic features of the offspring are controlled by the temperature of the water during gametogenesis of the broods thus showing the effect of temperature prior to fertilization.

In fishes, the growth pattern is also related with the environmental temperature and the maximum individual growth rate increases with temperature to an optimum value, and then it decreases rapidly at higher temperatures. The decreasing growth rate of fish at high temperature is probably related to many other factors, importantly the increasing temperature decreases dissolved oxygen in the water and this decreasing dissolved oxygen directly influencing on the fish growth. This function allows ectotherms to maintain thermal homeostasis within a wide range of temperatures. Temperature influences the rate of myogenesis (formation of muscular tissue), including composition of the functional structures within the cells, patterns of gene expression, number and size of muscle fibres, fibre diameters and myofibril areas (Wilkes et al., 2001).

Water temperature is the major environmental factor that controls food utilization at all levels and all stages of fish growth. Growth of fishes differs depending on whether they are grown at constant or fluctuating temperatures. Temperature affects growth, maturation, gonadal development, food intake, maintenance requirements, metabolic rates, enzyme and membrane function. Temperature affects the utilization of endogenous reserves during ontogenesis of teleosts. The amount of yolk consumed shows a dip with an elevation of water temperature. The utilization of endogenous nitrogen reserves in common carp was variably affected by temperature (Kaushik et al., 1982). For most species, the metabolic rate of embryos at hatching is correlated with temperature. Temperature also influences the rate of food consumption. Metabolic rates of fishes and subsequently the maintenance energy in several species of fish increases with temperature. During digestion, the rates of transit of food-stuffs through the different digestive compartments, digestibility of nutrients and activities of digestive enzymes are all influenced by water temperature. The proteolytic and amylolytic activities in the roach (*Rutilus rutilus*) and the rudd (*Scardinius erythrophthalmus*) were correlated with the seasonal temperature cycles (Hofer, 1979). In most teleosts, with increase in water temperature, there was an increase in gastric evacuation and transit rates (Fange and Grove, 1979). Most enzymes of intermediary metabolism viz., aspartate and alanine aminotransferase, glutamate dehydrogenase, glutaminase and uricase increased with an increase in water temperature and then remained unchanged. Protein synthesis rates are impacted by temperature and an increase in temperature results in an increase in protein synthesis and its re-utilization at the cellular level that lead to growth of a fish. The relative weights of various organs and tissues are also affected by water temperature. In Chinook salmon, maximum tissue weights varied inversely with temperature (Heming et al., 1982).

In general, it has been found that mean temperature has a close relationship with growth rate, metabolism and survival, especially during the juvenile phase. The alterations of pathways key to metabolism by enzymes are hypothesized to be the cause of the link between temperature and growth rate (Buentello et al., 2000). The rate of growth under conditions of optimal levels of nutrition is proportional to temperature in majority fish that inhabit warmer habitats (Talbot, 1993). Ficke et al. (2007) observed that each species of fish has an optimal temperature range for growth performance; for warm water or fish in tropical regions in general, optimal temperature for growth ranges from 20 to 32°C (Ficke et al., 2007; McLarney, 1998). Thus, several studies have focused on identifying optimum temperature for fish growth, which helps to avoid the culture of fish in unfavorable temperature condition and wasting the resources.

Salinity

Fish growth in water is greatly influenced by different environmental parameters. Salinity is one of the most exhaustively researched factors that have been subject to study. Fish lives in different salinities according to their origin such as freshwater, seawater and brackish water. Based on the salinity tolerance, the fish can be grouped in to euryhaline and stenohaline fishes. Osmoregulation that is driven by energy to maintain optimum levels of ions and osmotic pressures is used by teleost fish (Sampaio & Bianchini, 2002). Along similar lines, the overall outcome between the amount of food consumed and the energy lost in metabolism dictates fish growth (Jobbling, 1994). Therefore, by understanding the osmoregulatory cost as being proportional to the osmotic gradient existing between the fish body fluids and the external medium (Handeland et al., 1998), it is reasonable to expect that growth would be maximized when fish is reared in water that is with salinity near the isosmotic condition. By doing so, the energy utilized for osmoregulation could be reduced and diverted for growth. Thus, identifying the optimum salinity for fish culture would lead to improve the fish growth through salinity regulation. Influence of water salinity on fish development and growth has been studied well in most of the fish species. Salinity exerts control on several activities such as the fertilization of eggs followed by incubation, utilization of the yolk, the formation of embryos, inflation of bladder as well as development of larvae. The growth of big fish is controlled by salinity to a major extent in some cases. Changes in the growth rate due to salinity variation are result from an action on different factors like standard metabolic rate; food intake; food conversion; hormonal stimulation. Different studies showed that food intake and stimulation of food conversion are dependent on the environmental salinity. The association between salinity and temperature is multi-fold and complicated as the regulation of growth and osmoregulatory activities are simultaneously affected by certain hormones (Boeuf & Payan, 2001).

As euryaline fish can alter the protein transporters of salt depending on whether they inhabit fresh or salt water, they can inhabit areas of different salinity (Kidder III et al., 2006). Thus, such fish can be grown on waters across a range of environments. However, identifying the optimum salinity for better growth of the animal is needed to be standardized for different species. Study on indentifying optimum salinities for better growth was conducted for lebranche mullet *Mugil liza* in different range of salinities. As compared to the juveniles in freshwater, those reared in saline waters displayed an increase in growth of 24%. This implicated that in isosmotic environments, the amount of energy used for osmoregulatory mechanisms can be lowered in juvenile *M. liza* (Lisboa et al., 2015). An increase in energy employed in osmoregulation due to enhanced NKA in gills could serve a plausible contributor of the decreased levels of growth in freshwater. Studies in *Sparus sarba* that recorded better growth in isosmotic levels of salinity was attributed two reasons. First a decreased level of metabolism when fish were placed in isosmotic environments and second is an alteration in metabolism to use proteins for growth while other biomolecules are metabolized (Woo & Kelly, 1995).

Salinity effects have been studied extensively in marine fish embryos and larvae (Jørgensen and Hansen, 2010), but they are limited in freshwater species (Bonisławska, 2009). It has shown that some freshwater teleost eggs can be incubated and hatched in 5ppt salinity but fertility and eggs hatching rate of freshwater teleost decreases in saline water (Gbulubo and Erondu, 1998). In most marine species, optimum growth is observed at salinities ranging from 19 to 35 ppt. But, it appears that marine fish present higher developmental or growth rates at lower salinity and freshwater fish shows better growth at higher salinity. Many species, while in juvenile stage, opt for intermediary salinity conditions of the brackish water (8-16 ppt) and grow optimally in estuaries and costal systems. It is indicated that the salinity tolerance during early developmental stages depends on how internal fluids perform in various extents (Enayatmehr & Jamili, 2013).

Most of the developmental processes in marine fish are influenced by the salinity. The efficiency of yolk utilization is found to be linearly related to water salinity; for turbot, the efficiency was greater at 32 ppt (87%) than at 17 ppt in brackishwater condition (Quantz, 1985). In addition, the nutrients utilization in some of the fishes is also influenced by the salinity. Protein and lipid utilization in rainbow trout adapted to fresh water and sea water were found to vary and also observed that in brackish water it required higher protein (Lall and Bishop, 1979). Protein sparing action of fat was also found to differ in rainbow trout grown at different salinities (Jurss et al., 1985). Free amino acids are known to take an active part in osmoregulation and in rainbow trout, a significant increase in muscle free amino acid levels was observed at salinities above 29 ppt (Kaushik et al., 1977). The maintenance protein requirement of trout decreases with increasing salinity (Kaushik and Luquet, 1977). The endogenous nitrogen excretion in rainbow trout was lowest at the isotonic salinity of 10 ppt and higher both in fresh water and at 20 ppt (Roberts, 1977). Digestible energy of several feedstuffs was found to increase at sea water salinities. Absorption and transport of individual amino acids were also affected by water salinity in rainbow trout (Kaushik and Luquet, 1977).

Enzymes of intermediary metabolism in rainbow trout plasma were most active at salinities of 10 ppt (Sauer and Haider, 1979). The energy charge of rainbow trout was higher in sea water than in fresh water for rainbow trout, indicating lesser energy demands at higher salinities. Metabolic rates of juvenile carp were also found to decrease with increasing salinities (Maceina et al., 1980). Gluconeogenesis and lipogenesis decreased in sea water adapted salmonids. Seabass, transferred from sea water (38 ppt) to a lower salinity (4 ppt), resulted in an increase in oxygen uptake and in the respiratory quotient. Oxidative catabolism of glucose and acetate were higher at lower salinities. Gluconeogenesis and protein sparing effects were higher at lower salinities than in sea water.

Dissolved Oxygen

The efficiency of water as a respiratory medium depends to a great extent on the solubility of oxygen and carbon dioxide in a given body of water. The temperature of water and salinity play a role in controlling the solubility of oxygen. Hypoxia affects blood oxygen carrying capacity and under hypoxic conditions severe disturbances of energy metabolism have been recorded viz., increase in blood glucose, lactate and glycolysis. Tolerance towards hypoxic conditions is greater during early embryonic stages than in larval and post-larval stages. Protein synthesis is affected by ambient oxygen levels in *Fundulus* sp. (Jackim and Laroche, 1973). A decrease in oxygen saturation from 100% to 50% led to significant changes in protein metabolism of rainbow trout. Trout grown at an ambient oxygen concentration of 5 mg/l had decreased food intake, depressed growth and food conversions relative to those grown in normoxic water (8 mg/l).

The quantity of dissolved oxygen dictates life in aquatic milieu as it is essential for metabolism encompassing the growth and survival of aerobic organisms such as fish. Dissolved oxygen is influenced by several factors that include temperature, salinity, currents, wind effects, pressure as well as photosynthesis and respiration. For warm water fishes, the optimum level of dissolved oxygen required is 4 to 5 mg/l (Wedemeyer and Goodyear, 1984). The processes of metabolism and respiration depend on oxygen and in fish the level of oxygen in the cultivated habitats influence the metabolism. The critical oxygen tension is a point where fish undergo a transformation in the response of metabolism to the concentration of oxygen. At this point the rate of metabolism depends on the level of oxygen as opposed to an initial scenario of the metabolism not dependent on the gas. The consumption of food is oxygen dependent with a decrease in respiration and consequent food consumption as the dissolved oxygen decreases. A study by Randolph and Clemens showed that channel catfish displayed a drop in feeding as the level of oxygen decreased below 59% (Randolph and Clemens, 1976).Another 1995 study showed a drop of oxygen below 60% was accompanied by decrease in appetite in rainbow trout (*Oncorhynchus mykiss*) (Jobling,1995). In tilapia (*Oreochromis niloticus*), FCR was observed to be inversely proportional to the dissolved oxygen level (1.45 at higher dissolved oxygen level and 6.75 at lower dissolved oxygen level (Tsadik and Kutty, 1987). This results in depression in growth rate and an elevation in susceptibility to disease. This dictates saturated levels of dissolved oxygen for optimal growth. Reduction in growth of several fish has been observed when the levels of oxygen are lower such as juvenile turbot (Pichavant et al., 2001), channel catfish (*Ictalurus punctatus*) (Buentello et al., 2000), European sea bass (*Dicentrarchus labrax*) (Thetmeyer et al., 1999), common carp (*Cyprinus carpio*) and blue tilapia (*Oreochromis aureus*) (Papoutsoglou and Tziha, 1996).

pH

The level of acidity or basicity is reflected by pH. Fishes have an average blood pH of 7.4, thus the water with a pH close to this is considered optimum for growth and survival and well being of fish. The acceptable range of pH for different category of fish ranges from 6.5 to 9.0. Keeping the fish in the water in the extreme range of pH may affect the growth of the fish. Water pH affects metabolism and physiology of fishes. At low pH or as acidity increases, the level of elevated hydrogen ions induces the epithelial cells of gills to produce mucus that in turn cause an interference of gaseous exchange or blood acidosis or gill protein precipitation. Rapid acidification also causes necrosis of the integumental epithelium, and results in damage of the skin and high plasma electrolyte loss and impairment of the response of prolactin cells. At higher pH values, the elevated levels of hydroxyl ions induce hypertrophy of cells that produce mucus in the gills causing impairment of the epithelial cells of the gills and skin. Thus, the pH alterations are translated as hindrance to respiration or an imbalance of ions. Such stress is manifested when the pH of water is in the range of 4.0 to 6.5 and 9.0 to 11.0 with fish growth impaired when values are lesser than 6 or more than 9 (Boyd, 1982). Acid and alkaline stresses cause suppression of growth hormone secretion. At low pH, acid load through the gills is the source of acid-base disturbances and there is an increase in H^+ and NH_4^+ excretions in urine to compensate this problem. Acidic pH impairs growth in rainbow trout due to a decrease in food intake (D'Cruz and Wood, 1998), as was also observed in silver catfish (Copatti et al., 2005). Changes in Na^+ plasma levels require more energy to maintain the ion-regulatory balance, and explains the lower growth observed in the fish kept at pH 5.5. In acidic waters, high H^+ concentrations disrupt the tight junctions of gill epithelia, increasing ion loss by a paracellular route, and leading to whole body ion loss, as observed in silver catfish (Zaions and Baldisserotto, 2000). In very acidic waters fishes fail to regulate their internal ion content. Both influx and efflux of sodium and chloride through the gills and kidneys are affected by this. Sodium influx rates are largely dependent of pH. Empirical evidences have shown that fish blood, muscle and cellular parameters are altered by pH. Growth rate decreases significantly because it requires more energy to osmoregulate, and the fish are allocating more energy towards osmoregulation, and less towards growth. Aluminium ions (Al^{3+}) attached to minerals in nearby soil are released into water at lower pH and causes chronic stress that leads to lower body weight and smaller size and makes fish less able to compete for food and habitat. Higher pH reduces affinity for oxygen in the blood and thus decreases the oxygen carrying capacity of haemoglobin. Ammonia excretion increases at higher pH. Growth of fishes is less possibly because on its effects on food consumption and bacterial decomposition rate (involving nutrient availability). The reduction in growth at pH below 6.0 and above 8.0 is attributed to the imbalances in homeostasis. At pH values below 4.0 or higher than 10.0 mortality is ensured. The pH of water in a pond shows shifts as pH reaches a low at dawn due to the production of carbonic acid from CO2 from respiring plants.

Alkalinity and Hardness

The term alkalinity refers to the level of bases such as hydroxides, bicarbonates and carbonates and is the property of water to oppose shifts in pH as such basic compounds that bind acidic compounds to hence serve as buffers. The major contributors of alkalinity include carbonates and bi-carobonates and aid in buffering water. Excess CO_2 is stored as bicarbonates to avoid limiting levels of the gas that is a raw material in photosynthesis that in turn maintains steady levels of oxygen. Water that possesses an elevated level of alkalinity can resist changes in pH to a larger extent. Fish growth and production increases with an increase in total alkalinity and the value should be more than 20mg/l in any production system (Boyd, 1982).

The levels of alkaline earth elements like magnesium and calcium refer to a property of water called hardness. Such calcium and magnesium play vital roles in processes such as the formation of scales and bones. Greater the level of these metals, the water becomes "harder". Hardness can be temporary where magnesium and calcium are present as bicarbonates and carbonates or permanent where the metals associate with anions like sulfate. The pH of water is a reflection of both alkalinity and hardness through carbon dioxide cycle. Additionally, hardness also impacts osmoregulation that involves maintenance of a steady state of concentration of fluids in fish. The greater the difference in concentration between the fish body fluids and the surrounding water with the greater the osmotic effect. Due to the presence of salts, hard water has a higher concentration that leads to lower difference in the body fluids and water that leads to lower osmotic effects saving the fish of energy diverted in osmoregulation. It has been reported that fish in warm habitats achieve growth and health with a hardness level of 15 mg/l (Jhingran, 1988) while a level of 50 mg/l or more was optimal for tilapia (Lucas and Southgate, 2012).

Ammonia

Ammonia, the major product of teleost excretion represents the majority of nitrogenous waste in intensive aquaculture production. It is also the nitrogenous end-product of amino acid and protein oxidation. In fish, ammonia is produced predominantly in the liver through the transamination of amino acids, but it may also be produced in the kidneys, gills, and skeletal muscle. Among all the water quality parameters, which affect fish, ammonia is considered as one of the most important after oxygen. In intensive fish culture, high volumes of uneaten feed and feces lead to the accumulation of nitrogenous waste. When fishes are reared in a condition, where reuse of effluent is followed, the concentration of ammonia increases to concentrations that threaten the growth and life of fish. It was reported that culture of fish necessitates the level of ammonia to be lower than 0.1 mg/l (Boyd, 1998) while fish in warm habitats are healthy when the level of ammonia is 0.02 mg/l (Jhingran, 1988). The amount of ammonia in water is comprised of ionized NH_4^+ ions that are non-toxic and ammonia NH_3 that is toxic. The combination of the levels of both these forms yields the total ammonia nitrogen

(TAN). The temperature and pH of the environment influences the equilibrium between NH_4^+ and NH_3. Ammonium (NH_4^+) is regarded as the less toxic form of ammonia due to its larger size and charge, which makes it much less diffusive and thus is restricted to movement through water-filled channels. The solubility of NH_3 coupled with its absence of charge enables the molecule to diffuse across the gills and thus becomes several hundred folds more toxic than the ionized form. When ammonia accumulates to toxic levels, fishes cannot extract energy from the feed and hence growth is retarded fishes fall into a highly stressed condition and eventually dies. As the molecule causes blockage of gas exchange across the gills, the effect is rapid and lasts for a long time. The usual effects appear as a general thickening of the epithelial membrane with associated hyperplasia; sometimes there is a breakdown of the pillar cell structure of the secondary lamellae. This general disruption of branchial structure effectively reduces the surface area of the gill membrane, with resultant reduction in oxygen-diffusing capacity. Additionally, the molecule hampers the metabolic pathways in the cerebrum and other tissues such as spleen, kidney and liver. The growth and reproduction ability of the fish is severely compromised apart from death. Therefore, excessive ammonia in water causes a reduction of growth apart from impairment of the immune system, damage of tissues and death.

The growth of fish is inhibited by ammonia and thus the toxic effects can be reflected by the growth patterns. The detoxification of ammonia requires an investment of energy and hence compromises the growth of fish. In juvenile Nile tilapia, total ammonia in water had a negative effect on fish whole body growth (Lemarie et al., 2004). This was explained by the fact that ammonia toxicity disrupted the production of energy resources and resulted in a reduction of 68% of potential energy output. It followed that energy that was normally allocated to optimal growth was shunted towards the detoxification process. The negative effect of ambient ammonia concentrations on weight gain has been noted in several other species, such as seabass (Lemarie et al., 2004), juvenile turbot (Rasmussen and Korsgaard, 1996), lake trout (Beamish and Tandler, 1990) and channel catfish (Robinette, 1976). Exposure of fish to ammonia solutions generally raises the blood ammonia levels either due to ammonia retention in the fish or by entry of ammonia from the external medium resulting in a reduction in overall ammonia excretion.

Paradoxically recent studies have noted beneficial effects of low levels of total ambient ammonia (Wood 2004; Madison et al., 2009) and its potential to enhance fish growth. In juvenile rainbow trout, exposure to low levels of ammonia stimulated growth without an increase in food consumption. Furthermore, ammonia exposed fishes had an increased nitrogen and energy conversion efficiency with higher nitrogen retention at a lower metabolic cost. Fishes exposed to elevated ambient water ammonia were retaining greater overall levels of nitrogen in the form of protein, and utilizing it to enhance growth. Sub-lethal levels of water ammonia, elevated ammonia concentrations in fish plasma, triggering ammonia detoxification mechanisms, which in turn stimulated protein turnover rates and led to a higher retention of nitrogen. In another study, significant increases in mass, food conversion and protein content was recorded in juvenile rainbow trout exposed for a long period to low levels of ammonia (Wood, 2004). The effect of ammonia on growth is a product of the relative concentration of NH_3, which increases with temperature relative to NH_4^+. The increased protein accretion without an alteration in food consumption is the result of reduction of metabolic costs by the fish exposed in less ammonia concentration in water body.

Nitrite

The toxic effects of nitrite in fish are attributed to suppression of growth as well as disruption of physiology. Nitrite is a potential problem in aquatic environments. Increased levels of nitrite are attained as fish especially the ones that inhabit fresh waters imbibe the compound by their gills. Fish in marine environments are not that affected but reports indicate the intestine and gills absorbing nitrite. Freshwater fishes actively take up nitrite across the gills, leading to high internal concentrations. Sea water fish are less susceptible but do take up nitrite across intestine and gills. Nitrite has multiple physiological effects that damage regulation of ions, circulation, respiration, excretion, endocrine functions as well as dip in the levels of chloride against which the compound is taken up. In rainbow trout, *Oncorhynchus mykiss*, nitrite acted as a competitive inhibitor of chloride uptake in the gills. The treatment of nitrite on gills of carp (*Cyprinus carpio*) was observed as formation of vacuoles, hyperplasia and increase of chloride cells (Svobodova et al., 2005). The levels of potassium inside and outside the cells are also disrupted as nitrite facilitates the efflux of the metal from muscles and RBCs. A 1987 study observed a huge increase in the level of K^+ in carp outside the cells (Jensen et al., 1987). Additional studies in carp (Jensen, 1990) and trout (Stormer et al., 1996) revealed efflux of potassium from RNCs and muscles of the skeleton leading to increase in the ion in the body fluids. This increased level of K^+ outside the cell results in depolarisation of excitable tissues such as heart leading to damages to heart and nervous systems. The decrease in level of K^+ inside the cell can hamper the functioning of muscles. Nitrite transfer across the erythrocytic membrane leads to oxidation of haemoglobin to methemoglobin (metHb) that leads to an impairment of transport of oxygen, compromising blood O_2 transport. Other haem proteins are also oxidized. There are reports of observation of hyperventilation and increased levels of lactate due to the lack of oxygen. The methemoglobin levels that can impair growth depend across species and habitats. The amount of methemoglobin necessary to reduce growth varies with species and the environment conditions. Nitrite also takes a toll on cardiovascular systems as observed with an elevation of heart rate in rainbow trout studies (Aggergaard and Jensen, 2001). Rapid increases in heart rate suggest nitrite-induced vasodilation, possibly via nitric oxide generated from nitrite that is countered by increasing the cardiac pumping to re-establish the blood pressure. Disruptions and death are caused in lysosomes and mitochondria of fish by nitrite. Nitrite forms and/or mimics nitric oxide and thereby interferes with processes regulating the secretions of growth hormone. Reduced feeding and an elevation of the maintenance metabolism of the fish, through activation of the defence systems (NADH-reductase activity to reduce methemoglobin to haemoglobin) at high nitrite concentrations leads to lower growth rates, increased food conversion rates and higher susceptibility to diseases.

Crowding Stress

Another stress factor to the physiology of fish is crowding. This factor is translated as effects ranging from poor feeding, behavioral changes that is observed as hampered growth or death. Crowding stress indirectly reduces the growth through a negative effect on energy balance. High stocking density also causes a breakdown of resources as well as stress in fish. Crowding causes changes in immunology of fish as well as metabolic changes such as hampered thyroid. Studies in Amur sturgeon showed that the density of stocking was indirectly proportional to the levels of thyroid hormones in the serum (Li et al., 2012). Studies in brook charr (*Salvelinus fontinalis*) showed hampered growth with lower levels of circulating thyroid hormone levels (Vijayan et al., 1990). These altered levels were accompanied by reduction in haemoconcentration and immune functions such as alternative complement pathway. The levels of haemoglobin, erythrocytes and haematocrit were elevated in fish in crowded environments. This effect has been described as a strategy for increasing oxygen carrying capacity of blood during periods of high energy demand. The concentration of growth hormone in the plasma in rainbow trout was reduced due to stress (Pickering et al., 1991). The level of cortisol in plasma was increased with crowding that accounts for the depression in the immune functions. Elevation of plasma cortisol with high rearing density has been described for Atlantic salmon, *Salmo salar* production (Mazur and Iwama, 1993). Cortisol is known to cause reduced growth rates by direct action or by suppression of food intake. Crowding has an additional impact of elevating the levels of glucose in the plasma. High stocking density also produced an elevation of plasma glucose. It has been shown that plasma cortisol affected carbohydrate metabolism and high plasma cortisol levels activated gluconeogenesis stimulating glucose production. High plasma cortisol levels produced an immunosuppressive effect in fish in reducing the circulating lymphocytes. The levels of key enzymes such as lipase and trypsin in Japanese flounder, *Paralichthys olivaceus* (Bolasina et al., 2006) or hepatic aminotransferase in channel catfish, *Ictalurus punctatus* (Davis et al., 1985) was affected by the cortisol induced by stress. The functioning of the thyroid was hampered by higher amount of cortisol in Nile tilapia, *Oreochromis niloticus* due to overcrowding (Walpita et al., 2007). Changes in fatty acid constituents of the liver and reduced hepatosomatic index were also effects of overcrowding. This reduction is related to higher hepatic lipid utilization, associated with an increased lipid mobilization. This reduction in hepatic lipid content could also be associated with a decrease in the activity of enzymes related to lipid synthesis, such as glucose-6-phosphate dehydrogenase, and to an increase in the activity of enzymes associated with lipid oxidation, such as hydroxyacyl-Coenzyme A-dehydrogenase. Studies have shown the metabolism of triglycerides is altered in brook charr, *Salvelinus fontinalis* due to overcrowding (Vijayan et al., 1990). The mobilization of triglycerides helped to meet the increased energy demand associated with high rearing densities, suggesting an enhancement of gluconeogenic capacity from glycerol. Overcrowding was translated as lower levels of monounsaturated fatty acids such as oleic acid in the liver in gilthead sea bream, *Sparus aurata* (Montero et al., 1999). This fatty acid is considered one of the most important lipid energy sources in situations that impose a high energy demand such as larval development or starvation. Decreased growth of salmonids at high stocking density had been reported to be caused by decreased food consumption (Vijayan and Leatherland, 1988), social interactions (Wedemeyer, 1997) and decreased water quality (Pickering and Stewart, 1984). Fish growth is compromised by social interactions such as competition. The increased metabolic demands and additional expenditure of energy occurs at the expense of growth. A negative relation between stocking densities and growth performance and survival rates of the Mozambique tilapia (*Oreochromis mossambicus*) was reported (Shubha and Reddy, 2013). Growth performance of African sharp tooth catfish (*Clarias gariepinusis*) decreased along with an increase in stocking density (van de Nieuwegiessen et al., 2008). The amount of protein in plasma was influenced by the chronic stress due to overcrowding. The secretion of growth hormones in fish was suppressed during periods of stress. At high population densities, the activity of liver aspartate aminotransferase (AST) was not significantly affected. Serum lactate dehydrogenase (LDH) activity showed a significant increase at high stocking density due to increased formation of lactic acid production in the muscle.

SUMMARY

The growth in the fish is continuous in nature, and it is different from terrestrial animal, which means that fish grows larger with its longevity. The growth across different fish species is differs due to physiological function and environmental factors and is much more dependent on external environmental conditions. In general, the factors that affect water quality and in turn growth rate of fish can be chemical or biological or physical. Several factors such as temperature and level of solids in water constitute the physical aspects while pH, hardness and alkalinity constitute chemical factors. Other than the water conditions, the crowding of the animal in the culture system also affect the growth through its impact on several physiological parameters. Thus, understanding the optimum requirement of different water quality parameters for different commercially important fish species is most vital. This helps to take decision on correctly identifying suitable species in different culture environment or changing the possible water quality parameters to the requirement of the target fish species to get enhanced growth. Added to this, growth improvement in fishes through environmental manipulation is comparatively easier and low-cost method when compared to the methods such as feed manipulation, genetic manipulation and use of hormones for growth improvements.

REFERENCES

1. Aggergaard, S.; Jensen, F. B. Cardiovascular changes and physiological response during nitrite exposure in rainbow trout. *Journal of Fish Biology*. 2001, 59, 13-27.
2. Beamish, F. W. H.; Tandler, A. Ambient ammonia, diet and growth in lake trout. *Aquatic Toxicology*. 1990, 17, 155-156.
3. Boeuf, G.; Payan, *P.* How should salinity influence fish growth? *Comparative Biochemistry and Physiology Part C: Toxicology & Pharmacology*. 2001, 130 (4), 411-423.
4. Boeuf, G.; Payan, P. How should salinity influence fish growth? *Comparative Biochemistry and Physiology, Part C*. 2001, 130, 411-423.
5. Bolasina, S.; Tagawa, M.; Yamashita, Y.; Tanaka, M. Effect of stocking density on growth, digestive enzyme activity and cortisol level in larvae and juveniles of Japanese flounder, *Paralichthys olivaceus*. Aquaculture, 2006, 259, 432-443.
6. Boyd, C. E. Water Quality for Pond Aquaculture. Research and Development Series No. 43. International Center for Aquaculture and Aquatic Environments, Alabama Agricultural Experiment Station, Auburn University, Alabama, 1998.
7. Boyd, C. E. Water quality management for pond fish culture. *Elsevier Scientific Publishing Company,* 1982, 318 p.

8. Buentello, J. A.; Gatlin, D. M. and Neill, W. H. Effects of water temperature and dissolved oxygen on daily feed consumption, feed utilization and growth of channel catfish (*Ictalurus punctatus*). *Aquaculture*, 2000, 182, 339-352.
9. Clarke, W. C.; Shelbourn, J. E.; Brett, J. R. Effect of artificial photoperiod cycles, temperature and salinity on growth and smelting in underyearling coho (*Oncorhynchus kisutch*), chinook (*O. tshawytscha*) and sockeye (*O. nerka*) salmon. *Aquaculture*, 1981, 22: 105 -116.
10. Copatti, C. E.; Coldebella, I. J.; Radunz Neto, J.; Garcia, L. O., Rocha, M. C. and Baldisserotto, B. 2005. Effect of dietary calcium on growth and survival of silver catfish juveniles, *Rhamdia quelen* (Heptapteridae), exposed to different water pH. *Aquaculture Nutrition*. 11(5), 345-350.
11. D'Cruz, L. M.; Wood, C. M. The influence of dietary salt and energy on the response to low pH in juvenile rainbow trout. *Physiol. Zool.* 1998, 71, 642- 657.
12. Davis, K. B.; Torrance, P.; Parker, N. C.; Suttle, M. A. Growth, body composition, and hepatic tyrosine aminotransferase activity in cortisol fed channel catfish, *Ictalurus punctatus*. *Journal of Fish Biology*. 1985, 27, 177-184.
13. Enayatmehr. M.; Jamili, S. Effects of Salinity on Growth, Hormonal and Enzymatic Status in Fish: A Review. *International Journal of Science and Research (IJSR)*. 2015, 1(7), 805-813.
14. Fange, R.; Grove, D. 1979. Digestion. In: Fish Physiology, Vol VIII, Bioenergetics and Growth (Hoar, W. S., Randall, D. J. and Brett, J. R., eds.), Academic Press, New York; pp. 163-260.
15. Ficke, A.D.; Myrick, C.A.; Hansen, L.J. Potential impacts of global climate change on freshwater fisheries. *Reviews in Fish Biology and Fisheries*. 2007, 17, 581-613.
16. Handeland, S. O.; Berge, A.; Th. Bjornsson, B.: Stefansson, S. O. Effect of temperature and salinity on osmoregulation and growth of Atlantic salmon (*Salmo salar* L.) smolts in seawater. *Aquaculture*, 1998, 168, 289-302.
17. Heming, T. A.; McInerney, J. E. and Alderdin, D. F. Effect of temperature on initial feeding in alevins of chinook salmon

(*Oncorhynchus tshawytscha*). *Can. J. Fish. Aquat. Sci.* 39, 1554-1562.

18. Hofer, R. The adaptation of digestive enzymes to temperature, season and diet in roach, *Rutilus rutilus* L. and rudd, *Scardinius erythrophthalmus* L. *J. Fish. Biol.* 1979, 15: 373-379.
19. Hutchings, J. A. Life histories of fish. In Handbook of Fish Biology and Fisheries; Hart, P.J.B.; Reynolds, J.D., Ed.; 1st edn., Blackwell Publishing, London; pp. 149-174.
20. Jackim, E.; Laroche, G. Protein synthesis in *Fundulus heteroclitus* muscle. *Comp. Biochem. Physiol.* 1973, 44, 851-866.
21. Jensen, F. B. Nitrite and red cell function in carp: control factors for nitrite entry, membrane potassium ion permeation, oxygen affinity and methaemoglobin formation. *Journal of Experimental Biology*, 1990, 152, 149-166.
22. Jensen, F. B., Andersen, N. A. and Heisler, N. 1987. Effects of nitrite exposure on blood respiratory properties, acid base and electrolyte regulation in the carp (*Cyprinus carpio*). *Journal of Comparative Physiology.* 1987, 157, 533-541.
23. Jhingran, V. G. 1988. Fish and fisheries of India. Hindustan Publishing Corporation, Delhi, India, pp. 79-100.
24. Jobling, M. 1994. Fish Bioenergetics. Springer Netherlands, 309 p.
25. Jobling, M. 1995. Environmental biology of fishes. Chapman and Hall Fish and Fisheries Series-16.
26. Johnston, I. A.; Goldspink, G. Thermodynamic activation parameters of fish myofibrillar ATPase enzyme and evolutionary adaptations to temperature. *Nature. 1975,* 257, 620-622.
27. Jurss, K.; Bittorf, K. H.; Volker, T. H. Influence of salinity and ratio of lipid to protein in diets on certain enzyme activities in rainbow trout (*Salmo gairdneri* Richardson). *Comp. Biochem. Physiol.* 1985. 81: 73-79.
28. Kaushik, S. J. and Luquet, P. Study of free amino acids in rainbow trout in relation to salinity changes. *Ann. Hydrobiol.* 1977, 8, 135-144.
29. Kaushik, S. J., Dabrowski, K. and Luquet, P. Patterns of nitrogen excretion and oxygen consumption during ontogenesis of common

carp (*Cyprinus carpio*). *Can. J. Fish. Aquat. Sci.,* 1982, 39, 1095-1105.

30. Kaushik, S. J., Harache, Y. and Luquet, P. Variations in the total free amino acids level in rainbow trout muscle and blood during its adaptation to seawater. *Ann. Hydrobiol.* 1977, 8, 145-151.
31. Kidder III, G. W.; Petersen, C. W.; Preston, R. L. Energetics of osmoregulation: II. Water flux and osmoregulatory work in the euryhaline fish, *Fundulus heteroclitus*. *Journal of Experimental Zoology, Part A*. 2006, 305: 318-327.
32. Lall, S. P.; Bishop, F. J. 1979. Studies on the nutrient requirements of rainbow trout, *Salmo gairdneri*, growth in seawater and freshwater. In: Advances in aquaculture (Pillai, T. V. R. and Dill, W. A., eds.), Fishing News Books, Surrey England, pp. 580-584.
33. Lemarié, G.; Dosdat, A.; Covès, D.; Dutto, G.; Gasset, E.; Ruyet, J. P. 2004. Effect of chronic ammonia exposure on growth of European seabass (*Dicentrarchus labrax*) juveniles, *Aquaculture*. 2004, 229, 479-491.
34. Li, D.; Liu, Z.; Xie, C. Effect of stocking density on growth and serum concentrations of thyroid hormones and cortisol in Amur sturgeon, *Acipenser schrenckii*. *Fish Physiology and Biochemistry.* 2012, 38 (2), 511-520.
35. Lisboa, V.; Barcarolli, I.F.; Sampaio, L.A., Bianchini, A. Effect of salinity on survival, growth and biochemical parameters in juvenile Lebranch mullet *Mugil liza* (Perciformes: Mugilidae). *Neotropical Ichthyology*. 2015, 13(2), 447-452.
36. Lucas, J. S.; Southgate, P. C. Aquaculture-Farming aquatic animals and plants. *Aquacult. Int*. 2012, 20 (4), 807-809.
37. Maceina, M. J.; Nordlie, F. G. and Shireman, J. V. The influence of salinity on oxygen consumption and plasma electrolytes in grass carp, *Ctenopharyngodon idella*. *J. fish. Biol.* 1980, 16, 613-619.
38. Madison, B. N.; Dhillon, R. S.; Tufts, B. L. and Wang, Y. X. Moderate ammonia promotes growth in walleye (*Stizostedion vitreum*). *Journal of Fish Biology*. 2009, 74, 872-890.
39. Mazur, C. F.; Iwama, C. K. Handling and crowding stress reduces number of plaque-forming cells in Atlantic salmon. *J. Aquat. Anim.*

Health. 1993, 5, 98-101.

40. McLarney, W. Freshwater aquaculture. Hartley and Marks Publishers, Point, Roberts, 1998, WA.
41. Montero, D., Izquierdo, M. S., Tort, L., Robaina, L. and Vergara, J. M. High stocking density produces crowding stress altering some physiological and biochemical parameters in gilthead seabream, *Sparus aurata*, juveniles. *Fish Physiology and Biochemistry*. 1999, 20, 53-60.
42. Papoutsoglou, S. E. and Tziha, G. Blue tilapia (*Oreochromis aureus*) growth rate in relation to dissolved oxygen concentration under recirculated water conditions. *Aquacultural Engineering*. 1996, 15, 181-192.
43. Pichavant, K.; Person-Le-Ruyet, J.; Le Bayou, N.; Severe, A.; Le Roux, A.; Bcouf, G. Comparative effects of long-term hypoxia on growth, feeding and oxygen consumption in juvenile turbot and European Sea Bass. *J. Fish Biol*. 2001, 59, 875-883.
44. Pickering, A. D.; Pottinger, T. G.; Sumpter, J. P.; Carragher, J. F.; Le Bail, P. Y. Effects of acute and chronic stress on the levels of circulating growth hormone in the rainbow trout, *Oncorhynchus mykiss*. *General and Comparative Endocrinology*. 1991, 83, 86-93.
45. Pickering, A. D.; Stewart, A. Acclimation of the interrenal tissue of the brown trout, *Salmo trutta* L., to chronic crowding stress. *J. Fish Biol*. 1984, 24, 731-740.
46. Quantz, G. Use of endogenous energy sources by larval turbot (*Scopthalmus maximus* L.). *Trans. Am. Fish. Soc*. 1985, 114, 558-563.
47. Randolph, K. N.; Clemens, H. P. Some factors influencing the feeding behavior of channel catfish in culture ponds. *Transactions of the American Fisheries Society*. 1976, 105(6), 718-724.
48. Rasmussen, R. S.; Korsgaard, B. The effect of external ammonia on growth and food utilization of juvenile turbot (*Scopthalmus maximus* L.). *Journal of Experimental Marine Biology and Ecology*. 1996, 205, 35□48.
49. Roberts, J. K. 1977. The metabolism and growth of rainbow trout, *Salmo gairdneri* in fresh and saline waters. Ph. D. Thesis, Univ.

Aston, Birmingham, UK.

50. Robinette, H. R. Effects of selected sublethal levels of ammonia on the growth of channel catfish *Ictalurus punctatus*. *Progressive Fish Culturist*. 1976, 38, 26-29.
51. Sampaio, L. A.; Bianchini. A. Salinity effects on osmoregulation and growth of the euryhaline flounder *Paralichthys orbignyanus*. *Journal of Experimental Marine Biology and Ecology*, 2002, 269, 187-196.
52. Sankian, Z.; Khosravi, S.; Kim, Y-O.; Lee, S-M. Effect of dietary protein and lipid level on growth, feed utilization, and muscle composition in golden mandarin fish *Siniperca scherzeri*. *Fish. Aquat. Sci.* 2017, 20, 7.
53. Sauer, D. M.; Haider, G. Enzyme activities in the plasma of rainbow trout (*Salmo gairdneri*): the effects of nutritional status and salinity. *J. Fish. Biol.* 1979, 14, 407- 412.
54. Shubha, M.; Reddy, S. R. Effect of stocking density on growth, maturity, fecundity, reproductive behaviour and fry production in the mouth brooding cichlid *Oreochromis mossambicus* (Peters). *African Journal of Biotechnology*. 2013, 10, 9922 -9930.
55. Stormer, J.; Jensen, F. B.; Rankin, J. C. Uptake of nitrite, nitrate, and bromide in rainbow trout, *Oncorhynchus mykiss*: effects on ionic balance. *Canadian Journal of Fisheries and Aquatic Sciences*. 1996, 53, 1943-1950.
56. Sumpter, J. P. Control of growth of rainbow trout (*Oncorhynchus mykiss*). *Aquaculture, 1992,* 100, 299-320
57. Svobodova, Z.; Machova, J.; Drastichova, J.; Groch, L.; Luskova, V.; Poleszczuk, G.; Velisek, J.; Kroupova, H. Haematological and biochemical profile of carp blood following nitrite exposure at different concentration of chloride. *Aquaculture Research*. 2005, 36, 1177-1184.
58. Thetmeyer, H.; Waller, U.; Black, K. D.; Inselmann, S.; Rosenthal, H. Growth of European Sea Bass (*Dicentrarchus labrax* L.) under hypoxic and oscillating oxygen conditions. *Aquaculture*, 1999, 174, 355-367.
59. Tsadik, G. and Kutty, M. N. 1987. Influence of ambient oxygen on

feeding and growth of tilapia, *Oreochromis niloticus*. United Nation Development Programme. Food and Agriculture Organization of the United Nations, Nigerian Institute for Oceanography and Marine Research Project RAF/87/009.

60. Van De Nieuwegiessen, P. G.; Boerlage, A. S.; Verreth, J. A. J.; Schrama, J. Assessing the effects of a chronic stressor, stocking density, on welfare indicators of juvenile African catfish, *Clarias gariepinus* Burchell. *Applied Animal Behaviour Science*. 2008, 115, 233-243.
61. Viadero, R. C. Factors Affecting Fish Growth and Production. *Water Encyclopedia*. 2005, 3, 129-133.
62. Vijayan, M. M.; Ballantyne, J. S.; Leatherland, J. F. High stocking density alters the energy metabolism of brook charr, *Salvelinus fontinalis*. *Aquaculture*. 1990, 88, 371-381.
63. Vijayan, M. M.; Leatherland, J. F. Effect of stocking density on the growth and stress-response in brook charr, *Salvelinus fontinalis*. *Aquaculture*. 1988, 75, 159-170.
64. Walpita, C. N.; Grommen, S. V. H.; Darras, V. M.; Van der Geyten, S. The influence of stress on thyroid hormone production and peripheral deiodination in Nile tilapia (*Oreochromis niloticus*). *General and Comparative Endocrinology*. 2107, 150, 18-25.
65. Wedemeyer, G. A; Goodyear, C. P. 1984. Diseases caused by environmental stressors. In: Diseases of marine animals (Kinne, O. ed.), Vol. IV, Part 1, Pisces. Biologische Anstalt Helgoland, Hamburg, p. 424-434.
66. Wedemeyer, G.A. 1997. Effects of rearing conditions on the health and physiological quality of fish in intensive culture. In: Fish Stress and Health in Aquaculture (Iwama, G. K., Pickering, A. D., Sumpter, J. P. and Schreck, C. B., eds.), Cambridge University Press, Cambridge, pp. 35-71.
67. Wilkes.D.; Xie, S.Q.; Stickland, N.C.; Alami-Durante, H.; Kentouri, M.; Sterioti, G.; Koumoundouros,G.; Fauconneau, B.; Goldspinki, G. Temperaure and myogenic factor transcript levels during early development determines muscle growth potential in rainbow trout (*Oncorhynchus mykiss*) and sea bass (*Dicentrarchus labrax*). *The*

Journal of Experimental Biology. 2001, 204, 2763-2771.

68. Woo, N. Y. S.; Kelly, S. P. Effects of salinity and nutritional status on growth and metabolism of *Sparus sarba* in a closed seawater system. *Aquaculture*. 1995, 135, 229-238.
69. Wood, C. M. Dogmas and controversies in the handling of nitrogenous wastes: Is exogenous ammonia a growth stimulant in fish? *Journal of Experimental Biology*, 2004, 207, 2043-2054.
70. Zaions, M. I.; Baldisserotto, B. Na^+ and K^+ Body levels and survival of juveniles of *Rhamdia quelen* (Siluriformes, Pimelodidae) exposed to acute changes of water pH. *Ciencia Rural*. 2000, 30 (6), 1041-104.

CHAPTER

5

GROWTH RELATED HORMONES AND GROWTH FACTORS IN CRUSTACEANS

E. Suresh and M. Kavitha

[1]Dr. E. Suresh is working as Assistant Professor in the Dept. of Aquaculture in Fisheries College and Research Institute constituent of Tamil Nadu Fisheries University since 2015. Earlier he was working as Technical Assistant in National Bureau of Fish Genetic Resources, Lucknow for 18 months. He has obtained his PhD in Fish Genetics and MFSc in Inland Aquaculture from ICAR – Central Institute of Fisheries Education, Mumbai and BFSc from Tamil Nadu Veterinary and Animal Science University, Chennai. He has also gained research experience in working projects sponsored by DBT and ICAR-Norway collaborative project. In addition to research, he involved in the establishment of Shrimp brood bank in Kanyakumari Dist, Tamil Nadu funded by National Agricultural Development Project (NADP), Govt. of India. He has published more than 15 articles in National and International Journal. He was recipient of ICAR – CIFE fellowship for pursuing Master and Doctoral degree.

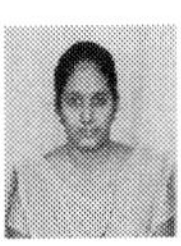

[2]Ms M. Kavitha obtained her M.F.Sc in Aquatic Environment and Management, qualified Agricultural Research Service (ARS) in Aquaculture discipline and joined as a Scientist in ICAR-CMFRI during 2014 in Molluscan Fisheries Division. Currently, she works on molluscan resource management and molluscan breeding. Ms Kavitha is also involved in hatchery rearing of squid and cuttlefish from egg masses and sea ranching of young ones in the natural fishing grounds at Tuticorin coast, Gulf of Mannar.

INTRODUCTION

Crustaceans belong to phylum Arthropoda, are ancient, biologically diverse organisms that are distributed throughout the globe. Since the past few decades, several decapod crustaceans in general and brachyurans in particular, have been the subject of study. Aquatic organisms belonging to more than 200 taxa (including molluscs, crustaceans, fish, and some algae) are now considered resources to be exploited for human consumption and profit. Accordingly, if current fishing practices and rates of wild fish population exploitation continue, the future of most marine ecosystems and wild fisheries will be uncertain due mainly to overfishing and habitat degradation (Worm et al., 2006). Thus, aquaculture has developed rapidly to meet the rising global demand for aquatic animal food, and is now practised worldwide. In a resource-constrained world where the redirection of edible foods as feedstock for animal production is coming under increased scrutiny, achieving maximum productivity in aquaculture systems will be an imperative as the global community strives to achieve sustainable food security. Over the past 50 years, the role of genetics has been vital in fisheries management (aquaculture, breed development programmes and hatchery production) enabling some industries to develop into production systems that are intensive, high-input and supply large markets internationally (e.g. Atlantic salmon in Norway). This has led to an increase in understanding and importance of marine and freshwater environments as well as the species therein (Hauser and Carvalho, 2008). As a result of expanding human populations, there has been an increase in demand for seafood that comes from aquaculture, indicating that in future, genetically improved breeds will play a major role in addressing food security issues and contribute to economic growth (Hulata, 2001; Hutchings & Fraser, 2008). Important aquaculture species constitute a number of decapod crustaceans, including lobsters, crabs, crayfish, shrimps and prawns contributing nearly US$60 billion to global aquaculture industry (FAO, 2009). Since several cultured crustaceans are high value species, research is now focussed towards programmes involving efficient stock enhancement and improving their culture performance (Hamasaki and Kitada, 2008). To date, however, a lack of genomic resources for most aquatic species, and associated poor understanding of molecular and biochemical processes underpinning growth and other economically important traits, has hindered efforts to increase the productivity of crustacean species in aquaculture.

A wide range of species are being investigated using genomic approaches to elucidate genetic factors that contribute to phenotypes and/or traits that are economically important and to manage genetic diversity in crop and livestock species (Wheeler et al., 2010; Morrell et al., 2012). New technologies that allow rapid cost-effective high-throughput sequencing, often referred to as next generation sequencing (NGS), offer significant opportunities to improve our understanding of the domestication process, contribute to the evolution of modern quantitative genetic theory, and can accelerate stock improvement of target populations in agriculture and aquaculture (Morrell et al., 2012). Next generation sequencing has enabled the possibility of fast and cost-effective collation of genome wide sequence data sets, also in non-model organisms (Ekblom and Galindo, 2011). There are reviews that extensively discuss the technical aspects of NGS in both model and non-model species (Gilad et al., 2009; Ekblom and Galindo, 2011). The chapter will concentrate primarily on information relevant to growth and growth-related traits in crustacean species.

Understanding gene function and genetic effects on expressed phenotypes of an organism is now important for future production systems. Modern DNA-sequencing technologies have replaced the once laborious hunt for individual genes associated with specific phenotypes (Stillman et al., 2008; Clark et al., 2011; Jung et al., 2011; Kawahara-Miki et al., 2011) and these advancements in genomic sequencing techniques will possibly revolutionize genetic improvement programmes for cultured crustaceans (Jung et al., 2011; Kawahara-Miki et al., 2011). Application of genomic approaches to stock improvement in aquaculture species has already resulted in some significant production gains in some aquatic organisms including carp, rainbow trout, Atlantic salmon, tilapia and catfish (McAndrew and Napier, 2011). This is, however, a new horizon for crustacean species and to date studies have been confined mostly to only a small number of penaeid species (Leekitcharoenphon et al., 2010) and to a few non-penaeids (Jung et al., 2011; Kawahara-Miki et al., 2011). Although genomic analysis of crustaceans through NGS technologies is still just beyond infancy, it is certain that there will be a rapid growth in the amount and quality of data generated. Studies will mainly focus on the identification and characterization of genes and markers that bring about variations in traits that are economically important and then strategically apply the knowledge obtained to speed up genetic improvement programmes for cultured crustacean species via marker assisted selection (MAS) programmes. Despite the limited genetic information that is available, a number of putative candidate genes related to growth and muscle development have already been identified in crustacean species.

In an effort to stimulate more research on identification of growth-related genes in crustacean species, this review will address the following topics:

Recent genome and functional genomics (transcriptomics) projects that have been undertaken since 2008 on five major crustacean groups (crabs, crayfish, lobsters, shrimps and prawns) (Stillman et al., 2008),

- Associations between individual genes and growth traits in crustaceans,
- Growth-related genes including muscle build-up and degradation genes involved with moulting,
- DNA sequences that have confirmed growth trait effects in farmed terrestrial species and where homologous sequences also exist in crustacean species.

CANDIDATE GENES THAT AFFECT GROWTH

Most eukaryotic genes contain coding (exons, or amino acid translated regions) and non-coding regions (introns, 5′ and 3′ untranslated regions that include promoter regions). In any of these regions, allelic variation in DNA sequence has the potential to alter the structure and activity of the encoded protein, or induce changes in the regulation of mRNA transcription levels (De Santis and Jerry, 2007). Allele specific gene expression and genetic variants within a functional gene can determine phenotypic variation and influence the expression of complex traits, growth traits and genetic diseases via regulation of gene expression and mRNA stability (Palacios et al., 2009; Ciobanu et al., 2010b). While a variety of molecular markers and techniques have been developed to screen allelic polymorphisms in genes, single nucleotide polymorphisms (SNPs) and mRNA expression profiles are currently considered to be the marker and technique of choice, respectively for studying how allelic variation in a gene can affect its regulation by identifying its location in the genome (Palacios et al., 2009; Ciobanu et al., 2010b; Garvin et al., 2010; Seeb et al., 2011).

SNPs are inherited in a co-dominant fashion and represent type I markers and, they are simple to score and compared with some alternative genetic markers, they are relatively inexpensive to screen (Garvin et al., 2010; Kim et al., 2011; Seeb et al., 2011). The recent development of technologies that allow the simultaneous genotyping of thousands of SNPs in a single assay has resulted in a number of genome-wide association studies (GWAS) being completed for model species (McClure et al., 2010; Bolormaa et al., 2011; Pausch et al., 2011). This approach allows researchers to map regions with significant Quantitative Trait Loci (QTLs) association with the phenotype of interest more accurately (McClure et al., 2010; Pausch et al., 2011). Even with well annotated genomes, the next step in identifying genes and mutations that underpin phenotypic variation for many complex traits has proven to be a significant exercise. However, there are no annotated genomes for a crustacean, except that of Daphnia, making GWAS impossible in these organisms. Although many internal and external environmental factors can influence growth rate (e.g. diet, food ration, stocking density, temperature, salinity, dissolved oxygen, pathogens and water chemistry etc.), here we will focus the discussion on the roles of genetic factors (SNPs and mRNA expression) in specific candidate genes that affect growth rate in crustaceans. A recent study applied 418 SNP markers in Pacific white shrimp, *Litopenaeus vannamei*, to develop genetic maps and to identify QTLs for economically important traits (Du et al., 2010). A number of SNPs in anonymous genes were reported to show significant associations with weight gain, growth rate, survival and/or pathogen resistance in these populations (Ciobanu et al., 2010a), demonstrating the value of having markers located within functional genes. The coming sections will highlight specific candidate genes in crustacean species that potentially may affect their growth rate based on previous studies on them, or in model species. In addition, the functional roles of these genes or gene families on growth, muscle production, moulting and degradation will also be discussed.

5-Hydroxytryptamine Receptor

It is known that serotonin (5-hydroxytryptamine, 5-HT), a biogenic amine neurotransmitter found in the nervous systems of all organisms, can affect a diverse array of behavioural, cognitive and physiological, functions (Vazquez-Acevedo et al., 2009; Zhang et al., 2011b). In crustaceans, serotonin has been shown to interact with various receptor subtypes and stimulates the release of several neurohormones including the red pigment dispersing hormone, crustacean hyperglycaemic hormone (CHH), moult-inhibiting hormone (MIH) and neuro-depressing hormone (Gerhardt and Van Heerikhuizen, 1997; Ongvarrasopone et al., 2006). A fourteen subtypes of 5-HT receptor have been identified, cloned and then classified into seven families (5-HT1 to 5-HT7) according to their amino acid sequence homology, pharmacological binding properties, and coupling to secondary messengers in invertebrates (Escamilla-Chimal et al., 2002; Ongvarrasopone et al., 2006). A number of 5-HT receptors have been cloned and characterized for their function in crustaceans (Clark et al., 2004; Komali et al., 2005). They show pharmacological and signalling properties similar to their mammalian counterparts and include 5-HT1, 5-HT2 and 5-HT7-like receptors (Tierney, 2001). Two exonic synonymous SNPs in the 5-HT1 receptor of the Pacific white shrimp, *L. vannamei*, have been linked to high body weight (Martin-Marti et al., 2010). An up-regulation of 5-HT1 in the brain and eyestalk of *Metapenaeus ensis* has been observed and may be vital to the understanding the mechanism of 5-HT1 action and the role of its receptor in control of reproduction (Tiu et al., 2005). Receptor localization plays a critical role in regulating ovarian maturation and spawning in giant tiger shrimp, *Penaeus monodon* (Ongvarrasopone et al., 2006). Undoubtedly, a better understanding of the 5-HT receptor will provide important information about 5-HT regulation mechanisms in different aspects of crustacean physiology.

Alpha-Amylase

The maintenance of appropriate glucose levels in the haemolymph of crustaceans is essential for supporting several key physiological functions and for responding to a variety of environmental stress (Lorenzon, 2005; Asaro et al., 2011). On digestion, glycogenic carbohydrates (i.e. complex polysaccharides that include starch and other disaccharides) provide one of the main sources of haemolymph glucose (Verri et al., 2001). The starch digesting ability depends on key enzymes levels of amylase and maltase in the hepatopancreas, the main production site for digestive enzyme in crustaceans.

Amylase (AMY) hydrolyses starch molecules to produce products including dextrins, and smaller polymers composed of glucose units (Windish & Mhatre 1965). A member of the glycoside hydrolase family a-1, 4-Glucan-4-glucanohydrolase (a-AMY) catalyses the hydrolysis of a-(1,4) glycosidic linkages in starch and related compounds (Janecek, 1997; Van der Maarel et al., 2002). It is a major glucosidase associated with digestive function in the hepatopancreas (Van Wormhoudt and Sellos, 1996; Asaro et al., 2011). A number of studies have suggested that digestive capacity is closely linked to high growth rate, lower mortality rate and final body weight (Brito et al., 2000; Gamboa-Delgado et al., 2003). While a full range of digestive enzymes are present in the crustacean gut, AMY may be an important component that maximizes the assimilation rate and influences growth rate (Brito et al., 2000; Wang, 2007; Simon and Jeffs, 2011).

Preliminary analysis in the Pacific white shrimp, *L. vannamei* AMY gene has identified a number of exonic and intronic SNPs (Glenn et al., 2005). To date, association studies involving this gene have been limited to one intronic SNP correlated with individual body weight in small populations (Glenn et al., 2005). Scientists have suggested that there is less chances of identifying significant associations of the AMY gene variation and body weight (Glenn et al., 2005), although the role of AMY should not be discounted because it has a functional role in digestion and hence potential to influence growth. These enzymes have also been shown to be much more active in freshwater crustaceans (red-claw crayfish) than in the saltwater penaeid species, perhaps reflecting an increased utilization of plant matter in the diet of many freshwater species (Pavasovic et al., 2006). Since feed can contribute to more than 60% of production costs in aquatic species in culture, optimizing efficiency of digestive enzymes to deal with artificial feeds can in theory, contribute significantly to improving growth rates in crustacean species (Williams, 2007).

Cathepsin L

Cysteine proteinases (EC 3.4.22) in the papain super family, known as lysosomal cathepsins, include a large number of enzymes that play an essential role in several highly regulated life processes (McGrath, 1999). Turk et al. (2000) have reported the characterization of eleven cathepsins belonging to two families: the cathepsin L family and the cathepsin B family (Karrer et al., 1993; Berti and Storer, 1995). Conventionally, cathepsin L (CatL, EC 3.4.22.15) is regarded as a major component of the lysosomal proteolytic system. It is involved in intracellular protein turnover, act as a scavenger for unwanted proteins in the cell (McGrath, 1999; Turk et al., 2000) and plays diverse physiological roles in maturation of exopeptidase (Dahl et al., 2001), antigen presentation (Nakagawa et al., 1998), and matrix modelling (Yamada et al., 2000). Recent data reported from insects and crustaceans have revealed that CatL plays roles in food digestion (Hu and Leung, 2007), the nucleus of the oocyte (Hu and Leung, 2004), the innate immune system (Li et al., 2010) and moulting (Le Boulay et al., 1996; Liu et al., 2006). cDNA encoding CatL-like enzymes from crustaceans have been isolated from the digestive gland (hepatopancreas) of two shrimp species *Penaeus vannamei* (Le Boulay et al., 1996) and *M. ensis* (Hu and Leung, 2004), and American lobster (*Homarus americanus*) (Laycock et al., 1989). According to Aoki et al. (2004), an enzyme found in *Pandalus borealis* shares homology with mammalian cathepsins S, L and K, but shows differences in both enzymatic and structural properties. Although CatL is expressed ubiquitously in a shrimp (extremely high in the hepatopancreas) (Hu and Leung, 2004), indirect evidence for a food digestive role for crustacean CatL will require further analysis. In one sampled population, a single intronic SNP in CatL in shrimp correlated with high individual body weight (Glenn et al., 2005). In addition, genetic studies in cotton bollworm *Helicoverpa armigera* (Liu et al., 2006) and in a shrimp (Le Boulay et al., 1996) suggested an essential role for CatL in moulting. More studies of CatL could provide important information on regulation mechanisms in crustacean physiology.

Cyclophilin

Cyclophilins (Cyps) contain a single conserved peptidylprolyl cis-trans isomerase (PPIase domain). They are abundant and ubiquitous cytosolic proteins expressed constitutively in bacteria to vertebrates (Galat, 1999; Opiyo and Moriyama, 2009). Due to their PPIase activity, Cyps have been reported to possess diverse regulatory functions in protein folding (Schmid et al., 1993), receptor complex stabilization, apoptosis and receptor signalling (Lin and Lechleiter, 2001; Brazin et al., 2002). Cyclophilins have also been reported to be involved in pathogen responses (Fidantsef et al., 1999), immunological system activation (Hamilton and Steiner, 1998) and as modulators of oxidative stress response (Guedes et al., 2005). In addition, Cyps are involved potentially in promoting endothelial cell proliferation, migration, invasive capacity and tubulogenesis at low concentrations (Kim et al., 2004).

Cyclophilin studies in aquatic animals have suggested an important role in the innate immune system (especially isoform CypA). For example, CypA was identified as being involved in the early infection in channel catfish, *Ictalurus punctatus* (Yeh and Klesius, 2008) of the bacterium *Edwardsiella ictaluri*. Further, Qiu et al. (2009) have reported that the expression of CypA in the hepatopancreas in giant tiger shrimp *Penaeus monodon* was up-regulated after stimulation with lipo-polysaccharide, indicating that CypA probably contributed to the defence response against bacterial infections. In another example, an isoform of CypA was expressed in haemocytes of a clam (*Venerupis philippinarum*), suggesting its involvement in the defence response against bacterial infections (Chen et al., 2011). A negative correlation between cyclophilin-like expression and individual body weight as reported in *P. monodon* (Tangprasittipap et al., 2010). In a study of expression profiles in two size classes of Atlantic pink shrimp (*Farfantepenaeus paulensis*), cyclophilin, haemocyanin and myosin were strongly expressed in high weight individuals but were only weakly expressed or absent in low weight individuals (Kamimura et al., 2008). The results strongly suggest that these genes may have some effect on individual growth performance, and an improved understanding of the immune response and growth retardation in crustaceans as a result of pathogen challenge could help in developing new approaches for maximizing crustacean growth rate in culture.

Fatty Acid-Binding Protein

Fatty acid-binding proteins (FABPs) are cytosolic proteins of small size (approximately 15 kDa) that bind to hydrophobic ligands, primarily fatty acids non-covalently (Esteves and Ehrlich, 2006) and are important for the development of fatness traits (McArthur et al., 1999). The physiological roles of FABPs include; uptake and utilization of fatty acids, intracellular fatty acid transport to specific organelles and engagement in metabolic pathways, cell growth and differentiation, cellular signalling, gene transcription and protection of cellular structures from detergent effects of fatty acids (Ho et al., 2002; Zimmerman and Veerkamp, 2002; Corsico et al., 2004). In FABP families, 12 FABP isoforms have been identified in vertebrates (Liu et al., 2008) that can be divided into two main groups (Glatz and Van der Vusse, 1996): those associated with the plasma membrane (FABPPM) and those with intracellular or cytoplasmic proteins (FABPC). The FABP genes in livestock species have been the focus of many studies for finding economically important polymorphic sites. Several SNPs have been identified in both exonic and intronic regions in the adipocyte FABP gene (A-FABP) while performing association studies between polymorphisms in FABP orthologues and fat deposition traits in cattle (Wang et al. 2006) and heart FABP gene (H-FABP) (Wang et al., 2007). Both genes have significant effects on several growth traits including body weight (BW), intramuscular fat content (IMF), abdominal fat weight (AFW) and abdominal fat percentage (AFP) in chickens (Wang et al., 2006).

To date, approximately only 40 FABPs have been identified in invertebrate species (Zimmerman and Veerkamp, 2002; Esteves and Ehrlich, 2006). Evidently, several putative FABPs have been detected in crustaceans, for example, giant tiger prawn (*P. monodon*), the crayfish (*Pacifastacus leniusculus*), Kuruma prawn (*Marsupenaeus japonicus*), Chinese mitten crab (*E. sinensis*) and Chinese white shrimp (*F. chinensis*) (Lyons et al., 2007; Gong et al., 2010; Li et al., 2011). While the presence of Es-FABP transcripts in a variety of tissues was revealed through gene expression analysis of FABPs in *E. sinensis* using RT-PCR, Es-FABP expression analysis revealed participation of hepatopancreas, haemocytes, testis and ovary in lipid nutrient absorption and utilization processes. The expression of Es-FABP was dependent on the status of testis and ovary development as evidenced by a role for Es-FABP in lipid transport (Gong et al., 2010; Li et al., 2011). In *M. japonicus*, characterization of an amplified fragment length polymorphism (AFLP) band that resided under a previously identified QTL peak that influenced weight, total length and carapace length, revealed a gene homologous to the elongation of very long chain fatty acids-like (ELOVL) protein family (Lyons et al., 2007). Weight difference analyses showed poor growth phenotypes in individuals carrying isoform A (regardless of whether A- or AB), while a positive growth phenotype that contributed 16% to total phenotypic variation was observed in individuals without isoform A (B-). This gene represents a suitable candidate for growth studies in crustaceans (Lyons et al., 2007). ELOVLs have also been shown to have significant effects on growth in yeast (Rossler et al., 2003) and play an important role in lipid recruitment in adipose tissue in mice (Westerberg et al., 2006). It is not unreasonable, therefore, to predict that changes in FABP activity may have significant biochemical and physiological effects on growth in crustaceans and constitute potential gene candidates that influence growth traits in crustaceans.

Fibrillarin

Fibrillarin is a 40 kDa protein that is located in the dense fibrillar component of the nucleolus (Ochs et al., 1985) and has essential functions related to RNA splicing and ribosomal RNA processing (Stanek and Neugebauer, 2006; Deryusheva and Gall, 2009). Amino acid alignment studies in vertebrates and invertebrates fibrillarin-like ESTs indicate that these proteins are highly conserved for several cellular functions including an essential protein required for pre-rRNA cleavage, rRNA methylation, and ribosome assembly (Venema and Tollervey, 1999; Fatica and Tollervey, 2002; Brown et al., 2003). Fibrillarin may also control ribosome biosynthesis and p53 functions and be potentially involved in pre-rRNA processing and modification, regulation of cell proliferation and individual growth (Michael and Oren, 2002). When the fibrillarin in HeLa cells was knocked down using an RNAi technique, no effect on nuclear envelope assembly was observed, although it did cause abnormal (aberrant) nuclear morphology, reduced cell proliferation and growth (Amin et al., 2007). Fibrillarin has also been reported to stimulate HeLa cell growth, but high endogenous expression of a fibrillarin-like EST in the optic lobe of female shrimp showed a negative correlation with body weight (Tangprasittipap et al., 2010). These differences in effect may be a result of involvement different cell types, given that HeLa cells are tumorigenic while the shrimp cells are not (Tangprasittipap et al., 2010).

Glyceradehyde-3-Phosphate Dehydrogenase

Glyceraldehyde-3-phosphate dehydrogenase (GAPDH) converts glyceraldehyde-3-phosphate reversibly to 1,3-bisphosphoglycerate by coupling with the reduction of NAD+ to NADH (Sirover, 2011; Tristan et al., 2011), and is thus is a glycolytic enzyme. Initially, GAPDH was considered to be a simple housekeeping protein, but has been shown to be involved in many cellular processes, besides glycolysis (Sirover, 2011; Tristan et al., 2011). Thus, GAPDH has multifunctional role to play, including: DNA repair (Meyer-Siegler et al., 1991), tRNA export (Singh and Green, 1993), membrane fusion and transport (Tisdale, 2001), cytoskeletal dynamics (Kumagai and Sakai, 1983) and cell death (Hara et al., 2005). GAPDH is commonly used as a control gene for gene expression profiling studies in crustaceans (Dhar et al., 2009), but Wade et al. (2012) suggested that it cannot be used for gene expression normalization in prawns because GAPDH shows a strong spike in expression during the pre-moult stage. SNPs were identified in the GAPDH gene in *L. vannamei* populations, and some showed potential associations with pathogen resistance, survival, and weight gain (Ciobanu et al., 2010a). Such diverse functions of GAPDH may be important in crustacean species in association with reproduction, growth, and disease.

Growth Hormone and Insulin-like Growth Factor

The transforming growth factor superfamily and somatotrophic axis and have often been targeted as possible candidate genes because of their key physiological role in regulation of growth, development, metabolism and lactation in model taxa (De Santis and Jerry, 2007; Lucy, 2008; Duan et al., 2010). The somatotropic axis essentially consists of growth hormone inhibiting hormone (GHIH or somatostatin), growth hormone-releasing hormone (GHRH), insulin-like growth factors (IGF-I and -II), and growth hormone (GH) associated carrier proteins and receptors. In general, the insulin signalling pathway can modulate the uptake of glucose, fatty acids and amino acids and promotes the storage of nutrients in the form of glycogen, lipid and protein, respectively into adipose tissue, muscle and liver (Hwa et al., 1999; Adam et al., 2005). Evidence has accumulated that component of the IGF axis, IGFs, IGFRs and a superfamily of IGF binding proteins (IGFBPs), act together to control a number of crucial biological processes including cellular growth, cell proliferation, cell differentiation, survival from apoptosis and cell migration (Khandwala et al., 2000; Pollak et al., 2004). These processes are related to tissue formation and remodelling, bone growth, brain development and energy metabolism, which all ultimately influence organism growth and affect individual longevity. Together they are believed to stimulate anabolic processes including protein synthesis, cell proliferation, and skeletal growth (De Santis and Jerry, 2007). A number of studies have extensively addressed the range and roles of the many different IGFs and GFs, and their receptors in vertebrates (Hwa et al. (1999), Adam et al. (2005), Jiang et al. (2007), Ola et al. (2008), Kocour and Kohlmann (2011), and Mullen et al. (2011).

The transmembrane GH receptor (GHR) mediates the action of GH (Kopchick and Andry, 2000). Thus, for studies of genetic variation related to growth traits, the potential targets are GHR and GH genes. Growth hormone effects on growth have been observed in a number of livestock species (Gupta et al., 2009; Tahmoorespur et al., 2011; Zhang et al., 2011a; Zhao et al., 2011) and variation in GHR gene sequences have been associated with a number of performance traits particularly in cattle (Viitala et al., 2006; Waters et al., 2011) including milk yield (Blott et al., 2003; Viitala et al., 2006), as well as feed intake, feed conversation efficiency and body energy traits (Banos et al., 2008). The potential effect of GH and GHR genes on growth traits have also been reported in fish (Tao and Boulding, 2003) with the majority of polymorphisms identified in introns (De Santis and Jerry, 2007).

There is a lack of direct evidence for the role of GH and GHR genes in promoting growth, although they have been reported in crustaceans (Toullec and Van Wormhoudt, 1987). Subsequently, some commonplace practices involve direct administration of vertebrate GH to promote growth in crustaceans (Arenal et al., 2008; Santiesteban et al., 2010). Studies have shown that administration of vertebrate GH in crustaceans can promote growth (Charmantier et al., 1989; Xu et al., 2000; Arenal et al., 2008). An investigation reported using recombinant tilapia GH for enhancement of growth in *L. vannammei* showed that compared to controls, shrimp post larvae treated with tilapia GH were 42.4% heavier and 5.2% longer (Santiesteban et al., 2010). This indicates the GH activity may be recognised and mediated in presence of GH receptor-like proteins. Till date little work has been done to explore the pathways or identity of a number of ecdysteroidal neuropeptides including MIH and CHH, that have been reported in crustaceans. Much more work is needed to identify homologous gene in crustaceans and to elucidate these pathways.

The insulin-like growth factor (IGF) system is a complex network of cell surface receptors, circulating binding proteins and peptide hormones (IGF1 and IGF2). IGF1 and IGF2 are well documented in mammalian species and fish as a promising candidate gene for selection of growth traits that are marker-assisted. Several genetic polymorphisms in the IGF1 gene associated with growth traits have been reported in chickens (Bennett et al., 2006), in swine (Hao et al., 2011), bovines (Mullen et al., 2011) and in fish (Li et al. 2009). These studies suggest strong genetic control of IGF1 gene that is regulated at both transcriptional and translational level (Wang et al., 2003; Mullen et al,. 2011). There is also growing interest in the role(s) that IGF2, a peptide hormone well known for its mitogenic activity that can stimulate cell growth and differentiation in many tissues, plays in livestock and fish growth. A series of studies have shown that DNA sequence polymorphisms within the IGF2 gene in livestock and fish contribute to variation in complex traits notably in fat deposition, muscle mass, body weight, carcass traits, fertility and survival traits in pigs (Hou et al., 2010), bovines (Berkowicz et al., 2011) and in fish (Li et al., 2012).

In invertebrates, evidence for the GH and IGF axis also demonstrates that contain peptides share substantial sequence homologies and/or biological actions with those of mammalian genes and the presence of IGFRs, IGFBPs, insulin, and IGFs. Single insulin binding domain protein (SIBD) has been identified in invertebrates including in some crustacean species (Garofalo, 2002; Okuno et al., 2002; Wu and Brown, 2006; Gai et al., 2010). In a study that attempted to elucidate the functional significance of insulin-like peptides in the white shrimp *L. vannamei*, juveniles injected with a single dose of recombinant human IGF-1 or bovine insulin at the inter-moult stage showed a significant elevation in glucose and glycogen levels in the gills and haemolymph (Gutierrez et al., 2007). In addition, the presence of endogenous Penaeus insulin(s), just like their vertebrate counterparts, was probably involved in the carbohydrate metabolism regulation. Recently, a SIBD, a putative member of the IGFB super family, was identified in shrimp *L. vannamei* haemocyte cDNA library and was hypothesised to be involved in the shrimp immune response following challenge by bacterial. Another putative SIBD, identified at the mRNA level in Chinese mitten crab *E. sinensis* (Gai et al., 2010) is associated with the endocrine and immune systems, while an interesting study reports putative insulin-like peptides in the androgenic gland (AG) of crustaceans indicating their possible role in controlling sex differentiation (Manor et al., 2007; Ventura et al., 2011). A number of studies have indicated that AG can play an important role in sexual differentiation, development of the appendix masculine and secondary male sex characteristics in crustaceans (see Manor et al. 2007; Ventura et al., 2009, 2011; Rosen et al., 2010; Mareddy et al., 2011). Much more work will be required to understand the effect(s) of IGFs on crustacean growth traits and to elucidate how insulin-like peptides regulate sexual shifts (e.g. appropriate intervention) in crustaceans because sexually dimorphic growth between genders has been reported in a wide array of malacostracans (Gitterle et al., 2005; Buric et al., 2010; Gopal et al., 2010).

MOULTING REGULATED GROWTH IN CRUSTACEANS

Apart from the hormones involved in growth enhancement in crustaceans, the process of moulting in crustacean also helps for growth increment. In all crustaceans, the exoskeleton curbs their growth and sheds off their exoskeleton in a cyclic process known as moulting which allows them to grow both in length and weight. Mouting process is otherwise known as ecdysis. This is a multistage process divisible into post-moult, inter-moult, pre-moult and moult and each stage marked with distinguished physiological changes. During the post molt process development of 3 layers of cuticle takes place through the materials stored in the epidermis, as feeding does not begin until the crustaceans enter into the inter-moult stage (Dall et al., 1965). During the intermoult period, the water taken up during moulting is gradually replaced by tissue growth, by protein and mineral deposition resulting in growth of the crustaceans (Dall et al., 1965). Therefore, the volume and weight of the crustacean increases during this phase. The cuticle hardens appreciably within a few hours after moulting. Pre-moulting is the stage just before moulting that starts with an increase in the concentration of molting hormone in hemolymph (Chang, 1992). During this stage calcium is reabsorbed from the old exoskeleton and energy reserves are mobilized from midgut gland that leading to enlarged epidermal cells which makes clear separation of the old cuticle layers from the epidermis for moulting. The lipid content of hepatopancreas of the crustacean will increase appreciably during this phase to serve as energy reserve utilized immediately after moult, when the animal will not feed.

Hormones involved in molting

Ecdysone

The ecdysis or apolysis is a process that is essential for growth and during early phase of life cycle the frequency of moulting is relatively higher than the adult phase. Therefore, molting process control the growth in crustaceans through ecdysone, an ecdysteroid hormones from Y organ and moult inhibiting hormone (MIH) from X-organ sinus gland complex of eyestalk in the crustaceans (Passano, 1960; Aiken, 1980; Vijayan and Diwan, 1993). According to Blais et al. (1994) the Y-organ of the *Litopenaeus vannamei*, secrets 3-dehydroecdysone, but the major circulating ecdysteroid in haemolymph is 20-hydroxyecdysone. The active form of ecdysteroid hormone is 20-hydroxyecdysone and its concentration in the haemolymph found very low during postmoult stages and the surge in the level occurs during pre-moult process (Blais et al., 1994; Huberman, 2000). The alpheid shrimp, *Alpheus heterochaelis*, exposed to 20-hydroxyecdysone for 5 days, shortened the winter molt cycle by 65% and thereby growth increment was noticed (Mellon and Greer, 1987). The MIH curbs the ecdysteroid synthesis from Y organ. According to Vijayan et al. (1993), extirpation of Y organ from Indian with shrimp, *Fenneropenaeus indicus* inhibit the onset of premolting process which is the cue for growth. Cholesterol is the precursor for the synthesis of ecdysone in crustaceans. Therefore, supplementation of cholesterol in the diets of crustaceans is most warranted for ensued production of ecdysone and thereby promoting growth. Paulson and Skinner (1991) investigated the protein synthesis in the tissues of land crab *Gecarcinus lateralis* by treating their tissues from intermolt crabs with 20-hydroxyecdysone in vitro and showed changes in the incorporation of [^{35}S]-methionine into proteins. The sarcomeric protein synthesis takes place during premolt and postmolt stages and accretion of sarcomeric proteins in the leg and abdominal muscles during ecdysis in number of crustacean species (El Haj, 1999). El Haj (1999) found that during the process of ecdysis the new exoskeleton is being laid down, the muscle fibres are increasing their length and reattached with exoskeleton. In American lobster, *Homarus americanus,* during the premolt stage, protein synthesis increased in all three muscle groups, leg, abdominal and claw muscle at the same time the ecdysteroid level also found elevated (El Haj et al., 1995). Ecdysteroids have a direct effect on muscle turnover based on dose dependent manner. Regeneration of lost walking legs of fiddler crab *Uca pugilator* is takes place during the transition between intermolt and premolt is the crucial point where limb bud growth is influenced by the elevated level of 20-hydroxyecdysone during this phase (Hopkins, 1989). Mellon and Greer (1987) found that *Alpheus heterochaelis* exposed to 20-hydroxyecdysone in micromolar concentration accelerate claw transformation which indicated that ecdysteroids attributing in morphogenesis.

Molt Inhibiting Hormone (MIH)

The molt inhibiting hormone (MIH), a peptide hormone is a member of the crustacean hyperglycemic hormone (CHH) subfamily II. The eyestalks of crustaceans, consists neurosecretory system with a group of peptidergic neurons clustered in medulla terminalis X-organ and their bulbous axonic terminals that comprise sinus gland and this complex secretes and releases a number of peptide hormones into haemolymph, including the MIH (Huberman, 2000). The eyestalk ablation eliminates the MIH and increasing 20-hydroxyecdystrone production and up-regulating its physiological response (Choy, 1987). Therefore, eyestalk ablation in crustaceans experienced precocious moult cycles and thereby abrupt growth which are due to the absence of MIH that controls over the Y organ for the secretion of ecdysone (Aiken, 1980). Thus, it was inferred that MIH delay the onset of ecdysis and extend the interval between successive molts in crustaceans. The administration of MIH or eyestalk extract in the eyestalk ablated shrimp stopped its precocious moulting which is due to the inhibitory role of MIH on the moult promoting effect of Y organ (Diwan and Nagabhushanam, 1975). Eyestalk extract of grass shrimp, *Palaemonetes vulgaris* repress the moult stimulating effect of 20-hydroxyecdysone in striped barnacle, *Balanus Amphitrite* (Freeman and Costlow, 1979). According to Mattson and Spazianai (1987) MIH elicits intracellular cAMP which in turn stimulates a protein kinase that attributes inhibition of protein synthesis and thus steroidogenesis from Y-organ. Spaziani et al. (1994) also revealed that MIH release is also regulated by 5-HT (seratonin) that stimulates MIH secretion from X organ sinus gland complex and thus inhibits the steroidogenesis. Further MIH secretion is regulated by negative feedback of ecdysteroid level in haemolymph. According to Naya et al. (1989) in blue crabs (*Callinectes sapidus*) ecdysone biosynthesis inhibitor, 3-hydroxy-L-kynurenine is secreted from X-organ-sinus gland complex and released into the hemolymph. Thereafter, it will accumulate at the surface of the Y-organ and converted into the active form, xanthurenic acid which is profoundly suppress the ecdysteroidogenesis.

Methyl Farnesoate (MF)

Methyl farnesoate (MF) is a sesquiterpenoid compound found in crustaceans which is structurally almost similar to juvenile hormone (JH) of insects but the only difference is the juvenile hormone additionally possess epoxide moiety at the terminal end (Hui et al., 2010). The MF is synthesized from mandible organ and its secretion is negatively regulated by a peptide hormone, mandibular organ-inhibiting hormone (MOIH) produced from X-organ sinus gland complex of the eyestalk (Nagaraju et al., 2005). MF regulates moulting, reproduction (Laufer and Biggers, 2001), morphogenesis (Laufer et al., 2005) and protein synthesis (Paulson and Skinner, 1991) in crustaceans. Laufer et al. (2002) revealed that eyestalk ablation in male spider crabs, *Libinia emarginata* increase the ecdysteroid and MF levels that increased carapace size, but retained shorter non-allometric claws, with length shorter than the carapace. In case of non-ablated males the ecdysteroid and MF levels were lower than the ablated ones and showed large allometric claws but no changes in carapace size. Similarly, Laufer et al. (2005) found that level of methyl farnesoate controls adult male morphogenesis in the crayfish,

Procambarus clarkii. MF treatment causes similar physiological response as like 20 hydroxyecdysone and accelerate molting process ration in several crustacean species, including freshwater crab *Oziotelphusa senex senex* (Reddy et al., 2004) and in American whiteleg shrimp, *Litopenaeus vannamei* (Alnawafleh et al., 2014). Abdu et al. (2001) revealed that administration of MF in the crayfish *Cherax quadricarinatus* females during their winter reproductive arrest period did not show any improvement on reproduction but significantly accelerate the molting process which enhances growth during winter phase.

SUMMARY

Among the different groups of crustaceans, shrimps are the leading group in terms of world aquaculture production. Shrimp industry is a booming sector and the production is increased from <1 million metric tonnes during 1995 to about 5 million metric tonnes (projected value by GOAL 2015) during 2017 with nearly 75% of production contributed by *Litopenaeus vananmei.* Shrimp aquaculture production is elicited in the recent years even after a tremendous struggle caused by EMS/APHNS, is due to several factors such as specific pathogen free seeds, specific pathogen resistant seeds, quality feed, growth improvement through genetic interventions. The specific candidate genes that are potentially affect the growth rate in crustacean species may be influenced to improve the growth of crustaceans is the future perspective to induce them through dietary intervention or transgenic approach.

REFERENCES

1. Abdu, U.; Barki, A.; Karplus, I.; Barel, S.; Takac, P.; Yehezkel, G.; Laufer, H.; Sagi, A. Physiological effects of methyl farnesoate and pyriproxyfen on wintering female crayfish *Cherax quadricarinatus. Aquaculture*. 2001, 202, 163-175.
2. Adam, D.; Cosgrove, L.J.; Booker, G.W.; Wallace, J.C.; Forbes, B.E. Molecular interactions of the IGF system. *Cytokine Growth Factor Rev.* 2005, 16, 421-439.
3. Aiken, D.E. *Molting and growth*, In *The Biology and Management of Lobsters;* Cobb, J.S., Phillips, B.F., Eds.; Vol. 1; Academic Press: New York; 1980; pp. 91-163.

4. Alnawafleh, T.; Kim, B-K.; Kang, H-E.; Yoon, T-H.; Kim, H.W. Stimulation of molting and ovarian maturation by methyl farnesoate in the Pacific White Shrimp *Litopenaeus vannamei* (Boone, 1931). *Fish Aquat. Sci.* 2014, 17, 115-121.
5. Amin, M.A.; Matsunaga, S.; Ma, N.; Takata, H.; Yokoyama, M.; Uchiyama, S.; Fukui, K. Fibrillarin, a nucleolar protein, is required for normal nuclear morphology and cellular growth in HeLa cells. *Biochem. Biophys. Res. Commun.* 2007, 360, 320-326.
6. Aoki, H.; Ahsan, M.N.; Watabe, S. Molecular and enzymatic properties of a cathepsin L-like proteinase with distinct substrate specificity from northern shrimp (*Pandalus borealis*). *J. Comp. Physiol.* 2004, 174B, 59-69.
7. Arenal, A.; Pimentel, R.; Pimentel, E.; Martin, L.; Santiesteban, D.; Franco, R.; Aleström, P. Growth enhancement of shrimp (*Litopenaeus schmitti*) after transfer of tilapia growth hormone gene. *Biotechnol. Lett.* 2008, 30, 845-851.
8. Asaro, A.; Valle, J.C.D.; Mananes, A.A.L. Amylase, maltase and sucrase activities in hepatopancreas of the euryhaline crab *Neohelice granulata* (Decapoda: Brachyura: Varunidae): partial characterization and response to low environmental salinity. *Scientia Marina*. 2011, 75, 517-524.
9. Banos, G.; Woolliams, J.A.; Woodward, B.W.; Forbes, A.B.; Coffey, M.P. Impact of single nucleotide polymorphisms in leptin, leptin receptor, growth hormone receptor, and diacylglycerol acyltransferase (DGAT1) gene loci on milk production, feed, and body energy traits of UK dairy cows. *J. Dairy Sci.* 2008, 91, 3190-3200.
10. Bennett, A.K.; Hester, P.Y.; Spurlock, D.E. Polymorphisms in vitamin D receptor, osteoponin, insulin-like growth factor 1 and insulin, and their associations with bone, egg and growth traits in a layer-broiler cross in chickens. *Anim. Genet.* 2006, 37, 283-286.
11. Berkowicz, E.W.; Magee, D.A.; Sikora, K.M.; Berry, D.P.; Howard, D.J.; Mullen, M.P.; Evans, R.D.; Spillane, C.; MacHugh, D.E. Single nucleotide polymorphisms at the imprinted bovine insulin-like growth factor 2 (IGF2) locus are associated with dairy performance

in Irish Holstein- Friesian cattle. *J. Dairy Res.* 2011, 78, 1-8.

12. Berti, P.J.; Storer, A.C. Alignment/phylogeny of the papain superfamily of cysteine proteinases. *J. Mol. Biol.* 1995, 246, 273-283.
13. Blais, C.; Sefiani, M.; Toullec, J.Y.; Soyez, D. *In vitro* production of ecdysteroids by Y-organs of *Penaeus vannamei* (Crustacea, Decapoda), Correlation with haemolymph titers. *Invert. Reprod. Dev.* 1994, 26, 3-11.
14. Blott, S.; Kim, J.J.; Moisio, S.; Schmidt-Kuntzel, A.; Cornet, A.; Berzi, P.; Cambisano, N.; Ford, C., Grisart, B.; Johnson, D.; Karim, L.; Simon, P.; Snell, R.; Spelman, R.; Wong, J.; Vilkki, J.; Georges, M.; Farnir, F.; Coppieters, W. Molecular dissection of a quantitative trait locus: a phenylalanine-to-tyrosine substitution in the transmembrane domain of the bovine growth hormone receptor is associated with a major effect on milk yield and composition. *Genetics.* 2003, 163, 253-266.
15. Bolormaa, S.; Hayes, B.J.; Savin, K.; Hawken, R.; Barendse, W.; Arthur, P.F.; Herd, R.M.; Goddard, M.E. Genome-wide association studies for feedlot and growth traits in cattle. *J. Anim. Sci.* 2011, 89, 1684-1697.
16. Brazin, K.N.; Mallis, R.J.; Fulton, D.B.; Andreotti, A.H. Regulation of the tyrosine kinase Itk by the peptidyl-prolyl isomerase cyclophilin A. *Proc. Natl. Acad. Sci. USA.* 2002, 99, 1899-1904.
17. Brito, R.; Chimal, M.E.; Gaxiola, G.; Rosas, C. Growth, metabolic rate, and digestive enzyme activity in the white shrimp *Litopenaeus setiferus* early postlarvae fed different diets. *J. Exp. Mar. Biol. Ecol.* 2000, 255, 21-36.
18. Brown, J.W.; Echeverria, M.; Qu, L.H. Plant snoRNAs: functional evolution and new modes of gene expression. *Trends Plant Sci.* 2003, 8, 42-49.
19. Buric, M.; Kouba, A.; Kozak, P. Intra-sex dimorphism in crayfish females. *Zool.* 2010, 113, 301-307.
20. Chang, E.S. *Endocrinology,* In: *Marine Shrimp Culture: Principles and Practices;* Fast, A.W., Lester, L.J., Eds.; Elsevier: Amsterdam, 1992; pp. 53-91.

21. Charmantier, G.; Charmantier-Daures, M.; Aiken, D.E. Human somatotropin stimulates the growth of young American lobsters, *Homarus americanus* (Crustacea, Decapoda). *Comptes Rendus del Academie des Sciences-Series III.* 1989, 308, 21-26.
22. Chen, L.; Mu, C.; Zhao, J.; Wang, C. Molecular cloning and characterization of two isoforms of cyclophilin A gene from *Venerupis philippinarum. Fish Shellfish Immunol.* 2011, 31, 1218-1223.
23. Choy, S.C. Growth and reproduction of eyestalk ablated *Penaeus canaliculatus* (Olivier, 1811) (Crustacea: Penaeidae). *J. Exp. Mar. Biol. Ecol.* 1987, 112, 93-107.
24. Ciobanu, D.C.; Bastiaansen, J.W.M.; Magrin, J.; Rocha, J.L.; Jiang, D.H.; Yu, N.; Geiger, B.; Deeb, N.; Rocha, D.; Gong, H.; Kinghorn, B.P.; Plastow, G.S.; van der Steen, H.A.; Mileham, A.J. A major SNP resource for dissection of phenotypic and genetic variation in Pacific white shrimp (*Litopenaeus vannamei*). *Anim. Genet.* 2010a, 41, 39-47.
25. Ciobanu, D.C.; Lu, L.; Mozhui, K.; Wang, X.; Jagalur, M.; Morris, J.A.; Taylor, W.L.; Dietz, K.; Simon, P.; Williams, R.W. Detection, validation, and downstream analysis of allelic variation in gene expression. *Genetics.* 2010b, 184, 119-128.
26. Clark, M.C.; Dever, T.E.; Dever, J.J.; Xu, P.; Rehder, V.; Sosa, M.A.; Baro, D.J. Arthropod 5-HT2 receptors: a neurohormonal receptor in decapod crustaceans that displays agonist independent activity resulting from an evolutionary alteration to the DRY motif. *J. Neurosci.* 2004, 24, 3421-3435.
27. Clark, M.S.; Thorne, M.A.S.; Toullec, J.Y.; Meng, Y.; Guan, L.L.; Peck, L.S.; Moore, S. Antarctic krill 454 pyrosequencing reveals chaperone and stress transcriptome. *PLoS ONE* 2011, 6, e15919.
28. Corsico, B.; Liou, H.L.; Storch, J. The alphahelical domain of liver fatty acid binding protein is responsible for the diffusion- mediated transfer of fatty acids to phospholipid membranes. *Biochem.* 2004, 43, 3600-3607.
29. Dahl, S.W.; Halkier, T.; Lauritzen, C.; Dolenc, I.; Pedersen, J.; Turk, V.; Human recombinant pro-dipeptidyl peptidase I (cathepsin

C) can be activated by cathepsins L and S but not by autocatalytic processing. *Biochem.* 2001, 40, 1671-1678.

30. Dall, W. Studies on the physiology of a shrimp, *Metapenaeus* sp. (Crustacea: Decapoda: Penaeidae) II. Endocrines and control of molting. *Aust. J. Mar. Freshwater Res.* 1965, 16, 1-12.
31. De Santis, C.; Jerry, D.R. Candidate growth genes in finfish -where should we be looking? *Aquaculture* 2007, 272, 22-38.
32. Deryusheva, S.; Gall, J.G. Small Cajal body-specific RNAs of Drosophila function in the absence of Cajal bodies. *Mol. Biol. Cell* 2009, 20, 5250-5259.
33. Dhar, A.K.; Bowersb, R.M.; Licona, K.S.; Veazey, G.; Read, B. Validation of reference genes for quantitative measurement of immune gene expression in shrimp. *Mol. Immunol.* 2009, 46, 1688-1695.
34. Diwan, A.D.; Nagabhushanam, R. Neurosecretory cells of the central nervous system of the freshwater crab, *Barytelphusa cunicularis* (Westwood 1836). *Riv. Biol.* 1975, 68, 79-99.
35. Du, Z.Q.; Ciobanu, D.C.; Onteru, S.K.; Gorbach, D.; Mileham, A.J.; Jaramillo, G. A gene-based SNP linkage map for pacific white shrimp, *Litopenaeus vannamei. Ani. Gen.* 2010, 41, 286-294.
36. Duan, C.; Ren, H.; Gao, S. Insulin-like growth factors (IGFs), IGF receptors, and IGF-binding proteins: roles in skeletal muscle growth and differentiation. *Gen. Comp. Endocrinol.* 2010, 167, 344-351.
37. Ekblom, R.; Galindo, J. Applications of next generation sequencing in molecular ecology of non-model organisms. *Heredity*. 2011, 107, 1-15.
38. El Haj, A.J. Regulation of muscle growth and sarcomeric protein gene expression over the intermolt cycle. *Amer. Zool.* 1999, 39, 570-579.
39. El Haj, A.J.; Clarke, S.R.; Harrison, P.; Chang. E.S. *In vivo* muscle protein synthesis rates in the American lobster, *Homarus americanus,* during the molt cycle and in response to 20 Hydroxyecdysone. *J. Exp. Biol.* 1995, 199, 1-5.
40. Escamilla-Chimal, E.G.; Hiriart, M.; Sanchez-Soto, M.C.; Fanjul-Moles, M.L. Serotonin modulation of CHH secretion by isolated

cells of the crayfish retina and optic lobe. *General Comp. Endocrinol.* 2002, 125, 283-290.

41. Esteves, A.; Ehrlich, R. Invertebrate intracellular fatty acid binding proteins. *Comp. Biochem. Physiol.* 2006, 2C, 262-274.
42. Fatica, A.; Tollervey, D. Making ribosomes. *Curr. Opin. Cell Biol.* 2002, 14: 313-318.
43. Fidantsef, A.L.; Stout, M.J.; Thaler, J.S.; Duffy, S.S.; Bostock, R.M. Signal interactions in pathogen and insect attack: expression of lipoxygenase, proteinase inhibitor II, and pathogenesis related protein P4 in the tomato, *Lycopersicon esculentum. Physiol. Mol. Plant P.* 1999, 54, 97-114.
44. Freeman, J.A.; Costlow, J.D. Hormonal control of apolysis in barnacle mantle tissue epidermis, in vitro. *J. Exp. Zool.* 1979, 210, 333-345.
45. Gai, Y.; Wang, L.; Song, L.; Zhao, J.; Qiu, L.; Xing, K. A putative endocrine factor SIBD (single insulin binding domain protein) involved in immune response of Chinese mitten crab *Eriocheir sinensis. Fish Shellfish Immunol.* 2010, 28, 10-17.
46. Galat, A. Variations of sequences and amino acid compositions of proteins that sustain their biological functions: an analysis of the cyclophilin family of proteins. *Arch. Biochem. Biophys.* 1999, 371, 149-162.
47. Gamboa-Delgado, J.; Molina-Poveda, C.; Cahu, C. Digestive enzyme activity and food ingesta in juvenile shrimp *Litopenaeus vannamei* (Boone, 1931) as a function of body weight. *Aquac. Res.* 2003, 34, 1403-1411.
48. Garofalo, R.S. Genetic analysis of insulin signaling in Drosophila. *Trends Endocrinol. Metab.* 2002, 13, 156-162.
49. Garvin, M.R.; Saitoh, K.; Gharrett, A.J. Application of single nucleotide polymorphisms to non-model species: a technical review. *Mol. Ecol. Resour.* 2010, 10, 915-934.
50. Gerhardt, C.C.; Heerikhuizen, H.V. Functional characteristics of heterologously expressed 5-HT receptors. *Eur. J. Pharmacol.* 1997, 334, 1-23.
51. Gilad, Y.; Pritchard, J.K.; Thornton, K. Characterizing natural

variation using next-generation sequencing technologies. *Trends Genet.* 2009, 25, 463-471.

52. Gitterle, T.; Rye, M.; Salte, R.; Cock, J.; Johansen, H.; Lozano, C.; Rye, M. Genetic (co) variation in harvest body weight and survival in *Penaeus (Litopenaeus) vannamei* under standard commercial conditions. *Aquaculture.* 2005, 243, 83-92.
53. Glatz, J.F.C.; Van der Vusse, G. Cellular fatty acid-binding proteins: their function and physiological significance. *Prog. Lipid Res.* 1996, 3535, 243-282.
54. Glenn, K.L.; Grapes, L.; Suwanasopee, T.; Harris, D.L.; Li, Y., Wilson, K.; Rothschild, M.F. SNP analysis of AMY2 and CTSL genes in *Litopenaeus vannmei* and *Penaeus monodon* shrimp. *Anim. Genet.* 2005, 36, 235-236.
55. Gong, Y.N.; Li, W.W.; Sun, J.L.; Ren, F.; He, L.; Jiang, H.; Wang, Q. Molecular cloning and tissue expression of the fatty acidbinding protein (Es-FABP) gene in female Chinese mitten crab (*Eriocheir sinensis*). *BMC Mol. Biol* 2010, 11, 71.
56. Gopal, C.; Gopikrishna, G.; Krishna, G.; Jahageerdar, S.S.; Rye, M.; Hayes, B.J.; Paulpandi, S.; Kiran, R.P.; Pillai, S.M.; Ravichandran, P.; Ponniah, A.G.; Kumar, D. Weight and time of onset of female-Genes and growth performance in crustaceans superior sexual dimorphism in pond reared *Penaeus monodon. Aquaculture.* 2010, 300, 237-239.
57. Guedes, S.D.; Vitorino, R.; Domingues, R.; Tomer, K.; Correia, A.J.F.; Amado, F. Proteomics of immune-challenged Drosophila melanogaster larvae hemolymph. *Biochem. Biophys. Res. Commun.* 328, 106-115.
58. Gupta, N.; Pandey, A.; Malik, G.; Gupta, S.C. Single nucleotide polymorphism (SNP) in growth hormone gene of Jakhrana, a prominent milk goat breed in India. *Small Rumin. Res.* 2009, 81, 35-41.
59. Gutierrez, A.; Nieto, J.; Pozo, F.; Stern, S.; Schoofs, L. Effect of insulin/IGF-I like peptides on glucose metabolism in the white shrimp *Penaeus vannamei. Gen. Comp. Endocrinol.* 2007, 153, 170-175.
60. Hamasaki, K.; Kitada, S. Potential of stock enhancement for

decapods crustaceans. *Rev. Fish. Sci.* 2008, 16, 164-174.

61. Hamilton, G.S.; Steiner, J.P. Immunophilins: beyond immunosuppression. *J. Med. Chem.* 1998, 41, 5119-5143.
62. Hao, L.L.; Yu, H.; Zhang, Y.; Sun, S.C.; Liu, S.C.; Zeng, Y.Z.; Ai, Y.X.; Jiang, H.Z. Single nucleotide polymorphism analysis of exon 3 and 4 of IGF-1 gene in pigs. *Genet. Mol. Res.* 2011, 10, 1689-1695.
63. Hara, M.R.; Agrawal, N.; Kim, S.F.; Cascio, M.B.; Fujimuro, M.; Ozeki, Y.; Takahashi, M.; Cheah, J.H.; Tankou, S.K.; Hester, L.D.; Ferris, C.D.; Hayward, S.D.; Snyder, S.H.; Sawa, A. S-nitrosylated GAPDH initiates apoptotic cell death by nuclear translocation following Siah1 binding. *Nat. Cell Biol.* 2005, 7, 665-674.
64. Hauser, L.; Carvalho, G.R. Paradigm shifts in fisheries genetics: ugly hypotheses slain by beautiful facts. *Fish Fish.* 2008, 9, 333-362.
65. Ho, S.Y.; Delgado, L.; Storch, J. Monoacylglycerol metabolism in human intestinal Caco-2 cells: evidence for metabolic compartmentation and hydrolysis. *J. Biol. Chem.* 2002, 277, 1816-1823.
66. Hopkins, P.M. Ecdysteroids and regeneration in the fiddler crab *Uca pugilator*. *J. Exp. Zool.* 1989, 252, 293-299.
67. Hou, G.; Wang, D.; Guan, S.; Zeng, H.; Huang, X.; Ma, Y. Associated analysis of single nucleotide polymorphisms of IGF2 gene's exon 8 with growth traits in Wuzhishan pig. *Mol. Biol. Rep.* 2010, 37, 497-500.
68. Hu, K.J.; Leung, P.C. Food digestion by cathepsin L and digestion-related rapid cell differentiation in shrimp hepatopancreas. *Comp. Biochem. Physiol.* 2007, 146B, 69-80.
69. Hu, K.J.; Leung, P.C. Shrimp cathepsin L encoded by an intronless gene has predominant expression in hepatopancreas, and occurs in the nucleus of oocyte. *Comp. Biochem. Physiol.* 2004, 137B, 21-33.
70. Huberman, A. Shrimp endocrinology: A review. *Aquaculture* 2000, 191, 191-208.
71. Hui, J.H.L.; Hayward, A.; Bendena, W.G.; Takahashi, T.; Tobe, S.S. Evolution and functional divergence of enzymes involved in

sesquiterpenoid hormone biosynthesis in crustaceans and insects. *Peptides*. 2010, 31, 451-455.

72. Hulata, G. Genetic manipulations in aquaculture: a review of stock improvement by classical and modern technologies. *Genetica*. 2001, 111, 155-173.
73. Hwa, V.; Oh, Y.; Rosenfeld, R.G. The insulin-like growth factor-binding protein IGFBP superfamily. *Endocr. Rev.* 1999, 20, 761-787.
74. Janecek, S. α-Amylase family: molecular biology and evolution. *Prog. Biophys. Mol. Biol.* 1997, 67, 67-97.
75. Jung, H.; Lyons, R.E.; Dinh, H.; Hurwood, D.A.; McWilliam, S.; Mather, P.B. Transcriptomics of a giant freshwater prawn (*Macrobrachium rosenbergii*): de novo assembly, annotation and marker discovery. *PLoS ONE*. 2011, 6, e27938.
76. Kamimura, M.T.; Meier, K.M.; Cavalli, R.O.; Laurino, J.; Maggioni, R.; Marins, L.F. Characterization of growth-related genes in the south-western Atlantic pink shrimp *Farfantepenaeus paulensis* (Pe´rez-Farfante 1967) through a modified DDRT-PCR protocol. *Aquac. Res.* 2008, 39, 200-204.
77. Karrer, K.M.; Peiffer, S.L.; DiTomas, M.E. Two distinct gene subfamilies within the family of cysteine proteinase genes. *Proc. Natl. Acad. Sci. USA* 1993, 90, 3063-3067.
78. Kawahara-Miki, R.; Wada, K.; Azuma, N.; Chiba, S. Expression profiling without genome sequence information in a non-model species, Pandalid shrimp (*Pandalus latirostris*), by next-generation sequencing. *PLoS ONE* 2011, 6, e26043.
79. Khandwala, H.M.; McCutcheon, I.E.; Flyvbjerg, A.; Friend, K.E. The effects of insulin-like growth factors on tumorigenesis and neoplastic growth. *Endocrine Reviews*. 2000, 21, 215-244.
80. Kim, S.H.; Lessner, S.M.; Sakurai, Y.; Galis, Z.S. Cyclophilin A as a novel biphasic modiator of endothelial activation and dysfunction. *Am. J. Pathol.* 2004, 164, 1567-1574.
81. Kim, W.J.; Jung, H.T.; Gaffney, P.M. Development of Type I genetic markers from expressed sequence tags in highly polymorphic species. *Mar. Biotechnol.* 2011, 13, 127-132.

82. Kocour, M.; Kohlmann, K. Growth hormone gene polymorphisms in trench, *Tinca tinca* L. *Aquaculture*. 2011, 310, 298-304.
83. Komali, M.; Kalarani, V.; Venkatrayulu, C.H.; Reddy, D.C.S. Hyperglycaemic effects of 5-hydroxytryptamine and dopamine in the freshwater prawn, *Macrobrachium malcolmsonii*. *J. Exp. Zool.* 2005, 303A, 448-455.
84. Kopchick, J.J.; Andry, J.M. Growth hormone (GH), GH receptor, and signal transduction. *Mol. Genet. Metab.* 2000, 71, 293-314.
85. Kumagai, H.; Sakai, H. A porcine brain protein (35 K protein) which bundles microtubules and its identification as glyceraldehyde 3-phosphate dehydrogenase. *J. Biochem.* 1983, 93, 1259-1269.
86. Laufer, H.; Ahl, J.S.B.; Rotllant, G.; Baclaski, B. Evidence that ecdysteroids and methyl farnesoate control allometric growth and differentiation in a crustacean. *Insect Biochem. Mol. Biol.* 2002, 32, 205-210.
87. Laufer, H.; Biggers, W.J. Unifying concepts learned from methyl farnesoate for invertebrate reproduction and post-embryonic development. *Am. Zool.* 2001, 41, 442-457.
88. Laufer, H.; Demir, N.; Pan, X.; Stuart, J.D.; Ahl, J.S. Methyl farnesoate controls adult male morphogenesis in the crayfish, *Procambarus clarkia*. *J. Insect Physiol.* 2005, 51, 379-384.
89. Laycock, M.V.; Hirama, T.; Hasnain, S.; Watson, D.; Storer, A.C. Purification and characterization of a digestive cysteine proteinase from the American lobster (*Homarus americanus*). *Biochem. J.* 1989, 263, 439–444.
90. Le Boulay, C.; van Wormhoudt, A.; Sellos, D. Cloning and expression of cathepsin L-like proteinases in the hepatopancreas of the shrimp *Penaeus vannamei* during the intermolt cycle. *J. Comp. Physiol.* 1996, 166B, 310-318.
91. Leekitcharoenphon, P.; Taweemuang, U.; Palittapongarnpim, P.; Kotewong, R.; Supasiri, T.; Sonthayanon, B. Predicted subpopulations in a marine shrimp proteome as revealed by combined EST and cDNA data from multiple Penaeus species. *BMC Res. Notes*. 2010, 3, 295.
92. Li, W.W.; Jin, X.K.; He, L.; Gong, Y.N.; Jiang, H.; Wang, Q. Molecular cloning and tissue expression of the fatty acidbinding

protein (Es-FABP9) gene in the reproduction seasons of Chinese mitten crab, *Eriocheir sinensis*. *Mol. Biol. Rep.* 2011, 38, 5169-5177.

93. Li, W.W.; Jin, X.K.; He, L.; Jiang, H.; Gong, Y.N.; Xie, Y.N.; Wang, Q. Molecular cloning, characterization, expression and activity analysis of cathepsin L in Chinese mitten crab, *Eriocheir sinensis*. *Fish Shellfish Immunol.* 2010, 29, 1010-1018.
94. Li, X.; Bai, J.; Hu, Y.; Ye, X.; Li, S.; Yu, L. Genotypes, haplotypes and diplotypes of IGF-II SNPs and their association with growth traits in largemouth bass (*Micropterus salmoides*). *Mol. Biol. Rep.* 2012, 39, 4359-4365.
95. Li, X.; Bai, J.; Ye, X.; Hu, Y.; Li, S.; Yu, L. Polymorphisms in the 5′ flanking region of the insulin-like growth factor I gene are associated with growth traits in largemouth bass *Micropterus salmoides*. *Fish. Sci.* 2009, 75, 351-358.
96. Lin, D.T.; Lechleiter, J.D. Mitochondrial targeted cyclophilin D protects cells from cell death by peptidyl prolyl isomerization. *J. Biol. Chem.* 2001, 277, 31134-31141.
97. Liu, J.; Shi, G.P.; Zhang, W.Q.; Zhang, G.R.; Xu, W.H. Cathepsin L function in insect moulting: molecular cloning and functional analysis in cotton bollworm, *Helicoverpa armigera*. *Insect Mol. Biol.* 2006, 15, 823-834.
98. Liu, R.Z.; Li, X.; Godbout, R. A novel fatty acid-binding protein (FABP) gene resulting from tandem gene duplication in mammals: transcription in rat retina and testis. *Genomics*. 2008, 92, 436-445.
99. Lorenzon, S. Hyperglycemic stress response in Crustacea. *Invert. Surviv. J.* 2005, 2, 132-14.
100. Lucy, M.C. Functional differences in the growth hormone and insulin-like growth hormone and insulin-like growth factor axis in cattle and pigs: implications for post-partum nutrition and reproduction. *Reprod. Domest. Anim.* 2008, 43S, 31-39.
101. Lyons, R.E.; Dierens, L.M.; Tan, S.H.; Preston, N.P.; Li, Y. Characterization of AFLP markers associated with growth in the Kuruma prawn, *Marsupenaeus japonicus*, and identification of a candidate gene. *Mar. Biotechnol.* 2007, 9, 712-721.
102. Manor, R.; Weil, S.; Oren, S.; Glazer, L.; Aflalo, E.D.; Ventura,

T.; Sagi, A. Insulin and gender: an insulin-like gene expressed exclusively in the androgenic gland of the male crayfish. *Gen. Comp. Endocrinol.* 2007, 150, 326-336.

103. Mareddy, V.R.; Rosen, O.; Thaggard, H.B.; Manor, R.; Kuballa, A.V.; Aflalo, E.D.; Sagi, A.; Paterson, B.; Elizur, A. Isolation and characterization of the complete cDNA sequence encoding a putative insulin-like peptide from the androgenic gland of *Penaeus monodon*. *Aquaculture*. 2011, 318, 364-370.

104. Martin-Marti, S.; Onteru, S.K.; Du, Z.Q.; Rothschild, M.F. SNP analyses of the 5HT1R and STAT genes in Pacific white shrimp, *Litopenaeus vannamei*. *Span. J. Agric. Res.* 2010, 8, 53-55.

105. Mattson, M.P.; Spaziani, E. Demonstation of protein kinase C activity in crustacean Y-organs and partial definition of its role in regulation of steroidogenesis. *Mol. Cell. Endocrinol.* 1987, 49, 159-171.

106. McAndrew, B.; Napier, J. Application of genetics and genomics to aquaculture development: current and future directions. *J. Agr. Sci.* 2011, 149, 143-151.

107. McArthur, M.J.; Atshaves, B.P.; Frolov, A.; Foxworth, W.D.; Kier, A.B.; Schroeder, F. Cellular uptake and intracellular trafficking of long chain fatty acids. *J. Lipid Res.* 1999, 40, 1371-1383.

108. McClure, M.C.; Morsci, N.S.; Schnabel, R.D.; Kim, J.W.; Yao, P.; Rolf, M.M.; McKay, S.D.; Gregg, S.J.; Chapple, R.H.; Northcutt, S.L.; Taylor, J.F. A genome scans for quantitative trait loci influencing carcass, post-natal growth and reproductive traits in commercial Angus Cattle. *Anim. Genet.* 2010, 41, 597-607.

109. McGrath, M.E. The lysosomal cysteine proteinases. *Annu. Rev. Biophys. Biomol. Struct.* 1999, 28, 181-204.

110. Mellon, De.F. Jr.; Greer, E. Induction of precocius molting and claw transformation in alpheid shrimps by exogenous 20-hydroxyecdysone. *Biol. Bull.* 1987, 172, 350-356.

111. Meyer-Siegler, K.; Mauro, D.J.; Seal, G., Wurzer, J.; deRiel, J.K.; Sirover, M.A. A human nuclear uracil DNA glycosylase is the 37-kDa subunit of glyceraldehyde-3-phosphate dehydrogenase. *Proc. Natl. Acad. Sci. USA*. 1991, 88, 8460-8464.

112. Michael, D.; Oren, M. The p53 and Mdm2 families in cancer. *Curr.*

Opin. Genet. Dev. 2002, 12, 53-59.

113. Morrell, P.L.; Buckler, E.S., Ross-Ibarra, J. Crop genomics: advances and applications. Nature Reviews. *Genetics*. 2012, 13, 85-96.
114. Mullen, M.P.; Berry, D.P.; Howard, D.J.; Diskin, M.G.; Lynch, C.O.; Giblin, L.; Kenny, D.A.; Magee, D.A.; Meade, K.G.; Waters, S.M. Single nucleotide polymorphisms in the insulin-like growth factor 1 (IGF-1) gene are associated with performance in Holstein-Friesian dairy cattle. *Front. Genet.* 2011, 2, 1-9.
115. Nagaraju, G.P.C.; Prasad, G.L.V.; Reddy, P.S. Isolation and characterization of mandibular organ inhibiting hormone from the eyestalks of freshwater crab, *Oziotelphusa senex senex*. *Int. J. Appl. Sci. Eng.* 2005, 3, 61-68.
116. Nakagawa, T.; Roth, W.; Wong, P.; Nelson, A.; Farr, A.; Deussing, J.; Villadangos, J.A.; Ploegh, H.; Peters, C.; Rudensky, A.Y. Cathepsin L: critical role in Ii degradation and CD4 T cell selection in the thymus. *Science*. 1998, 280: 450-453.
117. Naya, Y.; Ohnishi, M.; Ikeda, M.; Miki, W.; Nakanishi, K. What is molt-inhibiting hormone? The role of an ecdysteroidogenesis inhibitor in the crustacean molting cycle. *Proc. Nati. Acad. Sci. USA* 1989, 86, 6826-6829.
118. Ochs, R.L.; Lischwe, M.A.; Spohn, W.H.; Busch, H. Fibrillarin: a new protein of the nucleolus identified by autoimmune sera. *Biol. Cell*. 1985, 54, 123-133.
119. Okuno, A.; Hasegawa, Y.; Nishiyama, M.; Ohira, T.; Ko, R.; Kurihara, M.; Matsumoto, S.; Nagasawa, H. Preparation of an active recombinant peptide of crustacean androgenic gland hormone. *Peptides*. 2002, 23, 567-572.
120. Ola, S.I.; Ai, J.S.; Liu, J.H.; Wang, Q.; Wang, Z.B.; Chen, D.Y.; Sun, Q.Y. Effects of gonadotrophins, growth hormone, and activin A on enzymatically isolated follicle growth, oocyte chromatin organization, and steroid secretion. *Mol. Reprod. Dev.* 2008, 75, 89-96.
121. Ongvarrasopone, C.; Roshorm, Y.; Somyong, S.; Pothiratana, C.; Petchdee, S.; Tangkhabuanbutra, J.; Sophasan, S.; Panyim, S. Molecular cloning and functional expression of the Penaeus monodon 5-HT receptor. *Biochim. Biophys. Acta*. 2006, 1759, 328-

339.

122. Opiyo, S.O.; Moriyama, E.N. Mining the Arabidopsis and rice genomes for cyclophilin protein families. *Int. J. Bioinformaics Res. Appl.* 2009, 5, 295-309.

123. Palacios, R.; Gazave, E.; Goni, J.; Piedrafita, G.; Fernando, O.; Navarro, A.; Villoslada, P. Allele-specific gene expression is widespread across the genome and biological processes. *PLoS ONE*. 2009, 4, e4150.

124. Passano, L.M. *Molting and its control.* In *The Physiology of Crustacea*; Waterman, T.H., Ed.; Vol. 1; Academic Press: New York; 1960; pp. 473-536.

125. Paulson, C.R.; Skinner, D.M. Effects of 20-hydroxyecdysone on protein synthesis in tissues of the land crab *Gecarcinus lateralis*. *J. Exp. Zool*, 1991. 257, 70-79.

126. Pausch, H.; Flisikowski, K.; Jung, S.; Emmerling, R.; Edel, C.; Gotz, K.U.; Fries, R. Genome-wide association study identifies two major loci affecting calving ease and growth-related traits in cattle. *Genetics*. 2011, 187, 289-297.

127. Pavasovic, A.; Richardson, N.A.; Mather, P.B.; Anderson, A.J. Influence of insoluble dietary cellulose on digestive enzyme activity, feed digestibility and survival in the red claw crayfish, *Cherax quadricarinatus* (von Martens). *Aquac. Res.* 2006, 37, 25-32.

128. Pollak, M.N.; Schernhammer, E.S.; Hankinson, S.E. Insulinlike growth factors and neoplasia. *Nat. Rev. Cancer*. 2004, 4, 505-518.

129. Qiu, L.; Jiang, S.; Huang, J.; Wang, W.; Zhu, C.; Su, T. Molecular cloning and mRNA expression of cyclophilin A gene in black tiger shrimp (*Penaeus monodon*). *Fish Shellfish Immunol.* 2009, 26, 115-121.

130. Reddy, P.R.; Nagaraju, G.P.C.; Reddy, P.S. Involvement of methyl farnesoate in the regulation of molting and reproduction in the freshwater crab *Oziotelphusa senex senex*. *J. Crust. Biol.* 2004, 24, 511-515.

131. Rosen, O.; Manor, R.; Weil, S.; Gafni, O.; Linial, A.; Aflalo, E.D.; Ventura, T.; Sagi, A. A sexual shift induced by silencing of a single insulin like gene in crayfish: ovarian upregulation and testicular degeneration. *PLoS ONE*. 2010, 5, e15281.

132. Rossler H.; Rieck, C.; Delong, T.; Hoja, U.; Schweizer, E. Functional differentiation and selective inactivation of multiple Saccharomyces cerevisiae genes involved in very-long-chain fatty acid synthesis. *Mol. Genet. Genomics*. 2003, 269, 290-298.

133. Santiesteban, D.; Martin, L.; Arenal, A.; Franco, R.; Sotolongo, J. Tilapia growth hormone binds to a receptor in brush border membrane vesicles from the hepatopancreas of shrimp *Litopenaeus vannamei*. *Aquaculture*. 2010, 306, 338-342.

134. Schmid, F.X.; Mayr, L.M.; Mucke, M.; Schonbrunner, E.R. Prolyl isomerases – role in protein-folding. *Adv. Protein Chem.* 1993, 44, 25-66.

135. Seeb, J.E.; Carvalho, G.; Hauser, L.; Naish, K.; Roberts, S.; Seeb, L.W. Single-nucleotide polymorphism (SNP) discovery and applications of SNP genotyping in nonmodel organisms. *Mol. Ecol. Res.* 2011, 11S, 1-8.

136. Simon, C.J.; Jeffs, A. The effect of dietary carbohydrates on the growth response, digestive gland glycogen and digestive enzyme activities of early spiny lobster juveniles, *Jasus edwardsii*. *Aquac. Nutr.* 2011, 17, 613-626.

137. Singh, R.; Green, M.R. Sequence-specific binding of transfer RNA by glyceraldehyde-3-phosphate dehydrogenase. *Science*. 1993, 259, 365-368.

138. Sirover, M.A. On the functional diversity of glyceraldehyde-3-phosphate dehydrogenase: biochemical mechanisms and regulatory control. *Biochim. Biophys. Acta*. 2011, 1810, 741-751.

139. Spaziani, E.; Mattson, M.P.; Rudolph, P.H. Regulation of crustacean molt-inhibiting hormone. *Perspect. Comp. Endocrinol.* 1994, 243-250.

140. Stanek, D.; Neugebauer, K.M. The Cajal body: a meeting place for spliceosomal snRNPs in the nuclear maze. *Chromosoma*. 2006, 115, 343-354.

141. Stillman, J.H.; Colbourne, J.K.; Lee, C.E.; Patel, N.H.; Phillips, M.R.; Towle, D.W.; Eads, B.D.; Gelembuik, G.W.; Henry, R.P.; Johnson, E.A.; Pfrender, M.E.; Terwilliger, N.B. Recent advances in crustacean genomics. *Integr. Comp. Biol.* 2008, 48, 852-868.

142. Tahmoorespur, M.; Taheri, A.; Gholami, H.; Ansary, M. PCRSSCP

variation of GH and STAT5A genes and their association with estimated breeding values of growth traits in Baluchi sheep. *Anim. Biotechnol.* 2011, 22, 37-43.

143. Tangprasittipap, A.; Tiensuwan, M.; Withyachumnarnkul, B. Characterization of candidate genes involved in growth of black tiger shrimp *Penaeus monodon*. *Aquaculture*. 2010, 307, 150-156.
144. Tao, W.J.; Boulding, E.G. Associations between single nucleotide polymorphisms in candidate genes and growth rate in Arctic charr (*Salvelinus alpinus* L.). *Heredity.* 2003, 91, 60-69.
145. Tierney, A.J. Structure and function of invertebrate 5-HT receptors: a review. *Comp. Biochem. Physiol.* 2001, 128A, 791-804.
146. Tisdale, E.J. Glyceraldehyde-3-phosphate dehydrogenase is required for vesicular transport in the early secretory pathway. *J. Biol. Chem.* 2001, 276, 2480-2486.
147. Tiu, S.H.K.; He, J.G.; Chan, S.M. Organization and expression study of the shrimp (*Metapenaeus ensis*) putative 5-HT receptor: up-regulation in the brain by 5-HT. *Gene*. 2005, 353, 41–52.
148. Toullec, J.Y.; Van Wormhoudt, A. Quantitative variations during the intermoult cycle of peptides related to human growth hormone in *Palaemon serratus* (Crustacea; Decapoda). *Comptes Rendus de l'Academie des Sciences-Series III*. 1987, 305: 265-269.
149. Tristan, C.; Shahani, N.; Sedlak, T.W.; Sawa, A. The diverse functions of GAPDH: views from different subcellular compartments. *Cell. Signal.* 2011, 23, 317–323.
150. Turk, B.; Turk, D.; Turk, V. Lysosomal cysteine proteinases: more than scavengers. *Biochim. Biophys. Acta*. 2000, 1477, 98-111.
151. Van der Maarel, M.J.E.C.; van der Veen, B.; Uitdehaag, J.C.M.; Leemhuis, H.; Dijkhuizen, L. Properties and applications of starch converting enzymes of the a-amylase family. *J. Biotechnol.* 2002, 94, 137-155.
152. Van Wormhoudt, A.; Sellos, D.Y. Cloning and sequencing analysis of three amylase cDNAs in the shrimp *Pennaeus vannamei* (Crustacae decapoda): evolutionary aspects. *J. Mol. Evol.* 1996, 42, 543-551.
153. Vazquez-Acevedo, N.; Reyes-Colon, D.; Ruiz-Rodriguez, E.A.; Rivera, N.M.; Rosenthal, J.; Kohn, A.B. Cloning and

immunoreactivity of the 5-HT1Mac and 5-HT2Mac receptors in the central nervous system of the freshwater prawn *Macrobrachium rosenbergii*. *J. Comp. Neurol.* 2009, 513, 399-416.

154. Venema, J.; Tollervey, D. Ribosome synthesis in Saccharomyces cerevisiae. *Annu. Rev. Genet.* 1999, 33, 261-311.

155. Ventura, T.; Manor, R.; Aflalo, E.D.; Weil, S.; Raviv, S.; Glazer, L. Temporal silencing of an androgenic gland-specific insulinlike gene affecting phenotypical gender differences and spermatogenesis. *Endocrinol.* 2009, 150, 1278-1286.

156. Ventura, T.; Rosen, O.; Sagi, A. From the discovery of the crustacean androgenic gland to the insulin-like hormone in six decades. *Gen. Comp. Endocrinol.* 2011, 173, 381-388.

157. Verri, T.; Mandal, A.; Zilli, L.; Bossa, D.; Mandal, P.K.; Ingrosso, L.; Zonno, V.; Vilella, S.; Ahearn, G.A.; Storelli, C. D-Glucose transport in decapod crustacean hepatopancreas. *Comp. Biochem. Physiol.* 2001, 130A, 585-606.

158. Viitala, S.; Szyda, J.; Blott, S.; Schulman, N.; Lidauer, M.; Maki-Tanila, A.; Georges, M.; Vilkki, V. The role of the bovine growth hormone receptor and prolactin receptor genes in milk, fat and protein production in Finnish Ayrshire dairy cattle. *Genetics*. 2006, 173, 2151-2164.

159. Vijayan, K.K.; Diwan, A.D. Studies on the physiology of moulting in the penaeid prawn *Penaeus indicus* H. Milne Edwards. *CMFRI Spl. Publ.* 1993, 56, 66-72.

160. Vijayan, K.K.; Mohamed, K.S.; Diwan, A.D. On the structure and moult controlling function of the Y-organ in the prawn *Penaeus indicus* H. Milne Edwards. *J. World Aqua. Soc.* 1993, 24, 516-521.

161. Wade, N.M.; Anderson, M.; Sellars, M.J.; Tume, R.K.; Preston, N.P.; Glencross, B.D. Mechanisms of colour adaptation in the prawn *Penaeus monodon*. *J. Exp. Biol.* 2012, 215, 343-350.

162. Wang, Q.; Li, H.; Li, N.; Leng, L.; Wang, Y.; Tang, Z. Identification of single nucleotide polymorphism of adipocyte fatty acid-binding protein gene and its association with fatness traits in the chicken. *Poult. Sci.* 2006, 85, 429-434.

163. Wang, Y.; Price, S.E.; Jiang, H. Cloning and characterization of the bovine class 1 and class 2 insulin-like growth factor-I mRNAs.

Domest. Anim. Endocrinol. 2003, 25, 315-328.

164. Wang, Y.; Shu, D.; Li, L.; Qu, H.; Yang, C.; Zhu, Q. Identification of single nucleotide polymorphism of H-FABP gene and its association with fatness traits in chickens. *Mol. Biol. Rep.* 2007, 20, 1812-1819.
165. Waters, S.M.; McCabe, M.S.; Howard, D.J.; Giblin, L.; Magee, D.A.; MacHugh, D.E.; Berry, D.P. Associations between newly discovered polymorphisms in the *Bos taurus* growth hormone receptor gene and performance traits in Holstein-Friesian dairy cattle. *Anim. Genet.* 2011, 42, 39-49.
166. Westerberg, R.; Mansson, J.E.; Golozoubova, V.; Shabalina, I.G.; Backland, E.C.; Tvrdik, P. ELOVL3 is an important component for early onset of lipid recruitment in brown adipose tissue. *J. Biol. Chem.* 2006, 281, 4958-4968.
167. Wheeler, M.B.; Monaco, E.; Bionaz, M.; Tanaka, T. The role of existing and emerging biotechnologies for livestock production: toward holism. *Acta Sci. Vet.* 2010, 38S2, 463-484.
168. Williams, K.C. Nutritional requirements and feeds development for post-larval spiny lobster: a review. *Aquaculture.* 2007, 263, 1-14.
169. Worm, B.; Barbier, E.B.; Beaumont, N.; Duffy, J.E.; Folke, C.; Halpern, B.S.; Jackson, J.B.C.; Lotze, H.K.; Micheli, F.; Palumbi, S.R.; Sala, E.; Selkoe, K.A.; Stachowicz, J.J.; Watson, R. Impacts of biodiversity loss on ocean ecosystem services. *Science.* 2006, 314, 787-790.
170. Wu, Q.; Brown, M.R. Signaling and function of insulin-like peptides in insects. *Annu. Rev. Entomol.* 2006, 51, 1-24.
171. Xu, B.; Zhang, P.J.; Mai, Y.L.; Miao, H.Z. Studies of the effects of recombinant fish growth hormone on survival and growth enhancement of Chinese prawn *Penaeus chinensis*. *Mar. Sci.* 2000, 24, 54-56.
172. Yamada, H.; Nishimura, F.; Naruishi, K.; Chou, H.H.; Takashiba, S.; Albright, G.M.; Nares,S.; Iacopino, V.; Murayama, Y. Phenytoin and cyclosporin A suppress the expression of MMP-1, TIMP-1 and cathepsin, but not cathepsin B in cultured gingival fibroblasts. *J. Periodontol.* 2000, 71, 955-960.
173. Yeh, H.Y.; Klesius, P.H. Channel catfish, *Ictalurus punctatus*,

cyclophilin A and B cDNA characterization and expression analysis. *Vet. Immunol. Immunop.* 2008, 121, 370-377.

174. Zhang, C.; Liu, Y.; Huang, K.; Zeng, W.; Xu, D.; Wen, Q.; Yang, L. The association of two single nucleotide polymorphisms (SNPs) in growth hormone (GH) gene with litter size and superovulation response in goat-breeds. *Genet. Mol. Biol.* 2011a, 34, 49-55.
175. Zhang, Y.; Benton, J.L.; Beltz, B.S. 5-HT receptors mediate lineage-dependent effects of serotonin on adult neurogenesis in *Procambarus clarkia. Neural Dev.* 2011b, 6, 2.
176. Zhao, W.M.; Zhao, R.X.; Qiao, N.; Xu, Q.; Huang, Z.Y.; Li, X. Association of GH polymorphism with growth traits in goose. *J. Anim. Vet. Adv.* 2011, 10, 692-697.
177. Zimmerman, A.W.; Veerkamp, J.H. New insights into the structure and function of fatty acid-binding proteins. *Cell. Mol. Life Sci.* 2002, 11, 1096–1116.

CHAPTER 6

USE OF FISH GROWTH HORMONE: AUGMENTING PRODUCTIVITY IN AQUACULTURE

J Raymond Jani Angel, Sekar Megarajan and Dhanasekaran Linga Prabu

Dr. J Raymond Jani Angel is a scientist in the Division of Genetics and Biotechnology, ICAR-Central Institute of Brackishwater Aquaculture, Chennai. He earned his PhD in Aquaculture from ICAR-Central Institute of Fisheries Education, Mumbai. His main research interests are applied to growth improvement of candidate brackishwater aquaculture species using molecular tools. Other research work includes molecular systematics of coastal decapod crustaceans especially shrimps of family Penaeidae and Palaemonidae. More recent research has focused on mangrove and ichthyo- faunal biodiversity of the remotely located Andaman and Nicobar group of Islands.

Dr. Sekar Megarajan graduated in fisheries science, and then obtained master degree in Fish Genetics and Biotechnology and PhD in Fish Nutrition and Biochemistry with specialization in fish molecular endocrinology from Central Institute of Fisheries Education (CIFE), India. He worked for two years with Transasia Bio-Medicals Ltd, Mumbai in the area of immunology for development of diagnostic kits for human infectious disease and joined Indian Agriculture Research Service (ARS) in 2013 as a scientist. He is currently working as scientist in mariculture division at Central Marine Fisheries Research Institute (CMFRI), Cochin, India. Dr Megarajan has been working on breeding, seed production and culture of high value marine finfishes. He has published more than 10 research papers in peer reviewed journals in the area of fish population genetics, molecular endocrinology and fish culture.

D. Linga Prabu, With Masters in Fish Nutrition and Biochemistry, he joined in Agricultural Research Service (ARS) as Scientist. Presently posted in Marine Biotechnology Division of

ICAR-Central Marine Fisheries Research Institute (CMFRI) Kochi, India. Before joining into ARS, Mr Prabhu worked with Growel Feeds Pvt Ltd., Andhra Pradesh, India, a leading fish and shrimp feed producer, as Assistant Manager- Fish Nutrition and Formulation for a period of two years since 2012 and gained rich experience in commercial aspects of aquafeed formulation and manufacture. Currently he is working in marine food fish nutrition with nutrigenomic approach, and also involving in the development of cost effective best performance feed for marine food fishes. Mr Prabhu has great interest in aquatic disease management in mariculture species through nutraceuticals and its effects on immune development.

INTRODUCTION

Natural growth rate of organisms has evolved to provide maximum fitness in wild environments and it matches with seasonal and spatial availability of food resources, variation in physical conditions, and associated with other correlated characteristics such as predation risk and disease resistance (Thorpe et al., 1998). Enhancement of natural growth rate of fish in aquaculture has been extensively explored, with gains arising from improvements in husbandry, nutrition, and genetic selection (Li et al., 2003). Growth enhancement can be advantageous for aquaculture by shortening production time, and enhancing feed conversion efficiency. Therefore, controlling growth through endocrine approaches has been extensively explored. The development and differentiation of seawater-type chloride cells of branchial epithelia are regulated by GH, IGF-I and cortisol. Gene transfer technology has produced a great impact in modern biology and biotechnology (Powers et al., 1998), principally in applications of somatotropins such as insulin like growth factor-1 (IGF-1), growth hormone, thyroid hormones, prolactin, placental lactogen, and sex steroids (McLean and Delvin, 2000). Significant growth gain in fish could be achieved through application of these growth factors with some practical difficulties. However, use of growth hormone for growth improvement has been the central focus than the other hormones in different studies. Growth Hormone (GH) otherwise known as "Somatotropin" is a single chain polypeptide of approximately with 22 kDa molecular weight containing about 190 amino acid residues, produced by the pituitary gland and responsible for number of anabolic processes among vertebrates. Besides regulating body growth, GH is involved in reproduction, immunity and osmoregulation and metabolic regulation such as lipolytic activity and protein anabolism in fishes

(Rousseau and Dufour, 2007). The different activities of growth hormone have been explored and used for both in reach and commercial purposes. In fish, the availability of naturally available fish growth hormone is meager, thus the recombinant growth hormone has been produced from different fishes for large scale production of hormones for commercial application. The prospect of using recombinant DNA technology for the production of vertebrate growth hormones has generated some interest for the potential use of recombinant growth hormone to stimulate growth and improve feed efficiency in aquaculture.

Similarly, growth enhancement in fish using molecular techniques is equally beneficial to aquaculture and is more effective than traditional breeding programmes to develop new fish strains. The technology which uses molecular technique is called as transgenic fish technology. In principle, the technology can be used to improve growth rate of the fish, control sexual maturation, sterility and sex differentiation, increasing disease resistance, cold resistance and modify the biochemical characteristics of the flesh to enhance the nutritional qualities. However, since fish can be readily improved by application of molecular techniques, it is timely to consider what genetically modified (GM) fish are likely to offer in the future, both in terms of benefits and disadvantages (Maclean, 2003). Growth hormone gene has been utilised extensively for construction of transgenic fishes to enhance growth and thus, species showing high growth rate is widely used to isolate growth hormone gene for the production of transgenic fish. The application of transgenic fish not only limited to enhance aquaculture production but also useful in biomedical sciences for various purposes. The transgenic fish produced through the incorporation of growth hormone transgene may have effects on feed conversion efficiency, color, body shape, behavior, carcass yield, body composition, muscle structure, hypoxia tolerance, disease resistance and reproductive traits. Though transgenic fish have lot of benefits to increase fish production of aquaculture industry, the regulatory agencies have yet to approve transgenic fish for commercial production. This chapter explains the effects of administration of GH on fish physiology including behaviour, immunity and growth and use of growth hormone gene in transgenic fish production for growth improvement.

EXOGENOUS APPLICATION OF GH ON FISH GROWTH

A common therapy that is used to enhance growth in domestic animals is using GH and it has also become increasingly popular in aquaculture for fin fishes and shell fishes. Initially, mammalian GH was found to accelerate the growth of fish (Pickford and Thompson, 1948). Thereafter, there have been many studies demonstrating the efficacy of mammalian GH in the acceleration of growth of various species. In fish, approaches to accelerate growth by GH administration have been studied and increased growth has been achieved by various modes of delivery like injection, and oral administration of hormones (Schulte et al., 1989). The effect of administering exogenous natural mammalian and piscine growth hormone has been studied in many fishes including, salmonids, tilapias, basses, eels, carps, mullets, and channel catfish (Donaldson et al., 1979). These studies suggest that the exogenous growth hormone stimulates appetite and improves feed conversion and thus enhances fish growth (Matty, 1986). Exogenous growth hormone stimulate appetite in fish by a direct action on centers in the hypothalamus that influence feed intake or by inducing a number of metabolic changes that feed back on the hypothalamic centers to affect appetite (Markert et al., 1977). Earlier experiments suggested that exogenous growth hormone may improve feed conversion by one or more possible mechanisms 1) stimulation of lipid mobilization and oxidation 2) an action on the rate of protein synthesis or breakdown 3) stimulation of insulin synthesis and release. However, the mechanism (s) underlying improvements of feed conversion efficiency in GH treated fish are not entirely understood. Earlier experiments reported that elevated GH has no effect on gut enzyme activity (Lemieux et al., 1997), as an alternative it stimulates intestinal Na+ dependent and Na+ independent amino acid transport (Sun, 1992), which may partially explain improved feed conversion efficiency in treated fish. Depending upon the dose, treatment duration, and growth stage, a more than double of individual weight relative to controls may be achieved (McLean et al., 1996).

Administration of growth hormone has shown to improve the growth satisfactorily in many of the fish species. Interestingly, in some of the fishes it is possible to increase the growth to double the size than the actual individual size of fishes, and it depends upon the factors like dose, duration and time of the administration of the hormone (McLean et al., 1997). A study on Atlantic salmon resulted that the fish treated with GH exhibited significant increase in the metabolic rate during parr and pre-smolt stages than the control fish (Sedikki et al., 1996). The biological activity of purified protein from recombinant growth hormone of *Pangasianodon hypophthalmus* (rPhGH) was determined by Sekar et al. (2015) in rohu (*Labeo rohita*) fingerlings to promote growth. The fingerlings of *L. rohita* were injected with the hormone at three different concentrations, and a significant increase in growth was observed in fingerlings administered with rPhGH at a dosage of 1.0 mg g^{-1} body weight. Similarly, growth promoting effect of growth hormone has been studies in different fishes such as trout (Sato *et al.*, 1989), eel (Sugimoto *et al.*, 1990), common carp (Fine *et al.*, 1993), flounder (Jeh *et al.*, 1998) and giant catfish (Promdonkoy *et al.*, 2004). Administration of the hormones in most of the teleost fishes has showed enhanced growth than the untreated fish in any mode of administrations such as injection (Guillén II et al., 1998; Li et al., 2003), immersion treatment (Schulte et al., 1989; Moriyama and Kawauch, 1990) and oral administration (Tsai et al., 1997; Ben-Atia et al., 1999).

Other than the major impact on growth improvement, the administration of exogenous growth hormone lead to increase in endocrine growth regulatory component, and changes in the tissue proximate composition in most of the studied fishes (Fleming et al., 2002). Rainbow trout when injected with bovine GH led to elevation of endogenous GH (Kieffer et al., 1994). The effect of exogenous GH administration on the serum IGF-I and IGF-II concentrations were studied by Shamblott et al. (1995). Also, elevated levels of T3 and hepatic 50-deiodinase that converts thyroxine to T3 was observed in salmonids injected with GH (MacLatchy et al., 1992). Additionally, predictable changes in body composition and conformation can be achieved over elevated levels of GH through shifts in lipid, protein and mineral metabolism. It was observed that, the body of salmonids appear less deep-bodied and slender, thus reducing the condition factor (k) invariably when GH is administered (McLean and Donaldson, 1993; Johnsson and Bjornsson, 1994). The changes in proximate composition of treated fish were extensively studied (Gill et al., 1985; McLean and Devlin, 2000). The ash content of the fish increased significantly owing to the augmented mineral deposition within bone leading to increased bone formation and desorption (Takagi et al., 1992). GH treatment is concomitant with rise in fatty acid and plasma glucose concentrations (Leatherland and Nuti, 1981) and also with elevated hepatic lipolysis (O'Connor et al., 1993). Protein synthesis in muscle and other tissues (Cheema and Matty, 1978; Foster et al., 1991; Fauconneau et al., 1996) and body moisture levels are elevated (McLean and Devlin, 2000) with exogenous GH treatment.

ADMINISTRATION OF GROWTH HORMONE AND FISH BEHAVIOUR

The effects of growth hormone (GH) on various types of behavior in teleost fish are well established but the underlying mechanism behind the behavioral change has not been fully understood. In salmonid fish, growth hormone increases locomotor activity, aggression, and feeding activity and at the same time reduces predator avoidance behavior (Johansson, et al., 2005). The impact of behaviour changes of growth hormone has been well demonstrated in mammals and fishes. However, it is not clear that the behavioural effect of growth hormone is mediated directly in the brain or by pheripheral changes (Johansson, et al., 2005). Growth hormone administration has shown to influence on the fish behaviour, it was understood that, GH administration increases energy demand and thereby elicit appetite which might reduce anti-predator response and promote competitive ability. This hypothesis was tested by Johnsson and Bjornsson (1994) and found that exogenous GH elevates feeding motivation, and thereby provide dominance status in juvenile rainbow trout, comparing the behaviour, growth rate, food conversion and hypo-osmoregulatory ability of GH-treated and control groups. GH-treated fish consumes twice as much as control fish over the 2 days of feeding and exhibited higher food conversion efficiency. Feeding competition trials involving size-matched and control fish showed GH-treated fish were significantly more dominant. Similarly, GH treatment promoted growth rate in brown trout and increased RNA levels, but had not observed effect on dominance (Johnsson et al., 1996). Jonsson et al. (1998) observed agonistic interactions to clarify the role of GH in social interactions, in size-matched pairs of juvenile rainbow trout consisting of two control fish, two GH-treated fish. In this observation GH did not appear to affect fighting ability or social status directly, though it is likely to do so indirectly through affecting body size. In another study, Both Rainbow trout and Brown trout treated with GH appear more willing to risk exposure to predators, whether it be simulated attacks using a model thereon (Jonsson et al., 1996a; 1996b) or the presence of a predator Brown trout (Johnsson et al., 1996) was not sure. The growth-promoting effect of GH, however, appears to be much more pronounced under conditions of unrestricted food supply in the hatchery than in the wild, where food availability is limited; suggesting that the payoff associated with increased growth investment is higher in the former case (Johnsson et al., 2000). Moreover, work with domesticated Salmonids, which show similar increases in growth, aggression and willingness to risk exposure to predators as observed in GH-treated fish (Einum and Fleming, 2001; Johnsson et al., 1996), has shown that survival consequences for greater growth in the wild depends on the predation environment, being expressed only when risk is high (Biro et al., 2004).

ROLE OF GH ON OSMOREGULATION

Besides the growth promoting role, growth hormone promotes the acclimation to seawater in most of the teleost fish through the aid of insulin-like growth factor I. GH, IGF-I and cortisol regulate the development and differentiation of seawater-type chloride cells of branchial epithelia (Sakamoto and McCormick, 2006). The administration of GH may increase the number and size of gill chloride cells, Na^+, K^+, $2Cl^-$ co-transporter (NKCC), Na^+, K^+-ATPase and ion transporters associated with salt secretion (McCormick, 2001). The role of GH on salinity tolerance and salt secretory mechanisms is reported in salmonids, tilapia and killifish (Sakamoto and McCormick, 2006). Several authors reported that growth hormone receptors found in the liver, gill, gut and kidney of several marine fishes and the numbers of GH receptors are increased in the osmoregulatory organs of rainbow trout after exposure to sea water (Sakamoto and Hirano, 1991). According to McCormick et al. (2002), plasma levels of GH and IGF-I increase during smolting of salmonids along with increase in gill chloride cell size and number, NKCC levels, Na^+,K^+-ATPase activity indicates salinity tolerance (Hoar, 1988). But, in marine eel and sea bream, (*Sparus sarba*) GH may not play an osmoregulatory role because of their marine origins or a limited capacity for hyper-osmoregulation (Deane and Woo, 2004).

ROLE OF GH ON FISH IMMUNITY

Activation of immune functions during the seawater acclimatization of some euryhaline fish species appears to be linked with increased level of GH in the circulation which improves seawater adaptability (Dominguez et al., 2005). In fish, the non-specific defense elements such as cytotoxic, phagocytic, hemolytic and lysozyme activities and specific defense mechanism mainly immunoglobulin levels were elicited by administration of GH (Yada, 2007). In higher vertebrates, the GH secreted in the lymphoid tissues has an effect on the proliferation of leucocytes (Jeay et al., 2002). The structural resemblances of GH and its receptors to cytokines and its receptors indicate the functional association of GH on immune system and its role (Venters et al., 2001). The *in vivo* administration of GH in fishes showed an increase in superoxide anion activity, lysozyme activity and phagoctytic activity (Narnaware et al., 1997; Yada et al., 2002). Administration of bovine GH in *Oreochromis mossambicus* and *Oncorhynchus mykiss* stimulates somatic growth and also enhances antibody production against this growth hormone (Leedom et al., 2002; Biga et al., 2005). Hypophysectomized fish exhibit plummeted level of immunoglobulins (Ig) in the circulation and administration of GH in hypophysectomized fish elevated and restore the level of Ig in comparison with normal fish which indicates the importance of GH in the specific immunity of fish (Yada and Azuma, 2002). The administration of GH stimulates the proliferation of leucocytes in the peripheral blood of Chum salmon (Sakai et al., 1996). The extra-pituitary expression of GH mRNA in the lymphoid tissues and phagocytic leucocytes was noticed in several fishes and not in some fishes which indicates the degree of variation in the endocrine regulation of the immune function in different fishes (Yada, 2007). According to Yada (2007), the paracrine secretion and regulation of GH is also an important mechanism in the peripheral defense of host against the invading pathogens.

GROWTH HORMONE –MYOSTATIN INTERACTION ON MUSCLE GROWTH AND QUALITY

During earlier times, the quality of any livestock meat is mainly considered on the sensory factors such as texture, flavor, color and visible fat. However, the quality is described on the basis of the composition of poly-unsaturated versus saturated fat and microbiological contamination mainly because of increasing health concern and food safety awareness from the consumers. Feed management strategies can preferably affect the biochemical composition, flavor and texture quality of carcass of livestock. Similarly, genetic variation also has a great influence on feed intake, metabolism and there by conversion of food into desirable ratio of fat and muscle, location of fat distribution, muscle fiber type and quantity as well as taste (Williams, 2008). Improved genetic tools helps to increase the resistance of animals against the potential pathogens through improved genetic makeup which in turn reduce the microbial contamination on carcasses and processed meat. Myostatin (MSTN) is a transforming growth factor β (TGF-β) family member and negative regulator of muscle growth (McPherron et al., 1997). Myostatin plays an essential role in the regulation of developmental pathway for the differentiation of progenitor cells into muscle and fat tissues. During the nonexistence of myostatin, the cells preferentially transform into muscle tissue instead of fat tissue. Myostatin and GH have negative relationship and expression patterns in several vertebrates (Marcell et al., 2001; Liu et al., 2003).

In MSTN gene knockout mice, the skeletal muscle mass is 2–3 folds larger than the wild-type mice, which accomplished through hyperplasia and hypertrophy of muscle fibers (McPherron et al., 1997). According to Jiang et al. (2002) variations in the myostatin gene results in reduction of myostatin expression and hence the muscular hypertrophy which yields higher average daily gain as well as improved muscling (Clop et al., 2006). Xu et al. (2003) found that over expression of MSTN pro-domain in zebrafish, binds with MSTN domain and inhibits its regular function and allow the increased number of muscle fibers development. The transgenic coho salmon produced through all-salmon GH/metallothionen (MT) gene construct, showed overexpression of GH and lesser expression of MSTN which is evident from 11-folds larger growth rate than the control coho salmon in 15 months (Devlin et al. 1994). According to Hill et al. (2000), the transgenic coho salmon have appreciably more numbers of small-diameter muscle fibers that indicates their larger body size which is achieved as a result of increased muscle hyperplasia. Similarly, Roberts et al. (2004) proved that the transgenic coho salmon, fast growing fish than its wild counterpart exhibit under expression of both MSTN transcripts and protein suggests that MSTN does act as a negative regulator of growth. Thus, this study provides evidence that the anabolic effects of GH could be attributed through MSTN. Hyperplasia and hypertrophy in transgenic salmon indicates its better muscle quality due to improved muscle texture than their wild counterparts (Roberts et al., 2004).

MOLECULAR TECHNIQUES FOR GROWTH ENHANCEMENT

There are many ways to enhance growth including inbreeding, gynogenesis, androgenesis, selection, intraspecific crossbreeding, interspecific hybridization, polyploidy, sex reversal and breeding, nuclear transplantation and transgenesis. Cloned populations have been produced via gynogenesis and androgenesis (Dunham, 2004), but direct cloning of an individual fish of interest has not yet been accomplished. A significant impact of gene transfer technology has been observed in modern biotechnology (Powers et al., 1998). The fish species used for gene transfer studies can be divided into two main groups: those which are used in aquaculture (Fletcher and Davies, 1991; Hew et al. 1995; Chen and Lu, 1998) and those which can be used as a model fish for basic research (Chen and Lu, 1998). Some of the major food fish species are tilapia (*Oreochromis* sp.), carp (*Cyprinus* sp.), channel catfish (*Ictalurus punctatus*), and salmon (*Salmo* sp., *Oncorhynchus* sp.), while gold fish (*Carassius auratus*), zebrafish (*Danio rerio*), and medaka (*Oryzias latipes*) are used in basic research. Genetic engineering of farm animals offers great potential for improvement of selected genetic traits of agricultural significance. Several species of fish have also been used to exploit this technology for commercial purposes and examples include induction of freeze resistance in transgenic salmon using an anti-freeze protein gene (Fletcher et al., 1988) and production of growth enhanced fish using novel growth hormone (GH) genes (Dunham et al., 1987; Brem et al., 1988; Penman et al., 1990) or an Insulin-like Growth Factor (IGF) gene (Chen et al., 1995). Although several species of fish have been used to produce lines of transgenic fish, in only a few cases has germline transmission and stable long-term transgene expression been satisfactorily demonstrated. The main problem with obtaining germline transmission has been the widespread mosaicism of the transgene due to delayed integration into the host genome during early embryonic development (Maclean and Rahman, 1994; Gong and Hew, 1995). Transgenic gene knockout technology (Wong and Van Eenennaam, 2008; Thresher et al., 2009) and gene editing technology that mutates the fish without inserting exogenous DNA has also been accomplished (Tan et al., 2013). Further, another biotechnological tool based on a bacterial CRISPR (Clustered Regularly Interspaced Short Palindromic Repeats)-associated protein-9 nuclease (Cas9) from *Streptococcus pyogenes* (Cong et al., 2013) has gained much importance in knocking down the myostatin gene, thereby augmenting muscular growth in mammals. CRISPR/Cas9 system has been widely used in plants and animals to trigger mutagenesis. Jao et al. (2013) opined that CRISPR/Cas9 have the potential to develop high rates of homozygous/biallelic mutagenesis and allow phenotypic evaluation in the transformed fish that obtain the mutation in every cell.

TRANSGENIC FISH FOR GROWTH ENHANCEMENT

Transgenic fish have been produced for numerous species of fish including non-commercial model species such as the loach, *Misgurnus anguilIicaudtis* (Maclean et al., 1987), Medaka, *Oryzias latipes* (Ozato et al., 1986), topminnows and zebra fish (Rainbow zebra fish) (Gong et al. 2002). Several experimental trials have also produced and evaluated transgenic farmed fish species including gold fish (Zhu et al., 1985), silver carp, common carp, mud loach, Atlantic salmon, rainbow trout (Chourrout, 1986), chinook salmon, coho salmon, Nile tilapia (Brem et al., 1988) and channel catfish (Dunham et al., 1987). Additionally, gene transfer has also been accomplished in a game fish, Northern pike (Gross et al., 1992). Growth performance of the transgenic salmon showed an average of 11-fold increase of growth compared to non-transgenic controls

(Devlin et al., 1995). Nam et al. (2001) found that 35-fold increases in growth of transgenic mud loach, *Misgurnus mizolepis* developed by transgene produced from the same species through microinjecting the mud loach growth hormone (mlGH) gene fused with the mud loach β-actin promoter than their non-transgenic siblings. Most of the farmed transgenic fish aimed for the growth improvement, and many of these fishes have used for studying and standardizing growth pattern. Result of growth evaluation of different studies has showed positive result in many of the fishes. But, some of the studied has proved that the technology is unsuccessful due to many unknown reasons. Some of the studies have been quoted in the following Table.

Table 6.1: Studies showing enhancement of growth achieved in different target organisms worldwide with citation

FAMILY AND SPECIES	CONSTRUCT	GROWTH	COUNTRY	SUPPORTING CITATION
Salmonidae				
Atlantic salmon, *Salmo salar*	opAFP-csGH	2–6-Fold	Canada	Du et al. (1992) and Fletcher et al. (2004)
Coho salmon, *Oncorhynchus kisutch*	ssMT-ssGH	Up to 11-fold	Canada	Devlin et al. (1994a,b)
Coho salmon *Oncorhynchus kisutch*	opAFP-csGH	3–10-Fold	Canada	Devlin et al. (1995)
Chinook salmon, *O. tshawhytscha*	opAFP-csGH	6-Fold	Canada	Devlin et al. (1995)
Rainbow trout, *O. mykiss*	opAFP-csGH	3.2-Fold	Canada	Devlin et al. (1995)
Cutthroat trout, *O. clarki*	opAFP-csGH	6-Fold	Canada	Devlin et al. (1995)

Arctic charr, *Salvelinus alpines*	Various constructs	Up to 14-fold	Finland	Pitkanen et al. (1999)
Rainbow trout *O. mykiss*	ssGH-ssGH	None	Finland	Pitkanen et al. (1999)
Cichlidae				
Nile tilapia, *Oreochromis niloticus*	opAFP-csGH	2–4-Fold	UK	Rahman et al. (1998; 2001) and Rahman and Maclean (1999)
Nile tilapia *Oreochromis niloticus*	ssMT-ssGH	None	UK	Rahman et al. (1998; 2001) and Rahman and Maclean (1999)
Tilapia, *O. hornorum* Hybrid	hCMV-tiGH	82%	Cuba	Martinez et al. (1996)
Ictaluridae				
Channel catfish, *Ictalurus punctatus*	RSVLTR-rtGH,	Up to 26%	USA	Dunham et al. (1992)
Channel catfish *Ictalurus punctatus*	mMT-hGH	None	USA	Dunham et al. (1987)
Heteropneustidae				
Heteropneustes fossilis	Zpb-ypGH	30–60%	India	Sheela et al. (1999)
Cyprinidae				
Goldfish, *Carassius auratus*	mMT-hGH	None	PR China	Zhu et al. (1985)
Common carp, *Cyprinus carpio*	mMT-hGH	None	PR China	Zhu et al. (1989)

Common carp *Cyprinus carpio*	cbA-gcGH	42–80%	PR China	Zhu (1992) and Wang et al. (2001)
Catla, *Catla catla*	RSVLTR-rtGH	None	India	Sarangi et al. (1999)
Common carp *Cyprinus carpio*	ccbA-ccGH	4-Fold	Israel	Hinits and Moav (1999)
Rohu *Labeo rohita*	CMV-roGH	4-Fold	India	Venugopal et al. (2004)
Rohu *Labeo rohita*	gcbA-roGH	4.5–5.8-Fold	India	Venugopal et al. (2004)
Esocidae				
Northern pike	RSVLTR-bGH	30%	USA	Gross et al. (1992)
Cobitidae				
Mud loach, *Misgurnus misolepis*	mlb-actin-mlGH	Up to 35-fold	Republic of Korea	Nam et al. (2001; 2002)

Techniques for Gene Transfer

Gene transfer research in fish began in mid 1980's using microinjection (Zhu et al., 1985; Dunham et al., 1987). Zhu et al. (1985) published the first report of transgene microinjection into the goldfish fertilized eggs. As pronuclei are extremely difficult to visualize in live one-cell fish embryos, in almost all fish gene transfer experiments, the foreign gene was microinjected into one-to- four cell embryo cytoplasm (Hayat, 1989). Ozato et al. (1986) used a slightly different approach and injected the transgenes in the oocytes of the Medaka, which had been removed from the ovaries nine hours before ovulation. The chicken delta-crystallin gene was injected and found in four of the eight Medaka embryos examined (Ozato et al. 1986). An average of about 5% of the surviving microinjected embryos possesses the foreign DNA. However, microinjection is a slow and tedious procedure (Powers et al. 1992), and can also result in high mortality of the eggs (Dunham et al. 1987). Following microinjection, new techniques such as retroviral integration, electroporation, and sperm mediated transfer, liposomal-reverse-phase-evaporation, and high velocity microprojectile bombardment were developed (Chen and Powers, 1990) for fast and efficient production of large numbers of transgenic individuals. The first successful gene transfer utilizing electroporation produced integration rates and survival similar to that for microinjection (Inoue et al., 1990). Further, Powers et al. (1992) successfully demonstrated that electroporation can be more efficient than microinjection with integration rates sometimes as high as 30-100%. Walker (1993) found that hatching rates were higher for electroporated embryos than for microinjected channel catfish embryos and post-fertilization electroporation treatments had higher hatching rates than electroporation of sperm and eggs prior to fertilization. Electroporation technique is quite easier and less skill required than microinjection and reasonably more numbers of first generation individuals can be produced rapidly. Further, electroporated embryos are more likely to keep alive than microinjected embryos during the initial phase of incubation (Dunham and Winn, 2014).

In most of the published data on transgenic fish, integration rate is rather low (<5%) although this figure varies widely (Hackett, 1993; Fletcher and Davies, 1991). Nuclei are not usually visible and DNA is injected into the cytoplasm; this could have an effect on the extent and timing of integration of the microinjected transgene. Integration of DNA constructs into the fish genome at stages beyond the one cell stage, results in mosaicism. Mosaicism may occur either in somatic or germ cells of the developing fish. Transfer of the transgene from the founder generation to their progeny in less than the expected Mendelian ratio may indicate that the transgenic fish have germline mosaicism (Guyomard et al., 1989; Penman et al., 1990; Culp et al., 1991). Other unexpected levels of transgene genetic transmittance may be due to transfer of unintegrated transgenes to the offspring (Stuart et al., 1988; Guyomard et al., 1989) or multiple transgene integration sites (Penman et al., 1990; Stuart et al., 1990; Tewari et al., 1992).

β-actin Gene Promoter and its Role in Growth Enhancement

The choice of promoter/enhancer could determine the success in obtaining expression of transgenes in transgenic fish. Tissue specificity of promoters and enhancers depends on tissue-specific transcription factors and their availability which may affect the expression of the transgene. A transgene has three components, a promoter sequence, a structural gene and a regulatory 3' sequence. The promoter is responsible for the transcription of the structural gene, which is then responsible for the desired phenotypic character. These two gene components can be isolated from different species or from the same species. Some researchers have suggested that constructs made of fish promoters and fish genes are expressed more efficiently than mammalian constructs and/or promoters in fish cells (Winkler et al., 1992; Betancourt et al., 1993). The aberrant or nil expression was sometimes attributed to the use of promoter sequences from more distantly related species (Chourrout, 1986; Brem et al., 1988; Penman et al. 1990).

Regulatory sequences of viral origin such as Rous Sarcoma Virus have been effectively used to drive GH transgene expression in Carp by Zhang et al. (1990) and 20% enhanced growth was reported in transgenics as compared to non-transgenic controls. However, the use of transgene constructs which are as near as possible homologous both with respect to regulatory and structural gene sequence is to be recommended, largely because they are more effective (Friedenreich and Schartl, 1990; Alam et al., 1996), but also because of the probability that use of the subsequent fish in aquaculture would have increased customer acceptance. This is especially true with respect to sequences of viral origin, which should surely be avoided because it may cause adverse effect if the progeny of the transgenic organism is ever likely to enter the human food chain. Moreover, the possible use of transgenic fish in the aquaculture industry has raised social and moral considerations which have encouraged the development of 'all-fish' gene constructs (Du et al., 1992; Hackett, 1993).

After achieving the isolation and characterization of the common carp (*Cyprinus carpio*) β-actin gene (Liu et al., 1990b) at first, a variety of fish gene constructs containing carp β-actin regulatory sequences (Liu et al., 1990a; Liu et al., 1990b) have been developed. These constructs were tested for their potency in fish tissue culture cells (Moav et al., 1992), and by transient expression in early fish development was accomplished (Moav et al., 1992; Moav et al., 1993). The β-actin gene promoter from medaka fish has been isolated and used to drive a lacZ reporter gene in embryos from the same species (Takagi et al., 1994). Strong Green Fluorescence Protein (GFP) expression has also been shown in medaka fish using a medaka β-actin gene promoter (Hamada et al., 1998). The β-actin gene promoter from zebrafish has been isolated by Higashijima et al. (1997) and these authors also showed that this sequence could drive the GFP gene efficiently in Zebrafish. The use of transgenic constructs comprising `all-fish' sequences has already been reported in several fishes. Du et al. (1992) have used a growth hormone gene from Chinook salmon (*Oncorhynchus tschawytscha*) driven by a promoter sequence from an antifreeze gene of the ocean pout (*Macrozoarrces americanus*) for the production of transgenic Atlantic salmon (*Salmo salar*) and found growth enhancement of 2 to 6-fold higher than controls. Similarly, Devlin et al. (1994b) reported extraordinary growth (up to 11-fold over controls) in Coho salmon (*Oncorhynchus kisutch*) following introduction and expression of an all-salmonid construct consisting of a sockeye salmon (*Oncorhynchus nerka*) GH gene spliced to a metallothionein promoter of the same origin.

Status of Transgenic Fish

Genetically modified fish species are not popularized or accepted for wide distribution because of the issues on food safety and environmental risk. The general public has little understanding of biology and the vagaries of how their food is grown and where it comes from, so public education of the positive and negative aspects of transgenic food and its effects is paucity and is most warranted (FAO, 2001). In order to study the impact and make public more aware, different researchers have studied the impact of transgenic fish on health. Berkowitz and Kryspin-Sorensen (1994) have elaborately discussed a food safety issue posed by genetically modified fish. Additionally, Dunham (2004) analyzed the theoretical food safety of GH transgenic salmon. The scientific organizations like FAO, WHO, US National Academy of Sciences etc., have stated that transgenic fish will not cause any human health risk in most of the consumers but rarely which may develop allergic reaction in sensible people. Added to health hazard, the other impact of transgenic fish is that it can cause risk to native gene pools and natural ecosystems. According to the detailed review by Dunham (2011) transgenic fish have relatively lower fitness than their wild counterparts. Hence, if the transgenic fish escape to natural environment, transgene from the fish would be likely eliminated due to its inferiority, which prevent the adverse environmental effects with due course of time (Dunham and Winn, 2014). In an observation, "GloFish", the ornamental transgenic fish developed were unable to establish in wild in the US due to their inferior fitness than normal zebrafish and hence, Glofish in not an environmental threat.

Commercialization of genetically growth enhanced transgenic fish has apparently taken place in some countries such as Chile, China, Cuba and New Zealand. However, in locations such as Europe and Japan, conservative approaches to the development of transgenic fish prevailing politically for many more years. Several attempts have been made to produce transgenic fish for improving growth in aquaculture, but most of the study remained at experimental level. However, some of the companies have succeeded and produce transgenic fish in commercial level. AquaBounty Technologies (Niiler, 2000) have developed a genetically modified (GM), triploid Atlantic salmon, and the AquAdvantage salmon. This transgenic fish was transformed with a construct with a promoter from an ocean pout and growth hormone-regulating gene from a Pacific Chinook salmon. This fish has been developed with aim of having year-round growth in the fish against the fish growth observed during summer and spring and also to achieve the fast growth than the native original salmon i.e attained marketable size in 16-18 months instead of three years observed in other fishes. This fish has shown positive result and the transgene allowed the fish to achieve the faster growth. In different experiment, the fish has been studied for different adverse effect caused by the fish and found that the most of the impact of the developed transgenic fish was not significantly different from the control (diploid salmon) salmon fish (Bodnar, 2010). Finally, in November, 2015, the Food and Drug Administration (FDA) approved AquaBounty Technologies' application after several trials to sell the AquAdvantage salmon to U.S. consumers. Thereafter, AquaBounty has potential licensees for their salmon in Scotland, New Zealand and the United, Canada, Chile and Europe. Added to this earlier, transgenic carps have been commercialized in China, and transgenic Nile tilapia in Cuba, although this is disputed. Moreover, the first commercial transgenic fish in the U.S. may be Zebrafish. These fish expressing various fluorescent pigments of jelly fish are marketed in the ornamental fish industry in early 2004 in the trade name of "Glofish". The fishes like medaka, zebrafish, mummichog and swordtails were widely used for transgenic fish production especially for the application as animal models for studies of human diseases, vertebrate development, toxicology, and genetics which attributed because of its small size, rapid generation time and cost-effective laboratory culture techniques (Dunham and Winn, 2014).

CONSTRAINTS

Although it is quite feasible to accomplish gene transfer in commercial fish species, research facilities adequate to produce and confine transgenic fish are relatively few. The basic disadvantages of transgenic fish production of commercially important aquaculture fish species include the large size, seasonality of their egg production, more expensive holding facilities and large generation intervals, at least 6 months for tilapia and 1-2 years for carps where as 2-6 years in case of salmonids (Dunham and Winn, 2014). Mosacism in the parent generation of transgenic fish lengthens and slows research projects. Many commercial species have relatively long generation intervals compared to species that serve as laboratory fish models and other experimental models such as fruit fly, medaka, zebrafish and mice. Not surprisingly, lack of control on where in the genome the transgenes are inserted and other possible genetic factors such genetic background and epistasis have lead to variable responses in individual transgenic families or lines. It dictates combining transgenesis with traditional techniques such as selection to identify the high performance transgenic lines and optimize their gene expression and phenotype. This makes gene transfer a medium to long term breeding program rather than a short term one. However, transplantation of primordial germ cells is now possible opening the avenue for new gene knockout technology. Early transgenic fish was also hindered by a lack of fish promoters and much of the early research was conducted with viral promoters. If commercialization is the objective, much of that early research needs to be reiterated using fish promoters which will likely receive much better public perception and acceptance. Despite these obstacles, in actuality, molecular technique in fish research has made very good progress and results have, in some cases, been dramatically successful.

SUMMARY

Absolutely no doubt that Growth Hormone incorporation is beneficial for the growth, meat quality and other physiological functions of the fish. The transgenic fish with Recombinant Growth Hormone transgene will the futuristic approach for increase the aquaculture production with limited and shrinking natural resources to supply feed the ever-increasing human population. Hence, it is need of the day to take a wise decision after considering all positive and negative impacts of transgenic fish by all the relevant scientific agencies and political bodies. Further, the transgenic fish production technology should be fine tuned able to produce the sterile transgenic fish population through more advanced technologies such as gene knockout and gene editing technologies. For getting approval of transgenic commercial fish production from all scientific agencies, there should be an absolute commitment from all the concerned people those who involved in the aquaculture farming to ensure infallible confinement in each stage of production cycle of fish to prevent the escapes of the fish.

REFERENCES

1. Alam, M.S.; Lavender, F.L.; Iyengar, I.; Rahman, M.A.; Ayad, H.H.; Lathe, R.; Morley, S.D.; Maclean, N. Comparison of the activity of carp and rat â-actin gene regulatory sequences in tilapia and rainbow trout embryos. *Mol. Reprod. Dev.* 1996, 45, 117-122.
2. Ben-Atia, I.; Fine, M.; Tandler, A.; Funkenstein, B.; Maurice, S.; Cavari, B.; Gertler, A. Preparation of recombinant gilthead seabream (*Sparus aurata*) growth hormone and its use for stimulation of larvae growth by oral administration. *Gen. Comp. Endocrinol.* 1999, 113, 155-164.
3. Berkowitz, D. B.; Kryspin-Sorensen, I. Transgenic fish: safe to eat? A look at the safety considerations regarding food transgenics. *Bio Technol.* 1994, 12, 247-252.
4. Betancourt, O.H.; Attal, J.; Theron, M. C.; Puissant, C.; Houdebine, L.M. Efficiency of introns from various origins in fish cells. *Mol. Mar. Biol. Biotechnol.* 1993, 2, 181-188.

5. Biga, P.R.; Peterson, B.C.; Schelling, G.T.; Hardy, R.W.; Cain, K.D.; Overturf, K.; Ott, T.L. Bovine growth hormone treatment increased IGF-I in circulation and induced the production of a specific immune response in rainbow trout (*Oncorhynchus mykiss*). *Aquaculture.* 2005, 246, 437-445.

6. Biro, P.A.; Abrahams, M.V.; Post, J.R.; Parkinson, E.A. Predators select against high growth rates and risk-taking behaviour in domestic trout populations. *Proc. R. Soc. Lond. Ser. B: Biol. Sci.* 2004, 271, 2233-2237.

7. Bodnar, A. *Risk Assessment and Mitigation of AquAdvantage Salmon.* Information systems for biotechnology news report. October, 2010, pp1-8

8. Brem, G.; Brenig, B.; Horstgen-Schwark, G.; Winnacker, E.L. Gene transfer in tilapia (*Oreochromis niloticus*). *Aquaculture.*1988, 68, 209-219.

9. Cheema, L.R.; Matty, A.J. Increased uptake of L-leucine-14C in the skeletal muscle of rainbow trout, *Salmo gairdneri*, after administration of growth hormone. *Pakistani J. Zool.* 1978, 10, 119-123.

10. Chen, T.T.; Lu, J.K. *Transgenic fish technology: Basic principles and its application in basic and applied research.* In *Gene transfer in aquatic organisms;* De la Fuente, J., Castro, F.O., Eds.; RG Landes Company and Germany: Springer-Verlag, Austin, Texas, USA.; 1998; pp. 45-73.

11. Chen, T.T.; Lu, J.K.; Shamblott, M.J.; Cheng, C.M.; Lin, C.M.; Burns, J.C.; Reimschuessel, R.; Chatakondi, N.; Dunham, R.A. Transgenic fish: ideal models for basic research and biotechnological applications. *Zool. Studies.* 1995, 344, 215-234.

12. Chen, T.T.; Powers, D.A. Transgenic fish. *Tren. Biotechnol.* 1990, 8, 209-214.

13. Chourrout, D. Techniques of chromosome manipulation in rainbow trout: a new evaluation with karyology. *Theor. Appl. Genet.* 1986, 72, 627-632.

14. Clop, A.; Marcq, F.; Takeda, H.; Pirottin, D.; Tordoir, X.; Bibe, B.; Bouix, J.; Caiment, F.; Elsen, J.M.; Eychenne, F.; Larzul,

C.; Laville, E.; Meish, F.; Milenkovic, D.; Tobin, J.; Charlier, C.; Georges, M. A mutation creating a potential illegitimate microRNA target site in the myostatin gene affects muscularity in sheep. *Nat. Genet.* 2006, 38, 813-818.

15. Cong, L.; Ran, F.A.; Cox, D.; Lin, S.; Barretto, R.; Habib, N., Hsu, P.D.; Wu, X.; Jiang, W.; Marraffini, L.A.; Zhang, F. Multiplex genome engineering using CRISPR/Cas systems. *Science*. 2013, 339, 819-823.
16. Culp, P.; Ntisslein-Volhard, C.; Hopkins, N. High-frequency germ-line transmission of plasmid DNA sequences injected into fertilized zebrafish eggs. *Proc. Nat. Acad. Sci. USA*. 1991, 88, 7953-7957.
17. Deane, E.E.; Woo, N.Y.S. Differential gene expression associated with euryhalinity in sea bream (*Sparus sarba*). *Am. J. Physiol.: Regul. Integr. Comp. Physiol.* 2004, 287, R1054-R1063.
18. Devlin, R.H.; Byatt, J.C.; McLean, E.; Yesaki, T.Y.; Krivi, G.G.; Jaworski, E.G.; Donaldson, E.M. Bovine placental lactogen is a potent stimulator of growth and displays strong binding to hepatic receptor sites of coho salmon. *Gen. Comp. Endocrinol.* 1994b, 95, 31-41.
19. Devlin, R.H.; Yesaki, T.Y.; Biagi, C.A.; Donaldson, E.M.; Swanson, E.M.P.; Chan, W.K. Extraordinary salmon growth. *Nature*. 1994a, 371, 209-210.
20. Devlin, R.H.; Yesaki, T.Y.; Donaldson, E.M.; Du, S.J.; Hew, C.L. Production of germline transgenic Pacific salmonids with dramatically increased growth performance. *Can. J. Fish. Aquat. Sci*. 1995, 52, 1376-1384.
21. Dominguez, M.; Takemura, A.; Tsuchiya, M. Effects of changes in environmental factors on the non-specific immune response of Nile tilapia, *Oreochromis niloticus* L. *Aquacult. Res.* 2005, 36, 391-397.
22. Donaldson, E.M.; Fagerlund, U.M.H.; Higgs, D.A.; McBride, J.R. Hormonal enhancement of growth in fish. In: *Fish physiology*; Hoar. W.S, Randall D.J. and Brett J.R. (ed): Academic press, New York, 1979., Vol 3, 455-597.
23. Du, S.J.; Gong, Z.Y.; Fletcher, G.L.; Shears, M.A.; King, M.J.; Idler, D.R.; Hew, C.L. Growth enhancement in transgenic Atlantic salmon

by the use of an all-fish chimeric growth hormone gene construct. *BioTechnol.* 1992, 10, 176-181.

24. Dunham, R.A. *Aquaculture and Fisheries Biotechnology Genetic Approaches;* CABI publishing: Wallingford , UK, 2004.
25. Dunham, R.A. *Aquaculture and Fisheries Biotechnology: Genetic Approaches, 2nd Edn.* CABI Publishing: Wallingford, UK, 2011.
26. Dunham, R.A.; Eash, J.; Askins, J.; Townes, T.M. Transfer of the metallothione in human growth hormone fusion gene into channel catfish. *Trans. Am. Fish. Soc.* 1987, 116, 87-91.
27. Dunham, R.A.; Ramboux, A.C.; Duncan, P.L.; Hayat, M.; Chen, T.T.; Lin, C.M.; Kight, K.; Gonzalez-Villasenor, I.; Powers, D.A. Transfer, expression and inheritance of salmonid growth hormone genes in channel catfish, *Ictalurus punctatus*, and effects on performance traits. *Mar. Mol. Biol. Biotechnol.* 1992, 1, 380-389.
28. Dunham, R.A.; Winn, R.N. *Production of Transgenic Fish.* In *Transgenic Animal Technology: A Laboratory Handbook,* 3rd Edn.; Pinkert, C.A., Ed.; Elsevier Inc: Amsterdom, 2014; pp. 306-334.
29. Einum, S.; Fleming, I.A. Implications of stocking: ecological interactions between wild and released salmonids. *Nordic J. Freshwater Res.* 2001, 75, 56-70.
30. Fauconneau, B.; Mady, M.P.; Le Bail, P.Y. Effect of growth hormone on muscle protein synthesis in rainbow trout (*Oncorhynchus mykiss*) and Atlantic salmon (*Salmo salar*). *Fish Physiol. Biochem.* 1996, 15, 49-56.
31. Fleming, I.A.; Agustsson, T.; Finstad, B.; Johnsson, J.I.; Bjornsson, B.T. Effects of domestication on growth physiology and endocrinology of Atlantic salmon (*Salmo salar*). *Can. J. Fish. Aquat. Sci.* 2002, 59, 1323-1330.
32. Fletcher, G.L.; Davies, P.L. Transgenic fish for aquaculture. *Gen. Eng.* 1991, 13, 331-369.
33. Fletcher, G.L.; Shears, M.A.; King, M.J.; Davies, P.L.; Hew, C.L. Evidence for antifreeze protein gene transfer in Atlantic salmon (*Salmo salar*). *Can. J. Fish. Aquat. Sci.* 1988, 45, 352-357.
34. Fletcher, G.L.; Shears, M.A.; Yaskowiak, E.S.; King, M.J.; Goddard, S.V. Gene transfer: potential to enhance the genome of Atlantic

salmon for aquaculture. *Aust. J. Exp. Agric.* 2004, 44, 1095-1100.

35. Foster, A.R.; Houlihan, D.F.; Gray, C.; Medale, F.; Fauconneau, B.; Kaushik, S.J.; LeBail, P.Y. The effects of bovine growth hormone on protein turnover in rainbow trout. *Gen. Comp. Endocrinol.* 1991, 82, 111-120.
36. Friedenreich, H.; Schartl, M. Transient expression directed by homologous and heterologous promoter and enhancer sequences in fish cells. *Nucleic Acid Res.* 1990, 18, 3299-3305.
37. Gill, J.A.; Sumter E.M.; Donaldson, H.M.; Dye, H.M.; Souza, L.M.; Berg, T.; Wypych, J.; Langley, K. Recombinant chicken and bovine growth hormone accelerate growth in aquacultured juvenile Pacific Salmon (*Oncorhynchus kisutch*). *Bio/Technol.* 1985. 3, 643-646.
38. Gong, Z.; Hew, C.L. Transgenic fish in aquaculture and developmental biology. *Curr. Tropics Dev. Biol.* 1995, 30, 177-214.
39. Gong, Z.; Wan, H.; Ju, B.; He, J.; Wang, X.; Yan, T. *Generation of living color transgenic zebrafish.* In *Aquatic Genomics: Steps Toward a Great Future*; Shimizu, N., Aoki, T., Hirono, I., Takashima, F., Eds.; Springer-Verlag: New York, 2002; pp 329-339.
40. Gross, M L.; Schneider, J.F.; Moav, N.; Moav, B.; Alvarez, C.; Myster, S.H.; Liu, Z.; Hallerman, E.M.; Hackett, P.B.; Guise, K.S.; Faras, A.J.; Kapuscinski, A.R. Molecular analysis and growth evaluation of northern pike (*Esox lucius*) microinjected with growth hormone genes. *Aquaculture.* 1992, 103, 253-273.
41. Guillén, I.; Lleonart, R.; Agramonte, A.; Morales. R.; Morales, A.; Hernández, C.A.; Vázquez, M.M.; Diaz, M.; Herrera, MT.; Alvarez-Lajonchere, L.; Hernández, O.; de la Fuente, J. Physiological changes in the juvenile euryhaline teleost, the tilapia *Oreochromis hornorum*, injected with *E. coli*-derived homologous growth hormone. *J Mar Biotechnol.* 1998, 6, 142-151.
42. Guyomard, R.; Chourrout, D.; Leroux, C.; Houdebine, L.M.; Pourrain, F. Integration and germline transmission of foreign genes microinjected into fertilized trout eggs. *Biochimie.* 1989, 71, 857-863.
43. Hackett, P.B. *The molecular biology of transgenic fish.* In *Biochemistry and Molecular Biology of Fishes;* Hochachka and, P.W., Mommsen, T.P., Eds.; Vol. 2*;* Elsevier: Amsterdam, 1993; pp.

207-239.

44. Hamada, K.; Tamaki, K.; Sasado, T.; Watai, Y.; Kani, S.; Wakamatsu, Y.; Ozato, K.; Kinoshita, M.; Kohno, R.; Takagi, S.; Kimra, M. Usefulness of the medaka β-actin promoter investigated using a mutant GFP reporter gene in transgenic medaka (*Oryzias latipes*). *Mol. Mar. Biol. Biotechnol.* 1998, 7, 173-180.
45. Hayat, M. Transfer, expression and inheritance of growth hormone genes in channel catfish (*Ictalurus punctatus*) and common carp (*Cyprinus carpio*). PhD Dissertation, Auburn University, AL, USA, 1989.
46. Hew, C.L.; Fletcher, G.L.; Davies, P.L. Transgenic salmon: tailoring the genome for food production. *J. Fish Biol.* 1995, 47, 1-19.
47. Higashijima, S.; Okamoto, H.; Ueno, N.; Hotta, Y.; Eguchi, G. High-frequency generation of transgenic zebrafish which reliably express GFP in whole muscles or the whole body by using promoters of zebrafish origin. *J. Dev. Biol*. 1997, 192, 289-299.
48. Hill, J.A.; Kiessling, A.; Devlin, R H. Coho salmon (*Oncorhynchus kisutch*) transgenic for a growth hormone gene construct exhibit increased rates of muscle hyperplasia and detectable levels of differential gene expression. *Can. J. Fish. Aquat. Sci.* 2000, 57, 939-950.
49. Hinits, Y.; Moav, B. Growth performance studies in transgenic *Cyprinus carpio. Aquaculture.* 1999, 173, 285-296.
50. Hoar, W.S. *The physiology of smolting salmonids.* In: *Fish Physiology, vol. XIB;* Hoar, W.S., Randall, D., Eds.; Academic Press: New York, 1988; pp. 275–343.
51. Inoue, K.; Yamashita, S.; Hata, J.I.; Kabeno, S.; Asada, S.; Nagahisa, E.; Fujita, T. Electrophoration as a new technique for producing transgenic fish. *Cell Differ. Dev.* 1990, 29, 123-128.
52. Jao, L.-E.; Wente, S.R.; Chen, W. Efficient multiplex biallelic zebrafish genome editing using a CRISPR nuclease system. *Proc. Natl. Acad. Sci. USA.* 2013, 110, 13904-13909.
53. Jeay, S.; Sonenshein, G.E.; Postel-Vinay, M.-C.; Kelly, P.A.; Baixeras, E. Growth hormone can act as a cytokine controlling survival and proliferation of immune cells: new insights into signaling pathways. *Mol. Cell. Endocrinol.* 2002, 188, 1-7.

54. Jiang, Y.L.; Li, N.; Du, L.X.; Wu, C X. Relationship of T–>A mutation in the promoter region of myostatin gene with growth traits in swine. *Yi Chuan Xue Bao*. 2002, 29, 413-416.

55. Johnsson, J.I.; Bjornsson, B.T. Growth hormone increases growth rate, appetite and dominance in juvenile rainbow trout, *Oncorhynchus mykiss. Anim. Behav*. 1994, 48, 177-186.

56. Johansson, V.; Winberg, S.; Bjornsson, T. Growth hormone-induced stimulation of swimming and feeding behaviour of rainbow trout is abolished by the D1 dopamine antagonist SCH23390. *General and Comparative Endocrinology*. 2005, 141, 58-65.

57. Johnsson, J.I.; Bjornsson, B.T. Growth-enhanced fish can be competitive in the wild. *Funct. Ecol*. 2001, 15, 654-659.

58. Johnsson, J.I.; Jonnson, E.; Petersson, E.; Jarvi, T.; Bjornsson, B.T. Fitness-related effects of growth investment in brown trout under field and hatchery conditions. *J. Fish Biol*. 2000, 57, 326-336.

59. Johnsson, J.I.; Petersson, E.; Jonsson, E.; Bjornsson, B.T.; Jarvi, T. Domestication and growth hormone alter antipredator behavior and growth patterns in juvenile brown trout, *Salmo trutta. Can. J. Fish. Aquat. Sci*. 1996, 53, 1546-1554.

60. Jonsson, E.; Johnsson, J.I.; Bjornsson, B.T. Growth hormone increases aggressive behavior in juvenile rainbow trout. *Horm. Behav*. 1998, 33, 9-15.

61. Jonsson, E.; Johnsson, J.I.; Bjornsson, B.T. Growth hormone increases predation exposure of rainbow trout. *Proc. R. Soc. Lond. Ser. B*. 1996a, 263, 647-651.

62. Jonsson, N.; Jonsson, B.; Fleming, I A. Does early growth cause a phenotypically plastic response in egg production of Atlantic salmon? *Funct. Ecol*. 1996b, 10, 89-96.

63. Kieffer, T.J.; Scieldrop, P.J.; McLean, E.; Donaldson, E.M.; Brown, J.C. A radioimmunoassay for onchorynchid growth hormone targeted at the physiological range. *Can. J. Physiol. Pharmacol*. 1994, 72, 1155-1161.

64. Leatherland, J.F.; Nuti, R.A. Effect of bovine growth hormone on plasma free fatty acid concentration and liver, muscle and carcass lipid content in rainbow trout (*Salmo gairdneri*. Richardson). *J. Fish Biol*. 1981, 19, 487-498.

65. Leedom, T.A.; Uchida, K.; Yada, T.; Richman III, N.H.; Byatt, J.C.; Collier, R.J.; Hirano, T.; Grau, E.G. Recombinant bovine growth hormone treatment of tilapia: growth response, metabolic clearance, receptor binding and immunoglobulin production. *Aquaculture.* 2002, 207, 359-380.

66. Lemieux, H.; Blier, P.; Dutil, J.D. Do digestive enzymes set a physiological limit upon growth rate and conversion efficiency in Atlantic cod (*Gadus morhua*)? Bull. *Aquac. Soc. Can.* 1997, 97, 54-56.

67. Li, Y.; Bai, J.; Jian, Q.; Ye, X.; Lao, H.; Li, X.; Luo, J.; Liang, X. Expression of common carp growth hormone in the yeast *Pichia pastoris* and growth stimulation of juvenile tilapia (*Oreochromis niloticus*). *Aquaculture*. 2003, 216, 329-341.

68. Liu, W.; Thomas, S.G.; Asa, S.L.; Gonzalez-Cadavid, N.; Bhasin, S.; Ezzat, S. Myostatin is a skeletal muscle target of growth hormone anabolic action. *J. Clin. Endocrinol. Metab.* 2003, 88, 5490-5496.

69. Liu, Z.; Moav, B.; Faras, A.J.; Guise, K.S.; Kapuscinski, A.R.; Hackett, P.B. Development of expression vectors for transgenic fish. *BioTechnol.* 1990a, 8, 1268–1272.

70. Liu. Z.; Moav, B.; Faras, A.J.; Guise, K.S.; Kapuscinski, A.R.; Hackett, P.B. Functional analysis of elements affecting expression of the β-actin gene of carp. *Mol. Cell. Biol.* 1990b, 10, 3432-3440.

71. MacLatchy, D.L.; Kawauchi, H.; Eales, J.G. Stimulation of hepatic thyroxine 50-deiodinase activity in rainbow trout (*Oncorhynchus mykiss*) by Pacific salmon growth hormone. *Comp. Biochem. Physiol.* 1992, 101A, 689-691.

72. Maclean, N. Genetically modified fish and their effects on food quality and human health and nutrition. *Tren. Food Sci. Technol.* 2003, 14, 242-252.

73. Maclean, N.; Penman, D.; Talwar, S. Introduction of novel genes into fish. *Biotechnol.* 1987, 5, 257-261.

74. Maclean, N.; Rahman, A.; *Transgenic fish*. In *Animals with novel genes;* Maclean, N., Ed.; Cambridge University Press: USA, 1994; pp. 63-105.

75. Marcell, T.J.; Harman, S.M.; Urban, R.J.; Metz, D.D.; Rodgers, B.D.; Blackman, M.R. Comparison of GH, IGF-I, and testosterone

with mRNA of receptors and myostatin in skeletal muscle in older men. *Am. J. Physiol.* 2001, 281, E1159-E1164.

76. Martinez, R.; Estrada, M.P.; Berlanga, J.; Guillen, I.; Hernandez, O.; Cabrera, E.; Pimentel, R.; Morales, R.; Herrera, F.; Morales, A.; Pina, J.C.; Abad, Z.; Sanchez, V.; Melamed, P.; Lleonart, R.; de la Fuente, J. Growth enhancement in transgenic tilapia by ectopic expression of tilapia growth hormone. *Mol. Mar. Biol. Biotechnol.* 1996, 5, 62-70.
77. Market, J.R.; Higgs, D.A.; Dye, H.M.; MacQuarrie, D.W. Influence of bovine growth hormone on growth rate, appetite, and food conversion of yearlings coho salmon (*Oncorhynchus kisutch*) fed two diets of different composition. *Can J. Zoo.l,* 1977, 55, 74-83.
78. Matty, A.J. Nutrition, hormones and growth. *Fish physiol. Biochem.,* 1986, 2, 141-150.
79. McCormick, S.D. Endocrine control of osmoregulation in teleost fish. *Am. Zool.* 2001, 41, 781-794.
80. McCormick, S.D.; Shrimpton, J.M.; Moriyama, S.; Bjornsson, B.T. Effects of an advanced temperature cycle on smolt development and endocrinology indicate that temperature is not a zeitgeber for smolting in Atlantic salmon. *J. Exp. Biol.* 2002, 205, 3553–3560.
81. McLean, E.; Devlin, R.H. *Application of biotechnology to enhance growth of salmonids and other fish.* In *Recent Advances in Marine Biotechnology: Aquaculture Part B. Fishes;* Fingerman, M., Nagabhushanam, R., Eds.; Science Publishers: Enfield, NH, vol. 4; 2000, 17-55.
82. McLean, E.; Devlin, R.H.; Byatt, J.C.; Clarke, W.C.; Donaldson, E.M. Impact of a controlled release formulation of recombinant growth hormone upon growth and seawater adaptation in coho (*Oncorhynchus kisutch*) and Chinook (*Oncorhynchus tshawytscha*) salmon. *Aquaculture.* 1997, 156, 113-128.
83. McLean, E.; Donaldson, E.M. *The role of somatotropin in growth poikilotherms.* In *The Endocrinology of Growth, Development and Metabolism in Vertebrates*; Schreibman, M.P., Scanes, C.G., Pang, P.K.T., Eds.; Academic Press: New York, USA, 1993; 43-71.
84. McPherron, A.C.; Lawler, A.M.; Lee, S.J. Regulation of skeletal muscle mass in mice by a new TGF-beta superfamily member.

Nature. 1997, 387, 83-90.

85. Moav, B.; Liu, Z.; Caldovic, L.D.; Gross, M. L.; Kapuscinski, A.R.; Faras, A.J.; Guise, K.S.; Hackett, P.B. *Expression of heterologous genes in transgenic fish*. In *Transgenic Fish;* Hew, C., Fletcher, G.L., Eds.; World Science Publisher: Singapore, 1992, 121-140.
86. Moav, B.; Liu, Z.; Caldovic, L.D.; Gross, M.L.; Hackett, P.B. Regulation of early expression of transgenes in developing fish. *Transgen. Res*. 1993, 2, 153-161.
87. Moriyama, S.; Kawauchi, H. Growth stimulation of juvenile salmonids by immersion in recombinant salmon growth hormone. *Nippon Suisan Gakkaishi*. 1990, 56, 31-34.
88. Nam, Y.K.; Cho, Y.S.; Cho, H.; Kim, D.S. Accelerated growth performance and stable germ-line transmission in androgenetically derived homozygous transgenic mud loach, *Misgurnus mizolepis*. *Aquaculture*. 2002, 209, 257-270.
89. Nam, Y.K.; Noh, J.K.; Cho, Y.S.; Cho, H.J.; Cho, K.N.; Kim, C.G.; Kim, D.S. Dramatically accelerated growth and extraordinary gigantism of transgenic mud loach *Misgurnus mizolepis*. *Transgenic Res*. 2001, 10, 353-362.
90. Narnaware, Y.K.; Kelly, S.P.; Woo, N.Y.S. Effect of injected growth hormone on phagocytosis in silver sea bream (*Sparus sarba*) adapted to hyper- and hypo-osmotic salinities. *Fish Shellfish Immunol.* 1997, 7, 515-517.
91. Niiler, E. Terminator technology temporarily terminated. *Nat. Biotechnol.* 1999, 17, 1054.
92. O'Connor, P.K.; Reich, B.; Sheridan, M.A. Growth hormone stimulates hepatic lipid mobilization in rainbow trout, *Oncorhynchus mykiss*. *J. Comp. Physiol.* 1993, 163B, 427-431.
93. Ozato, K.; Kondoh, H.; Inohara, H.; Iwamatsu, T.; Wakamatsu, Y.; Okada, T.S. Production of transgenic fish: introduction and expression of chicken delta-crystallin gene in medaka embryos. *Cell Differ. Dev.* 1986, 19, 237-244.
94. Penman, D.J.; Beeching. A.J.; Penn, S.; Maclean, N. Factors affecting survival and integration following microinjection of novel DNA into rainbow trout eggs. *Aquaculture*. 1990, 85, 35-50.

95. Pitkanen, T.I.; Krasnov, A.; Teerijoki, H.; Molsa, H. Transfer of growth hormone (GH) genes into Arctic charr (*Salvelinus alpinus* L.). I. Growth response to various GH constructs. *Genet. Anal.: Biomol. Eng*. 1999, 15, 91-98.
96. Pickford, G.E.; Thompson, E.F. The effect of purified mammalian growth hormone on the killifish, *Fundulus heteroclitus* (Linn.). *J. Exp. Zoo.l.* 1948, 109, 367-383.
97. Powers, D.A.; Cole, T.; Creech, K.; Chen, T.T.; Lin, C.M., Kight, K.; Dunham, R. Electroporation: a method for transferring genes into the gametes of zebrafish, *Brachydanio rerio*, channel catfish, *Ictalurus punctatus*, and common carp, *Cyprinus carpio*. *Mol. Mar. Biol. Biotech.* 1992, 1, 301-309.
98. Powers, D.A.; Gómez-Chiarri, M.; Chen, T.T.; Dunham, R. *Genetic Enginering of Finfish and shellfish.* In: *Gene transfer in aquatic organisms;* De la Fuente, J., Castro F.O., Eds.; RG Landes Company and Germany: Springer-Verlag, Austin, Texas, USA, 1998; 17-34.
99. Rahman, M.A.; Maclean, N. Growth performance of transgenic tilapia containing an exogenous piscine growth hormone gene. *Aquaculture*. 1999, 173, 333-346.
100. Rahman, M.A.; Mak, R.; Ayad, H.; Smith, A.; Maclean, N. Expression of a novel piscine growth hormone gene results in growth enhancement in transgenic tilapia (*Oreochromis niloticus*). *Transgenic Res*. 1998, 7, 357– 369.
101. Rahman, M.A.; Ronyai, A.; Engidaw, B.Z.; Jauncey, K.; Hwang, G.L.; Smith, A.; Roderick, E.; Penman, D.; Varadi, L.; Maclean, N. Growth and nutritional trials on transgenic Nile tilapia containing an exogenous fish growth hormone gene. *J. Fish Biol*. 2001, 59, 62-78.
102. Roberts, S.B.; McCauley, L.A.R.; Devlin, R.H.; Goetz, F.W. Transgenic salmon overexpressing growth hormone exhibit decreased myostatin transcript and protein expression. *J. Exp. Biol.* 2004, 207, 3741-3748.
103. Rousseau, K.; Dufour, S. Comparative aspects of GH and metabolic regulation in lower vertebrates. *Neuroendocrinol.* 2007, 86, 165-174.
104. Sakai, M.; Kobayashi, M.; Kawauchi, H. Mitogenic effect of

growth hormone and prolactin on chum salmon *Oncorhynchus keta* leukocytes in vitro. *Vet. Immunol. Immunopathol.* 1996, 53, 185-189.

105. Sakamoto, McCormick. Prolactin and growth hormone in fish osmoregulation. *Gen. Comp. Endocrinol.* 2006, 147, 24-30.
106. Sakamoto, T.; Hirano, T. Growth hormone receptors in the liver and osmoregulatory organs of rainbow trout: characterization and dynamics during seawater adaptation. *J. Endocrinol.* 1991, 130, 425-433.
107. Sarangi, N.; Mandall, A.B.; Bandyopadhyay, A.K.; Venugopal, T.; Mathavan, S.; Pandian, T.J. Electroporated sperm-mediated gene transfer in Indian major carps. *Asia-Pacific J. Mol. Biol. Biotechnol.* 1999, 7, 151-158.
108. Schulte, P.M.; Down, N.; Donaldson, E.M.; Souza, L.M. Experimental administration of recombinant bovine growth hormone to juvenile rainbow trout (*Salmo gairdneri*) by injection or by immersion. *Aquaculture*. 1989, 76, 145-156.
109. Sedikki, H.; Boeuf, G.; Maxime, V.; Peyraud, C. Effects of growth hormone treatment on oxygen composition and seawater adaptability in Atlantic salmon parr and pre-smolts. *Aquaculture.* 1996, 148, 49–62.
110. Sekar, M.; Singh, S.D.; Angel, J.R.J.; Meena, D.K.; Sivakumar, N.; Suresh, E.; Kathirvelpandian, A. Growth promoting activity of *Pangasianodon hypophthalmus* recombinant growth hormone expressed in *Escherichia coli. Indian J. Fish.* 2015, 62, 70-77.
111. Shamblott, M.J.; Cheng, C.M.; Bolt, D.; Chen, T.T. Appearance of insulin-like growth factor mRNA in liver and pyloric caeca of a teleost in response to exogenous growth hormone. *Proc. Natl. Acad. Sci. U.S.A*. 1995, 92, 6943-6946.
112. Sheela, S.G.; Pandian, T.J.; Mathavan, S. Electroporatic transfer, stable integration, and transmission of pZp beta ypGH and pZp beta rtGH in Indian catfish, *Heteropneustes fossilis* (Bloch). *Aquac. Res.* 1999, 30, 233-248.
113. Stuart, G.W.; McMurray, J.V.; Westerfield, M. Replication, integration, and stable germ-line transmission of foreign sequences

injected into early zebrafish embryos. *Development*. 1988, 103, 403-412.

114. Stuart, G.W.; Vielkind, J.R.; McMurray, J.V.; Westerfield, M. Stable lines of transgenic zebrafish exhibit reproducible patterns of transgene expression. *Development*. 1990, 109, 577-584.
115. Sun, L. Effect of bovine growth hormone on fish growth and intestinal amino acid transport. Ph.D. Dissertation, Rutgers, The State University of New Jersey, New Brunswick, NJ, 1990.
116. Takagi, S.T.; Sasado, G.; Tamiya, K.; Ozato, Y.; Wakamatsu, A.; Takeshita, M.; Kimura, M. An efficient expression vector for transgenic medaka construction, *Mol. Mar. Biol. Biotechnol*. 1994, 3, 192-199.
117. Takagi, Y.; Moriyama, S.; Hirano, T.; Yamada, J. Effects of growth hormones on bone formation and resorption in rainbow trout (*Oncorhynchus mykiss*), as examined by histomorphometry or the pharyngeal bone. *Gen. Comp. Endocrinol*. 1992, 86, 90-95.
118. Tan, W.; Carlson, D.F.; Lancto, C.A.; Garbe, J.R.; Webster, D.A.; Hackett, P.B.; Fahrenkrug, S.C. Efficient nonmeiotic allele introgression in livestock using custom endonucleases. *Proc. Natl. Acad. Sci. USA*. 2013, 110, 16526-16531.
119. Tewari, R.; Michard-Vanhee, C.; Perrot, E.; Chourrout, D. Mendelian transmission, structure and expression of transgenes following their injection into the cytoplasm of trout eggs. *Transgen. Res*. 1992, 1, 250-260.
120. Thresher, R.; Dunham, R.; Grewe, P.; Whyard, S.; Patil, J.; Templeton, C.M.; Chaimongkol, A.; Hardy, C.; Hinds, L. Development of repressible sterility to prevent the establishment of feral populations of exotic and genetically modified animals. *Aquaculture*. 2009, 290, 104-109.
121. Thorpe., J.E.; Mangel, M.; Metcalfe, N.B.; Huntingford, F.A. Modelling the proximate basis of salmonid life-history variation, with application to Atlantic salmon, *Salmo salar*. *Evol Ecol*. 1998, 12, 581-599.
122. Tsai, H.J.; Hsih, M.H.; Kuo, J.C. *Escherichia coli*-produced fish growth hormone as a feed additive to enhance the growth of juvenile black seabream (*Acanthopagrus schlegeli*). *J. Appl. Ichthyol*. 1997,

13, 79-82.

123. Venters, H.D.; Dantzer, R.; Freund, G.G.; Broussard, S.R.; Kelley, K.W. *Growth hormone and insulin-like growth factor as cytokines in the immune system*. In *Psychoneuroimmunology*, *3rd Edn*; Ader, R., Felten, D.L., Cohen, N., Eds.; Academic Press: San Diego, vol. 1, 2001; pp. 339-362.

124. Venugopal, T.; Anathy, V.; Kirankumar, S.; Pandian, T.J. Growth enhancement and food conversion efficiency of transgenic fish, *Labeo rohita*. *J. Exp. Biol.* 2004, 301A, 477-490.

125. Walker, D.S. Effect of electroporation and microinjection on survival of ictalurid catfish embryos. Master of Science Thesis, Auburn University, AL, 1993.

126. Wang, Y.; Hu, W.; Wu, G.; Sun, Y.; Chen, S.; Zhang, F.; Zhu, Z.; Feng, J.; Zhang, X. Genetic analysis of "all-fish" growth hormone gene transferred carp (*Cyprinus carpio* L.) and its F1 generation. *Chin. Sci. Bull.* 2001, 46, 1174-1177.

127. Williams, J.L. *Genetic Control of Meat Quality Traits*. In *Meat Biotechnology*; Toldra, F., Ed.; Springer Science: New York, 2008; pp 21-60.

128. Winkler, C.; Hong, Y.; Wittbrodt, J.; Schartl, M. Analysis of heterologous and homologous promoters and enhancers *in vitro* and *in vivo* by gene transfer into Japanese medaka (*Oryzias latipes*) and *Xiphophorus*. *Mol. Mar. Biol. Biotechnol.* 1992, 1, 326-337.

129. Wong, A.C.; Van Eenennaam, A.L. Transgenic approaches for the reproductive containment of genetically engineered fish. *Aquaculture*. 2008, 275, 1-12.

130. Xu, C.; Wu, G.; Zohar, Y.; Du, S.J. Analysis of myostatin gene structure, expression and function in zebrafish. *J. Exp. Biol.* 2003, 206, 4067-4079.

131. Yada, T. Growth hormone and fish immune system. *Gen. Comp. Endocrinol.* 2007, 152, 353-358.

132. Yada, T.; Azuma, T. Hypophysectomy depresses immune functions in rainbow trout. *Comp. Biochem. Physiol. C* 2002, 131, 93-100.

133. Yada, T.; Uchida, K.; Kajimura, S.; Azuma, T.; Hirano, T.; Grau, E.G. Immunomodulatory effects of prolactin and growth hormone

in the tilapia, *Oreochromis mossambicus. J. Endocrinol.* 2002, 173, 483-492.

134. Zhang, P.; Hayat, M.; Joyce, C.; Gonzalez-Villasenor, L.I.; Lin, C.M.; Dunham, R.A.; Chen, T.T.; Powers, D.A. Gene transfer, expression and inheritance of pRSV-rainbow trout GH-cDNA in the common carp, *Cyprinus carpio* (Linnaeus). *Mol. Reprod. Dev.* 1990, 25, 3-13.
135. Zhu, Z. *Generation of fast-growing transgenic fish: methods and mechanisms.* In *Transgenic Fish;* Hew, C.L., Fletcher, G.L., Eds.; World Publishing: Singapore, 1992; pp. 92-119.
136. Zhu, Z.; Li, G.; He, L.; Chen, S. Novel gene transfer into the fertilized eggs of goldfish (*Carassius auratus*, 1758). *J. Appl. Ichthyol.* 1985, 1, 31-33.
137. Zhu, Z.; Xu, K.; Xie, Y.; Li, G.; He, L. A model of transgenic fish. *Sci. Sin.* 1989, B2, 147-155.

CHAPTER 7

GENETIC AND STRUCTURAL CHARACTERIZATION OF THE GROWTH HORMONE GENE AND PROTEIN FROM TENCH, TINCA TINCA

R. Panicz[1], J. Sadowski[1], and R. Drozd[2]

[1]Department of Aquaculture, West Pomeranian University of Technology, Szczecin, Kazimierza Krolewicza 4, 71-550 Szczecin, Poland

[2]Department of Immunology, Microbiology and Physiological Chemistry, West Pomeranian University of Technology, Szczecin, Doktora Judyma 24, 71-466 Szczecin, Poland

ABSTRACT

The analysis of the tench growth hormone gene structure revealed a comparable organization of coding and non-coding regions than other from cyprinid species. Based on the performed mRNA and amino acid sequence alignments, *gh* tench is related to Asian than to European representatives of Cyprinidae family. Second aim of the work was to characterize and predict protein structure of the tench growth hormone. *Tinca tinca GH* share many common features with human GH molecule. The Tench GH protein binds to the growth hormone receptor (GHR) using two regions I and II that are situated at opposite sites of molecule. Binding site I is placed in the central part of *T. tinca* GH and H 189 amino acid in the middle region of the IV helix is crucial for GH–GHR interactions.

Keywords: Growth hormone, Growth rate, Protein structure modelling, Species diversity, Tench

INTRODUCTION

The Tench, *Tinca tinca* L., is a warm-water cyprinid fish commonly found in Europe and adjacent regions , although human activities are wide spread. Outside of Europe, the tench is found in the continents of North America (USA, Canada), South America (Chile), Africa (Tunisia, South Africa, Zimbabwe, Zambia) and Australia (Tasmania) as well as in Japan, Indonesia and New Zealand (Rowe 2004). The Tench generally prefers lower parts of stagnant waters (turbid, lentic) where they live on the muddy ground. During daytime, they tend to hide between plants, roots or other shelters. The Tench is a typical omnivorous species that is looking for food during the night-time using rather olfactory than visual cues (Herrero et al. 2003; Owen et al. 2010). Since centuries, the tench has been cultivated in polish fish farms mainly in polycultures with common carp (*Cyprinus carpio*) to improve its slow growth rate (Skrzypczak and Mamcarz 2006). Except one example, the Quolsdorf tench line this fish has always been considered as a non-profitable species, which needs much more time and care to gain an appropriate weight than other fish. The Quolsdorf tench was a fast-growing strain in early ages of XX century, which was established throughout long-term selection programme. However, after a few years of farming, this strain disappeared and never has been re-established despite huge demand for its meat (Milkau 1921; Müller 1961; Šestáková et al. 1989; Billard and Flajshans 1995). Since that time, many researchers have tried to find a solution to increase the low growth rate of tench and make its culture profitable. The idea of compensatory growth was tested as a possible way to achieve better results via feeding trials for juvenile stages. The initially obtained results were promising; however, it turned out that it was an insufficient attempt to solve this complex problem (Quirós and Alvariño 2000; Wolnicki et al. 2003, 2006; Celada et al. 2008). An approach experimenting with chromosomal manipulations, carried out by Flajshans et al. (1993, 2004, 2007), increased the specific growth rate of tench. However, according to authors' conclusions, additional analysis of alloenzymes and nuclear DNA must be performed.

The slow growth rate is affected by many factors and should be analysed using various methods. Nowadays, scientists use nutrigenomics to approach such problems because this is a multidisciplinary platform which analyse influence of nutrients on gene expression level (Leaver et al. 2008). In order to analyse and improve the slow tench growth, the most important factors in the pathway that is determining the somatotropic axis. Signals from the hypothalamus are carried to specific tissue receptors responsible for growth, osmoregulation, steroidogenesis or gametogenesis (Mommsen 2001; Sakamoto and McCormick 2006; Canosa et al. 2007). Growth hormone (GH) is regarded as a main component of this axis; however, insulin-like growth factors type I and type II, somatostatin, somatoliberin or specific receptors are equally important. Piscine components of somatotropic axis have been extensively studied and described by Reinecke et al. (2005), and the intricate interactions along this axis were characterized by Wood et al. (2005). A first detailed information for the tench growth hormone gene has been presented by Kocour and Kohlmann (2011). However, in order to study the specific molecular factors that might modulate interactions between hormones and their receptors, it is necessary to gain knowledge about three-dimensional structure (3D).

Considering the facts mentioned above, this study aims to analyse the tench growth hormone gene sequence and to predict the currently unknown, 3D structure of GH protein. Additionally, to define which amino acids in protein structure might be important for internalization of the GH-GHR complex. Information on the 3D structure will be used to emphasize to which extent the results help to improve the knowledge of aquaculture.

MATERIALS AND METHODS

Tench *gh* Sequence Determination

Tench (*T. tinca*) samples, by means of electrofishing, were collected during June of 2008 from Południewo Lake, part of the Masurian Lake District in NE, Poland. Południewo Lake has been chosen because this lake has never been subject to any fishery activities (fishing, fish introductions, annual fish catch, etc.). A total of 29 tench individuals were caught alive, and a small caudal fin-clip dissected from each fish before the fish was released. Total DNA from these samples was extracted using peqGOLD Tissue DNA Kit (PEQLAB Biotechnologie). Purity and concentration of DNA extracts were analysed on a 1 % agarose gel and Nanodrop ND-1000 (Thermo Fisher Scientific Inc.) spectrophotometer, then stored in −20 °C until PCR assays.

Primers were designed based on alignment of tench mRNA *gh* (NCBI Acc. Nr. DQ980027) and silver carp (*Hypophthalmichthys molitrix*) growth hormone gene (NCBI Acc. Nr. M94348) sequences. This step determined exon–exon boundaries (vertical dash lines) within tench mRNA sequence, and Primer 3 software generated 3 pairs of primers (Rozen and Skaletsky 2000). Figure 1 illustrates primer hybridization sites.

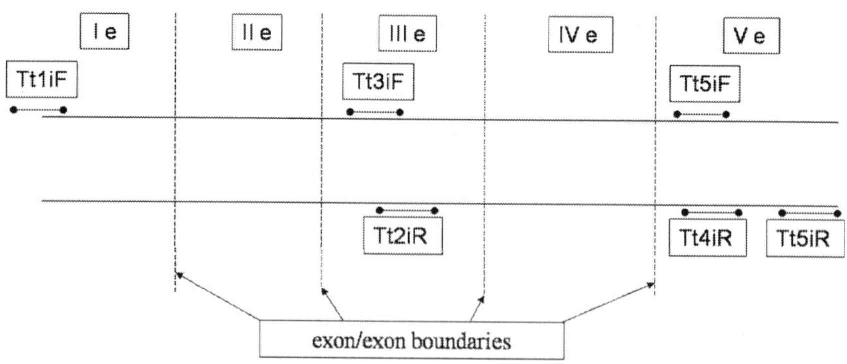

Figure 1: Primer hybridization sites along the *gh* mRNA of the tench. Ie–Ve—particular exons.

PCR cycling conditions were as follows: 94 °C/3 min followed by 35 cycles of 94 °C/30 s, depending on primer pair (I—60 °C, II—65 °C and III—57 °C) throughout 30 s, 72 °C/1 min and one cycle of final elongation 72 °C/7 min. Annealing temperatures proposed for each primer pair by Primer 3 software were verified and determined using gradient PCR.

Each PCR product without any additional bands was purified with peqGOLD Gel Extraction Kit (PEQLAB) and bidirectionally sequenced in a CEQ 8000 sequencer (Beckman Coulter). Conditions and reaction mixtures were done according to GenomeLab DTCS Quick Start Kit (Beckman Coulter). Subsequently, mRNA *gh* sequences of tench and other fish from Cyprinidae family were aligned to asses interspecies genetic similarities. DNA sequences were analysed by means of CEQ 8000 Sequence Analysis Software Package, BioEdit and MEGA5 software (Hall 1999; Tamura et al. 2011).

Tench GH Secondary Structure Prediction and Fold Recognition

After sequencing, the PCR product was translated to amino acid sequence in BioEdit software. Sequence was used as target for protein structure prediction and homologue searching in databases. Homologue searching was performed in the NR (non-redundant) database set by Psi-BLAST, 5 iteration and e value—0.005 (Altschul et al. 1997). For secondary structure prediction and fold recognition, full-length amino acid sequence was submitted to Meta-server Bioinfobank (Ginalski et al. 2003). Secondary structure was predicted by Psi-Pred (Jones 1999). Available profile–profile servers were used for structural homologue's searching: FFAS03 (Jaroszewski et al. 2005) and GRDB (Ginalski et al. 2003), and the threading servers FUGUE (Shi et al. 2001) and Sam-T02 (Karplus 2009).

Homology Modelling

Alignments from the fold recognition step with significance scores between target sequence and recognized templates were used for 3D structure modelling. Alignments were optimized manually in the Swiss-PdbViewer 4.0 (Guex and Peitsch 1997), with priority on reducing misalignments and gaps in the protein core regions. Finally, preliminary models were built using MODELLER 9v8 (Eswar et al. 2006). Models were evaluated by the ProQ and the MetaMQAPII server. For further analysis, 3D structures with the best overall score were chosen (Björn and Elofsson 2003; Pawlowski et al. 2008).

Model Analysis

The ConSurf server was used to map conserved regions on molecular surface a modelled GH (Ashkenazy et al. 2010). In order to calculate electrostatic potential on molecular surface, a model of the GH was submitted to PROPKA server to predict the protonation state at pH 7.0 (Li et al. 2005). The Adaptive Poisson–Boltzmann Solver (APBS) software was used for the electrostatic potential map calculation via PROPKA server (Baker et al. 2001). Structures and analysis were visualized using UCSF-Chimera 1.5 (Pettersen et al. 2004).

RESULTS

The Tench Growth Hormone Gene

The electrophoretic separation of 29 DNA isolates revealed a high purity and concentration of the samples; this was confirmed with the Nanodrop ND-1000. The purity of the extracted nucleic acids has been assessed on the basis of OD 260/280 ratio and ranged from 1.8 to 2.0 for all the samples. PCR and further bidirectional sequencing reactions yielded 174 high-quality reads, which after assembly gave 29 sequences representing the complete growth hormone gene sequence for each individual. Multiple alignment using ClustalW (via BioEdit software) suggested that the 29 growth hormone gene sequences were identical.

The nucleotide sequence of tench growth hormone gene that was determined in this study (1,776 bp) has been submitted in GenBank under accession number HM114351.1. The coding region of the sequence consists of four introns indicated by low letters (269, 424, 296 and 137 bp) and five exons shown by shaded capital letters (9, 141, 117, 162 and 201 bp). The open reading frame has 630 nucleotides, including a typical Kozak site (underlined sequence region) and ends with the TAG stop codon (Fig. 2).

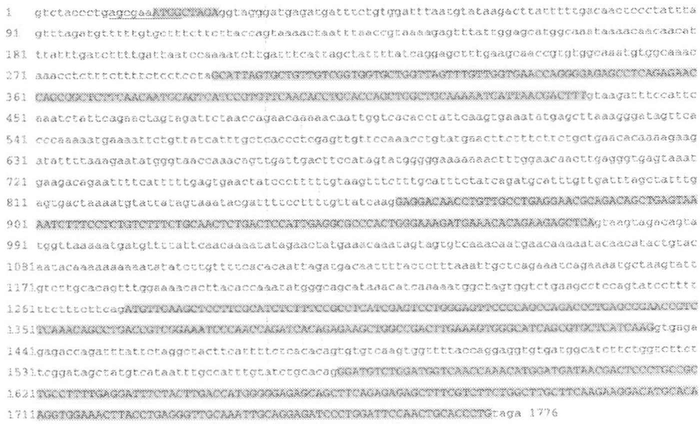

Figure 2: Nucleotide sequence of tench growth hormone sequence determined in this study (description in text).

Based on the alignment of the tench *gh,* mRNA sequence with six others species representing the Cyprinidae family (Fig. 3). The tench shared the highest similarity to bighead carp and grass carp. The similarity to those two species based on the nucleotide level was 97.6 and 97.1 %, respectively (Fig. 4). These results clearly suggest that the tench is more closely related to species that were imported from China and introduced into polish aquaculture than to native Polish (European) species.

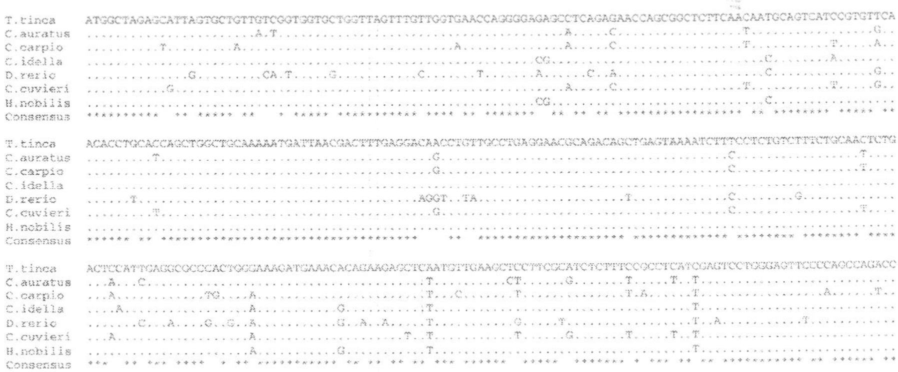

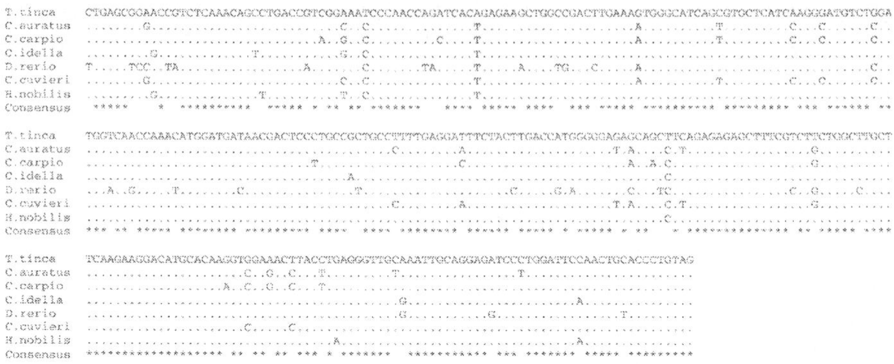

Figure 3: Alignment of seven ghmRNA sequences from members of the Cyprinidae family (Carassius auratus—AY265352.1, C. carpio—M27000.1, Ctenopharyngodon idella—AY616661.1, Danio rerio—AJ937858.1, Carassius cuvieri—AF389237.1, H. molitrix—EU157194.1 and Hypophthalmichthys nobilis—X60473.1).

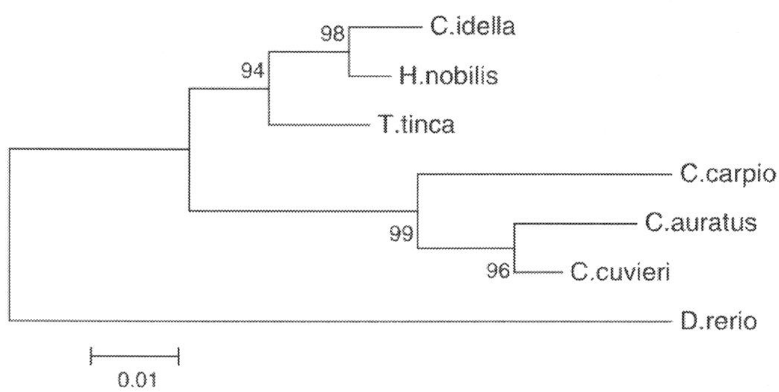

Figure 4: Phylogenetic tree prepared based on the Jukes-Cantor model shows the tench relationship to other species from the *Cyprinidae* family.

Tench GH Protein Homology Modelling

Analysis of the full-length sequence of the growth hormone of *T. tinca* revealed nearly one hundred piscine orthologs with an significant sequence similarity (with e value above 0.005). Based on the amino acid alignment, the tench was placed within the Cyprinidae family what correlated with results obtained based on DNA comparisons (Kocour and Kohlmann 2011). In the obtained set of protein sequences, any fish growth hormone structural homologue was found in the Protein Data Bank (Berman et al. 2000). However, the tench GH sequence also showed significant similarity to six human growth hormone sequences with described 3D structures, PDB ID: 1HWG, 1KF9, 1HWH, 1HHR, 1HGU and 1AXI. The secondary structure predictions by Psi-Pred revealed a characteristic α-α-α-α pattern (Fig. 5a) that is well known for human GH. This pattern consists of four helices arranged in an up-down bundle type of fold (CATH, 1.20) (Greene et al. 2007). The fold recognition performed on several servers suggested that the 1AXI, 1HUW, 1HGU and 1A22 human GH structures are the most useful templates (top positions in ranking) for homology modelling of the tench growth hormone protein.

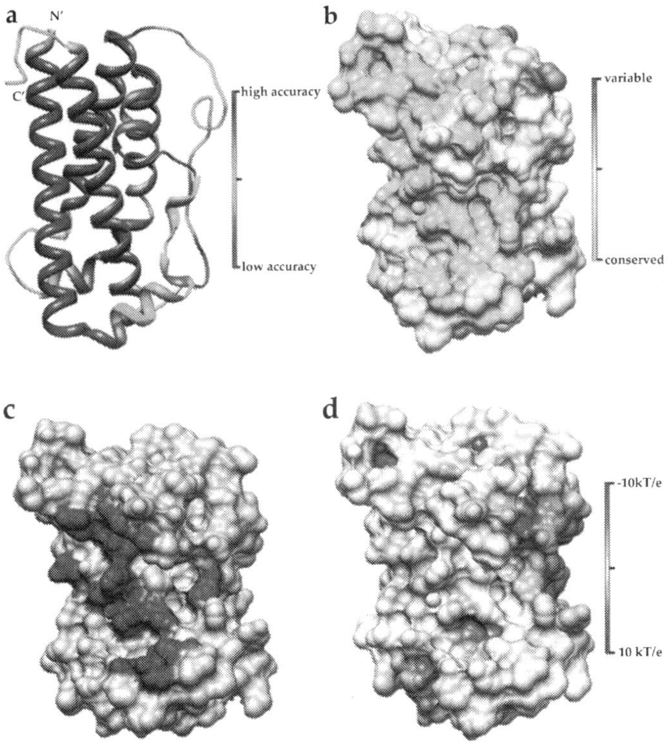

Figure 5: Analysis of the T. tinca structural GH model. a Results of the model quality evaluated by MetaMQAPII. b Surface conservation score of the GH model achieved in ConSurf based on the multiple alignment of 85 amino acid sequences of fish GH. c Localization of the predicted GH binding I site (red) for GHR. d The model highlighted with colours according to the distribution of electrostatic potential on the GH surface.

Homology Modelling

According to the 3D-jury scores, the most suitable reference template PDB ID: 1AXI chain A with the score 135.50 (FFAS3, score—81.5) was chosen for the homology modelling. The sequence similarity between the full target sequence (with the signal peptide) and the template calculated on alignment from the fold recognition was 29 %. The alignment covered almost the full length of the target sequence, and only one longer gap occurred in the region predicted by Psi-Pred, as loop, starting with K143 to F174. The best model was evaluated by the ProQ method as "very good" with LG score 3.309. The global accuracy was predicted as high by MetaMQAPII with RMSD (root-mean-square deviation) to native structure (currently unknown) 3.4 Å and GDT-TS score 54.8 (Fig. 5a).

DISCUSSION

Cultivating tench in Poland has a long tradition; however, the slow growth is a serious problem, which makes farming of this fish unattractive. The present study aims at the characterization of the tench growth hormone gene sequence and its protein. The information obtained in this work will be used for nutrigenomic studies in future where the influence of nutrition ingredients on the expression level of the somatotropic axis genes will be assessed. The structure of the tench *gh* resembles corresponding genes described for other species in the Cyprinidae family and comprises of 5 exons and 4 introns (Rajesh and Majumdar 2007). All analysed sequences were identical in length and base composition. This result likely depends on the sampling method where all fish were collected form one population. Such a bias is to be avoided in further analyses. In most cases, the sequences correlated with sequences in the NCBI database, and differences were only seen as a mispairing or short indels in intron regions. Non-coding *gh* regions that might have been mistakenly considered as useless may play an important role in regulation of growth hormone gene expression, as described for *Sparus aurata* (Almuly et al. 2000). Therefore, appropriate attention should be devoted to asses any possible correlation between polymorphisms and growth performance (Kocour and Kohlmann 2011).

Since decades, the wild stocks of tench populations are shrinking, and different approaches have been applied to bring them again to appropriate levels. Many artificial breading techniques have been developed (Kucharczyk et al. 2007), different food compositions were prepared for fry (Wolnicki et al. 2006) or tench density experiments were conducted (Celada et al. 2007) to find the best solution how effectively breed and culture this fish.

Therefore, the second aim of our work was to characterize and predict protein structure of tench growth hormone. GH protein binds to GHR (growth hormone receptor) by two regions I and II that are placed at the opposite sites of the analysed molecule (Wells et al. 1993; Walsh et al. 2003). The identification of binding sites (I, II) in the tench GH model was achieved by the superimposition of a known human growth hormone (PDB ID: 1A22) structure (RMSD Cα 0.56 Å) (Clackson et al. 1998). The recognized binding site I was placed in the central part of *T. tinca* GH (Fig. 5c). Structural alignments revealed that despite of the low overall sequence homology, the amino acid determinants binding site I of tench GH were similar to the human GH. Conservation mapping applied for the molecular surface of the GH model revealed that site I represents more conservative properties than the region located on the other side of analysed molecule (Fig. 5b). Most of the conserved residues were deployed in cluster at the C-terminus of the IV helix in the middle of the site I (Fig. 6). The predicted determinants in the adjacent helices I and III had distinctly higher variability.

The specificity of hormone–receptor binding greatly depends on the electrostatic properties of the molecules (Demchuk et al. 1994). Most determinants predicted in the present study were the polar amino acid residues that might form weak hydrogen bonds and the more stable salt bridges with oppositely charged residues of the receptor (Fig. 6). The presented results clearly indicate that the predicted binding site (positively charged) area is surrounded by a negatively charged rim (Fig. 5d). This is a defining characteristic of the transient interactions between proteins which are, inter alia, impacts between the hormone and the receptor (Nooren and Thornton 2003).

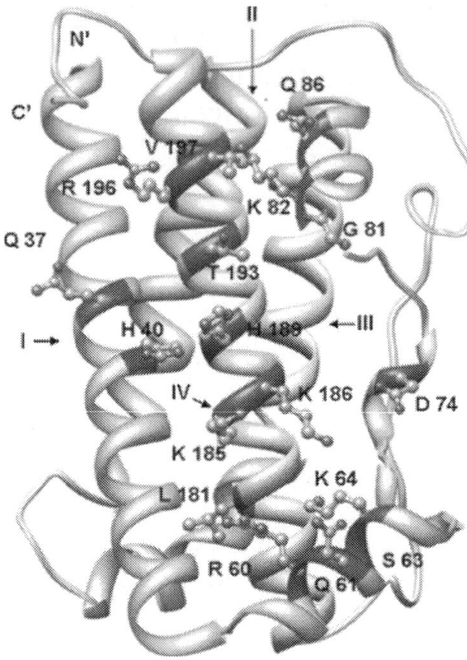

Figure 6: Predicted amino acid residue determinants of the binding site I for the *T. tinca* growth hormone.

The middle region of the IV helix is H 189, which occurs in all aligned GH fish amino acid sequences used in this study. The corresponding position of the human growth hormone is a conserved aspartic acid (D 171, PDB ID: 1A22). This residue was described as the major amino acid responsible for the binding specify of hGH by the formation salt bridge with R 43 of the hGHR (Behncken et al. 1997). Substitution of this residue in the hGH by a histidine decreases its affinity to the receptor hGHr that was described as species specifying factor (Souza et al. 1995). An alignment of the known fish protein GHr sequences (results not presented) and the human GHr showed that in complementary position to hGHR arginine 43 in the fish sequence, a glutamine that is surrounded by two glutamic acids can be found. This constellation can form a strongly negatively charged cluster on the receptor surface. The major affinity specifying factor between the tench GH and GHR can depends on the interaction between histidine 189 and glutamic acid residues. Another predicted GH-GHR binding determinates are well retained in all compared in this study fish GH protein sequences. However, this suggests that the problem of the slow growth tench in aquacultures compare to another species from Cyprinidae is not directly related with hormone structure. Probably, process may impact on another point in somatotropic axis.

ACKNOWLEDGEMENTS

We would like to thank Piotr Fenicki, director of the fish farm "Bartoły Wielkie" for information concerning the history, localization and the great help with the tench electrofishing.

REFERENCES

1. Almuly R, Cavari B, Ferstman H, et al. Genomic structure and sequence of the gilthead seabream (*Sparus aurata*) growth hormone encoding gene: identification of minisatellite polymorphism in intron I. Genome. 2000;43:836–845.
2. Altschul SF, Madden TL, Schäffer AA, et al. Gapped BLAST and PSI-BLAST: a new generation of protein database search programs. Nucleic Acids Res. 1997;25:3389–3402. doi: 10.1093/nar/25.17.3389.
3. Ashkenazy H, Erez E, Martz E, et al. ConSurf 2010: calculating evolutionary conservation in sequence and structure of proteins and nucleic acids. Nucleic Acids Res. 2010;38:W529–W533. doi: 10.1093/nar/gkq399.
4. Baker NA, Sept D, Joseph S, et al. Electrostatics of nanosystems: application to microtubules and the ribosome. PNAS. 2001;38:10037–10041. doi: 10.1073/pnas.181342398.
5. Behncken SN, Rowlinson SW, Rowland JE, et al. Aspartate 171 is the major primate-specific determinant of human growth hormone. Engineering porcine growth hormone to activate the human receptor. J Biol Chem. 1997;272:27077–27083. doi: 10.1074/jbc.272.43.27077.

6. Berman HM, Westbrook J, Feng Z, et al. The protein data bank. Nucleic Acids Res. 2000;28:235–242. doi: 10.1093/nar/28.1.235.
7. Billard R, Flajshans M. The current state of research and culture of tench, *Tinca tinca* Linnaeus (1758). Concluding remarks on the workshop. Pol Arch Hydrobiol. 1995;42:219–225.
8. Björn W, Elofsson A. Can correct protein models be identified. Prot Sci. 2003;12:1073–1086. doi: 10.1110/ps.0236803.
9. Canosa LF, Chang JP, Peter RE. Neuroendocrine control of growth hormone in fish. Gen Comp Endocrinol. 2007;151:1–26. doi: 10.1016/j.ygcen.2006.12.010.
10. Celada JD, Carral JM, Rodríguez R, et al. Tench (*Tinca tinca* L.) larvae rearing under controlled conditions: density and basic supply of *Artemia nauplii* as the sole food. Aquacult Int. 2007;15:489–495. doi: 10.1007/s10499-007-9116-z.
11. Celada JD, Aguilera A, Carral JM, et al. Rearing tench (*Tinca tinca* L.) larvae on live feed (*Artemia*) and on two transition schedules from live to dry diets. J Appl Ichthyol. 2008;24:595–600. doi: 10.1111/j.1439-0426.2008.01078.x.
12. Clackson T, Ultsch MH, Wells JA, Vos AM. Structural and functional analysis of the 1:1 growth hormone:receptor complex reveals the molecular basis for receptor affinity. J Mol Biol. 1998;277:1111–1128. doi: 10.1006/jmbi.1998.1669.
13. Demchuk E, Mueller T, Oschkinat H, et al. Receptor binding properties of four-helix-bundle growth factors deduced from electrostatic analysis. Prot Sci. 1994;3:920–935. doi: 10.1002/pro.5560030607.
14. Eswar N, Marti-Renom MA, Webb B et al (2006) Comparative protein structure modeling with MODELLER. Curr Protoc Bioinform, John Wiley and Sons, Inc., Supplement 15:5.6.1–5.6.30. doi:10.1002/0471140864.ps0209s50
15. Flajshans M, Linhart O, Kvasnicka P. Genetic studies of tench (*Tinca tinca* L.). Induced triploidy and tetraploidy and first performance data. Aquaculture. 1993;113:301–312. doi: 10.1016/0044-8486(93)90401-J.

16. Flajshans M, Kocour M, Gela D, Piackova V. The first results on relationships among amphimictic diploid, diploid gynogenic and triploid tench, *Tinca tinca* L. under communal testing. Aquacult Int. 2004;12:103–118. doi: 10.1023/B:AQUI.0000017191.52711.53.
17. Flajshans M, Kohlmann K, Rab P. Autotriploid tench *Tinca tinca* (L.) larvae obtained by fertilization of eggs previously subjected to postovulatory ageing in vitro and/or in vivo. J Fish Biol. 2007;71:868–876. doi: 10.1111/j.1095-8649.2007.01557.x.
18. Ginalski K, Elofsson A, Fischer D, Rychlewski L. 3D-Jury: a simple approach to improve protein structure predictions. Bioinformatics. 2003;19:1015–1018. doi: 10.1093/bioinformatics/btg124.
19. Greene LH, Lewis TE, Addou S, et al. The CATH domain structure database: new protocols and classification levels give a more comprehensive resource for exploring evolution. Nucleic Acids Res. 2007;35:D291–D297. doi: 10.1093/nar/gkl959.
20. Guex N, Peitsch MC. SWISS-MODEL and the Swiss-PdbViewer: an environment for comparative protein modeling. Electrophoresis. 1997;18:2714–2723. doi: 10.1002/elps.1150181505.
21. Hall TA. BioEdit: a user-friendly biological sequence alignment editor and analysis program for Windows 95/98/NT. Nucleic Acids Symp Ser. 1999;41:95–98.
22. Herrero MJ, Madrid JA, Sanchez-Vazquez FJ. Entrainment to light of circadian activity rhythms in tench (*Tinca tinca*) Chronobiol Int. 2003;20:1001–1017. doi: 10.1081/CBI-120025246.
23. Jaroszewski L, Rychlewski L, Li Z, et al. FFAS03: a server for profile–profile sequence alignments. Nucleic Acids Res. 2005;33:W284–W288. doi: 10.1093/nar/gki418.
24. Jones DT. Protein secondary structure prediction based on position-specific scoring matrices. J Mol Biol. 1999;292:195–202. doi: 10.1006/jmbi.1999.3091.
25. Karplus K. SAM-T08, HMM-based protein structure prediction. Nucleic Acids Res. 2009;37:W492–W497. doi: 10.1093/nar/gkp403.

26. Kocour M, Kohlmann K. Growth hormone gene polymorphisms in tench, *Tinca tinca* L. Aquaculture. 2011;310:298–304. doi: 10.1016/j.aquaculture.2010.10.006.

27. Kucharczyk D, Kujawa R, Mamcarz A, et al. Artificial spawning of common tench (*Tinca tinca* L.) collected from wild populations. Pol J Nat Sci. 2007;19:37–45.

28. Leaver M, Bautista J, Bjorn T, et al. Towards fish lipid nutrigenomics: current state and prospects for fin-fish aquaculture. Rev Fish Sci. 2008;16:71–92. doi: 10.1080/10641260802325278.

29. Li H, Robertson AD, Jensen JH. Very fast empirical prediction and interpretation of protein pKa values. Proteins. 2005;61:704–721. doi: 10.1002/prot.20660.

30. Milkau V. Die Resultate der Quolsdorfer Schleienzucht, ein Ansporn für die Forellenzucht. Fischerei-Zeitung (Neudamm) 1921;24:261–263.

31. Mommsen TP. Paradigms of growth in fish. Comp Biochem Physiol. 2001;129B:207–219.

32. Müller W. Schlechtes Schleienwachstum bei intensiver Karpfenteichwirtschaft. Deutsche Fisch. 1961;8:256.

33. Nooren IM, Thornton JM. Diversity of protein–protein interactions. EMBO. 2003;22:3486–3492. doi: 10.1093/emboj/cdg359.

34. Owen MG, Davies SJ, Sloman KA. Light colour influences the behaviour and stress physiology of captive tench (*Tinca tinca*) Rev Fish Biol Fish. 2010;20:375–380. doi: 10.1007/s11160-009-9150-1.

35. Pawlowski M, Gajda MJ, Matlak R, Bujnicki JM. MetaMQAP: a meta-server for the quality assessment of protein models. BMC Bioinform. 2008;29:403. doi: 10.1186/1471-2105-9-403.

36. Pettersen EF, Goddard TD, Huang CC, et al. UCSF Chimera—a visualization system for exploratory research and analysis. J Comput Chem. 2004;25:1605–1612. doi: 10.1002/jcc.20084.

37. Quirós M, Alvariño JMR. Growth and survival of tench larvae fed under different feeding strategies. J Appl Ichthyol. 2000;16:32–35. doi: 10.1046/j.1439-0426.2000.00155.x.

38. Rajesh R, Majumdar KC. A comparative account of the structure of the growth hormone encoding gene and genetic interrelationship in six species of the genus *Labeo*. Fish Physiol Biochem. 2007;33:311–333. doi: 10.1007/s10695-007-9164-3.

39. Reinecke M, Björnsson BT, Dickhoff WW, et al. Growth hormone and insulin-like growth factors in fish: where we are and where to go. Gen Comp Endocrinol. 2005;142:20–24. doi: 10.1016/j.ygcen.2005.01.016.

40. Rowe DK (2004) Potential effects of tench (*Tinca tinca*) in New Zealand freshwater ecosystems. NIWA Client Report HAM2004-005. National Institute of Water and Atmospheric Research, Hamilton, New Zealand

41. Rozen S, Skaletsky HJ. Primer3 on the WWW for general users and for biologist programmers. In: Krawetz S, Misener S, editors. Bioinformatics methods and protocols. Totowa, NJ: Met Mol Biol Humana Press; 2000. pp. 365–386.

42. Sakamoto T, McCormick SD. Prolactin and growth hormone in fish osmoregulation. Gen Comp Endocrinol. 2006;147:24–30. doi: 10.1016/j.ygcen.2005.10.008.

43. Šestáková I, Kouřil J, Faina R, et al. Age dependence of tench (*Tinca tinca* L.) larvae on food selectivity in experimental conditions. VURH Vodnany. 1989;25:3–11.

44. Shi J, Blundell TL, Mizuguchi K. FUGUE: sequence-structure homology recognition using environment-specific substitution tables and structure-dependent gap penalties. J Mol Biol. 2001;310:243–257. doi: 10.1006/jmbi.2001.4762.

45. Skrzypczak A, Mamcarz A. Changes in commercially exploited populations of tench, *Tinca tinca*(L.) in lakes of Northeastern Poland. Aquacult Int. 2006;14:179–193. doi: 10.1007/s10499-005-9024-z.

46. Souza S, Frick G, Wang X, Kopchick J, Lobo R, Goodman HM (1995) A single arginine residue determines species specificity of the human growth hormone receptor. Proc Natl Acad Sci USA 92:959–963 [PMC free article]

47. Tamura K, Peterson D, Peterson N, et al. MEGA5: molecular evolutionary genetics analysis using maximum likelihood, evolutionary distance, and maximum parsimony methods. Mol Biol Evol. 2011 [PMC free article]
48. Walsh ST, Jevitts LM, Sylvester JE, et al. Site2 binding energetics of the regulatory step of growth hormone-induced receptor homodimerization. Prot Sci. 2003;12:1960–1970. doi: 10.1110/ps.03133903.
49. Wells JA, Cunningham BC, Fuh G, et al. The molecular basis for growth hormone-receptor interactions. Recent Prog Horm Res. 1993;48:253–275.
50. Wolnicki J, Myszkowski L, Kamiński R. Effect of supplementation of a dry feed with natural food on growth, condition and size distribution of juvenile tench *Tinca tinca* (L.) J Appl Ichthyol. 2003;19:157–160. doi: 10.1046/j.1439-0426.2003.00468.x.
51. Wolnicki J, Myszkowski L, Korwin-Kossakowski M, et al. Effects of different diets on juvenile tench *Tinca tinca* (L.) reared under controlled conditions. Aquacult Int. 2006;14:89–98. doi: 10.1007/s10499-005-9017-y.
52. Wood AW, Duan CM, Bern HA. Insulin-like growth factor signalling in fish. Int Rev Cytol. 2005;243:215–285. doi: 10.1016/S0074-7696(05)43004-1.

CHAPTER

8

GENOTYPE-TEMPERATURE INTERACTION IN THE REGULATION OF DEVELOPMENT, GROWTH, AND MORPHOMETRICS IN WILD-TYPE, AND GROWTH-HORMONE TRANSGENIC COHO SALMON

Mare Lõhmus[1,2], L. Fredrik Sundström[1], Mats Björklund[3], Robert H. Devlin[1]

[1] Centre for Aquaculture and Environmental Research, Fisheries and Oceans Canada, West Vancouver, British Columbia, Canada

[2] Section for Environment and Biosecurity, Department for Chemistry, Environment and Feed Hygiene, National Veterinary Institute, Uppsala, Sweden

[3] Animal Ecology/Department of Ecology and Evolution, Evolutionary Biology Centre, Uppsala University, Uppsala, Sweden

ABSTRACT

Background

The neuroendocrine system is an important modulator of phenotype, directing cellular genetic responses to external cues such as temperature. Behavioural and physiological processes in poikilothermic organisms (e.g. most fishes), are particularly influenced by surrounding temperatures.

Methodology/Principal Findings

By comparing the development and growth of two genotypes of coho salmon (wild-type and transgenic with greatly enhanced growth hormone production) at six different temperatures, ranging between 8° and 18°C, we observed a genotype-temperature interaction and possible trend in directed neuroendocrine selection. Differences in growth patterns of the two genotypes were compared by using mathematical models, and morphometric analyses of juvenile salmon were performed to detect differences in body shape. The maximum hatching and alevin survival rates of both genotypes occurred at 12°C. At lower temperatures, eggs containing embryos with enhanced GH production hatched after a shorter incubation period than wild-type eggs, but this difference was not apparent at and above 16°C. GH transgenesis led to lower body weights at the time when the yolk sack was completely absorbed compared to the wild genotype. The growth of juvenile GH-enhanced salmon was to a greater extent stimulated by higher temperatures than the growth of the wild-type. Increased GH production significantly influenced the shape of the salmon growth curves.

Conclusions

Growth hormone overexpression by transgenesis is able to stimulate the growth of coho salmon over a wide range of temperatures. Temperature was found to affect growth rate, survival, and body morphology between GH transgenic and wild genotype coho salmon, and differential responses to temperature observed between the genotypes suggests they would experience different selective forces should they ever enter natural ecosystems. Thus, GH transgenic fish would be expected to differentially respond and adapt to shifts in environmental conditions compared with wild type, influencing their ability to survive and interact in ecosystems. Understanding these relationships would assist environmental risk assessments evaluating potential ecological effects.

INTRODUCTION

Growth of organisms arises as a net outcome of numerous behavioural and physiological functions and is influenced by variables such as food intake, digestion, absorption, and assimilation, as well as metabolic expenditure and excretion. The physiological limitations to growth are in turn strongly influenced by both biotic factors, such as the size of the individual and the availability of nutrition, and by abiotic factors, such as day length and temperature [1].

Poikilotherms, such as most fish, have limited biological means to control and regulate body temperature. Consequently they are particularly affected by thermal conditions which influence their metabolic rate and oxygen consumption, growth and other physiological characteristics [2], [3], [4], [5]. Since ambient temperatures vary both daily and seasonally, poikilotherms are often within a temperature regime that is not optimal for all functions. Hence these organisms have developed mechanisms to survive various thermal conditions both above and below optimal ranges [3], [6].

The nervous and endocrine systems are major signalling pathways between external cues, such as the ambient temperature, and internal physiology responding to environmental changes [7]. Hormones, the chemical messengers of the endocrine system, exert profound effects on organisms' physiology and behaviour and are able to act simultaneously on many target tissues, including the brain [8]. Thus, it is not surprising that the neuroendocrine machinery, more than any other physiological system, is critically involved in the evolution of entire suites of complex adaptive traits [9]. In general, hormonal control systems are rather complicated, including multiple levels of hierarchical control, negative and positive feedback loops and numerous signal substances. Consequently, an alteration in the secretion of one hormone not only affects the parameters of one control system but also others, and can have pleiotropic effects on many bodily functions [10].

Growth hormone (GH) is a principal regulator of somatic growth in vertebrates, being produced in the pituitary gland and regulated by neuroendocrine controls integrating external environmental (e.g. increased daylength and temperature in the spring) and internal metabolic (e.g. energy status) signals. GH has major metabolic effects on lipid mobilization and protein accretion, increases gluconeogenesis, and enhances feed intake and conversion during growth [11], [12], in part mediated by insulin-like growth factors (IGF). In addition to the direct effect on growth regulation, GH has been shown to increase feeding behaviour in fish [11], however is not known whether this occurs by direct action on the central nervous system or indirectly through metabolic or downstream endocrine targets [12], [13]. It has been suggested that GH may pass through the blood-brain barrier and stimulate its own receptor in the CNS, affecting neuroendocrine secretion of appetite by regulating peptides such as neuropeptide Y, bombesin, and cholecystokinin [14].

Transgenic animals can be useful models for studying long-term functional effects of neuroendocrine systems without applying invasive procedures to the test organism. In this way, transgenic model organisms increase the possibility to examine the costs and benefits of increased production of hormones without treatment effects such as repeated hormone injections [15], [16]. Studying GH-transgenic organisms also provides an opportunity to investigate how changes in growth hormone axis alter fish development, growth, and morphometrics.

Growth hormone transgenic salmon constitutively expressing GH in non-pituitary tissues show elevated plasma GH levels, earlier fry emergence dates and increased daily specific growth rates compared to wild-type under hatchery-conditions [15], [17], [18], [19], [20], [21]. The significantly increased growth rate of GH transgenic fish is associated with strongly elevated appetite and feed intake, and feed conversion efficiency relative to wild-type [19], [21], [22], [23], [24]. Consequently, when fed to satiation, transgenic individuals of the same age as wild-type are much larger [17], [20]. GH-transgenic salmon also demonstrate greater general activity and, because of the increased drive to forage, are more willing to take risks [22], [25], [26], [27].

Many aquatic habitats are presently experiencing environmental shifts which may influence the rate of growth and maturation of animals, with consequences for fitness and further effects on ecosystems.

Studying growth in different thermal conditions provides insights into the possible changes in and effects on physiological parameters (such as GH production and physiological effects) under changing ecological conditions. In the present study, the developmental rate and growth rates of wild-type and GH-transgenic coho salmon at 6 different temperatures ranging between 8° and 18°C has been assessed and modelled. Further, a morphometric analyses of juvenile salmon has been undertaken to detect differences in body shape of fish reared at different temperatures.

METHODS

Experimental Animals and Conditions

This study was performed at the DFO/UBC Centre for Aquaculture and Environmental Research, West Vancouver, Canada which houses a contained aquatic system designed to prevent the escape of transgenic fish. Our research was approved by and conducted according to guidelines of the Department of Fisheries and Oceans Pacific Region Animal Care Committee (AUP 08-003). Coho salmon of wild genotype were the offspring of wild-caught parents from the Chehalis River, BC, Canada. Transgenic coho salmon were originally produced by microinjecting eggs from wild parents with the gene construct OnMTGH1 with a metallothionein-B promoter driving the over-expression of the type-I GH gene from sockeye salmon (*O. nerka*) [17]. The transgenic strain utilized (M77) was maintained through crosses with wild salmon and thus contain on average the same genetic background as the wild-type fish except for the presence of the OnMTGH1 transgene. Experimental transgenic fish were the offspring of wild caught females from the Chehalis River and homozygous M77 transgenic males reared at the experimental facility. Half-sib wild-type experimental fish originated from the same females that were crossed to wild males obtained from the Chehalis River.

Hatching and Development of Eggs and Alevin

In March 2006, 24 Whitlock-Vibert hatching boxes (www.fedflyfishers.org) were filled with 80 eyed eggs each of either transgenic or wild-type genotype (previously incubated at 10°C from fertilization). The boxes were placed into twelve 200 L tanks, which were divided into six groups. Fresh well water was either heated or chilled and mixed in a flow-through system to gradually change tank water temperatures from 10°C to constant 8°, 10°, 12°, 14°, 16° and 18°C (with 2 replicate tanks at each temperature). Temperatures of 8° and 12°C where changed over a period of 18 hours. 14°C, 16° and 18° C were reached after 28, 38 and 48 hours respectively. Two boxes, one with transgenic and the other with wild-type eggs were placed into each tank so both types experienced the same temperature change. Artificial light was kept on a 10 h light: 14 h dark photoperiod regime.

The hatching boxes were checked every second day and any dead eggs or alevin were recorded and removed. Dates for reaching specific stages such as hatching or complete absorption of the yolk sack were observed for each genotype and temperature.

Growth of Fry and Juveniles

In May 2006, two floating incubation containers (15 L each, with bottom and side mesh to allow water flow), containing 30 first-feeding fry of either wild-type or transgenic genotype previously incubated at 10°C, were placed into each 200 L tank. The water temperatures were gradually changed to constant at 8°, 10°, 12°, 14°, 16° and 18°C (see above), the artificial daylight was constant at 10 h light:14 h dark per day. Thereafter, experimental fish were hand-fed to satiation, by throwing small amounts of pellets into the tank until fish lost interest in eating, with commercial fish food (Skretting Inc.,) from 6 (younger fry) to 2 (juveniles) times every day. The size of food pellets was chosen throughout the experiment to be appropriate according to fish size. Once every second week for 14 weeks, fish were either weighed in groups for an average group weight, or were weighed individually, resulting in seven data points for each group. After 14 weeks the replicate groups of the same genotype and temperature were pooled and released into the twelve 200 L tanks (genotypes separate) for an additional 35 days of growth.

Growth/Survival Statistics and Analyses

No significant differences were detected between the two replicates groups for the same treatment (at egg, fry or juvenile stages). Consequently, the data were pooled for further analyses.

The probability and time of egg hatching and survival of alevin to complete absorption of the yolk sack at different temperatures were tested with Kaplan-Meier survival analysis. Differences in alevin weight at the time when 50% of the individuals in a specific treatment group had absorbed their yolk sac completely, and data on the general weight differences between fry/juveniles of two genotypes at different temperatures, were tested with two-way ANOVAs with mass in grams as the dependent variable and the genotypes and temperatures as fixed factors.

The growth coefficient (Gc) for groups of juvenile coho salmon was calculated from a simple mathematical model presented by Iwama and Tautz [28], [29]. Gc was calculated based the following formula: $W_f^{1/3} = W_i^{1/3} + (T/1000\times Gc) \times Time$; there $W_f^{1/3}$ is the cube root of the final weights in grams; $W_i^{1/3}$ is the cube root of the initial weight in grams; T is the temperature in °C and Time is in days between measurements of W_f and W_i. A value 1 of Gc suggests that the fish are growing according to the model whereas values above or below indicate higher or lower growth rates, respectively, than what is predicted from the model. Additionally we fit our data into a classic exponential growth function: $W= a \times e^{b \times time}$, where *a* (intercept), and *b*(slope) were estimated constants.

Morphometrics

Differences in body shape among genotypes and rearing temperatures were examined by geometric morphometrics methods [30], [31]. We digitized 11 landmarks (Fig. 1) using the software program tpsDig2 [32]. Variation in shape was small enough to allow statistical analysis to be performed as assessed by TpsSmall [33]. Landmarks were analysed in Tps-Relw which uses the Generalized orthogonal least-squares Procrustes (GPA) procedure to produce both affine (uniform) and non-affine (non-uniform) partial warp scores (representing morphological deformations from a consensus individual) [34]. Differences in body shapes due to genotype and rearing temperature (while controlling for difference in size) were tested with a two-way MANCOVA where centroid size (similar to body size) was the covariate and the affine and non-affine partial warp scores as response variables (reported as Pillai›s trace). Because the discriminant analysis used to describe groups differences was applied to the combined groups of genotype and temperature (hence n=15 per group), partial warp scores could not be used. Instead, we extracted relative warps (RW) using TPSRelw [34]. An initial discriminant analysis was then applied to all relative warps and centroid size as predictor variables. The test of equality of group means was then used to assess which RW would be included in the final analysis since each group only had 15 individuals and hence not all 18 RW could be used for classification [35]. This strategy resulted in RW 3, 5, 8, 17, and 18 being excluded and the other 13 RW being included in subsequent analysis. This procedure allowed us to take advantage of the most information initially obtained from the partial warps scores considering both the number of landmarks and the sample size.

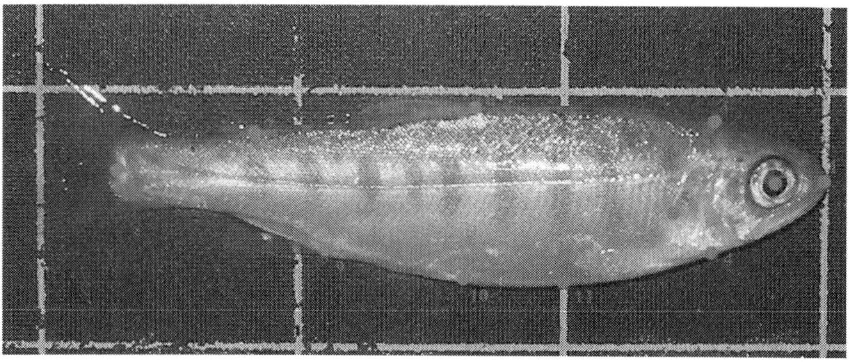

Figure 1: Locations of the 11 landmarks on a digital photograph.

Red dots are marking the 1) tip of the nose, 2) centre of the eye, 3) dorsal dividing-line of head and body, 4) ventral dividing-line of head and body, 5) posterior point of operculum, 6) anterior end of dorsal fin, 7) anterior end of adipose fin, 8) central caudal dividing line of body and tail, 9) anterior end of anal fin, 10) anterior end of pelvic fin, 11) lowermost part of the stomach.

https://doi.org/10.1371/journal.pone.0009980.g001

RESULTS

Development

There was an effect of both genotype and temperature on the probability and timing of hatching of coho salmon eggs (Fig. 2; left side of Table 1) with typically shorter time to hatch with increasing temperature, and transgenic genotypes hatching sooner than wild-type. At temperatures 8°, 10°, 12° and 14°C, a significant difference in hatching timing was found between wild-type and transgenic fish (χ^2=10; 30; 11; 11 respectively for each temperature, p<0.01; 0.0001; 0.001; 0.001; Kaplan-Meier; pair-wise comparison Log Rank/Mantel-Cox). There was no difference in timing between genotypes at 16° and 18° C (χ^2=0.23 and 0.28; p=0.63; 60 respectively). The magnitude of difference in hatch timing between genotypes increased as temperature decreased.

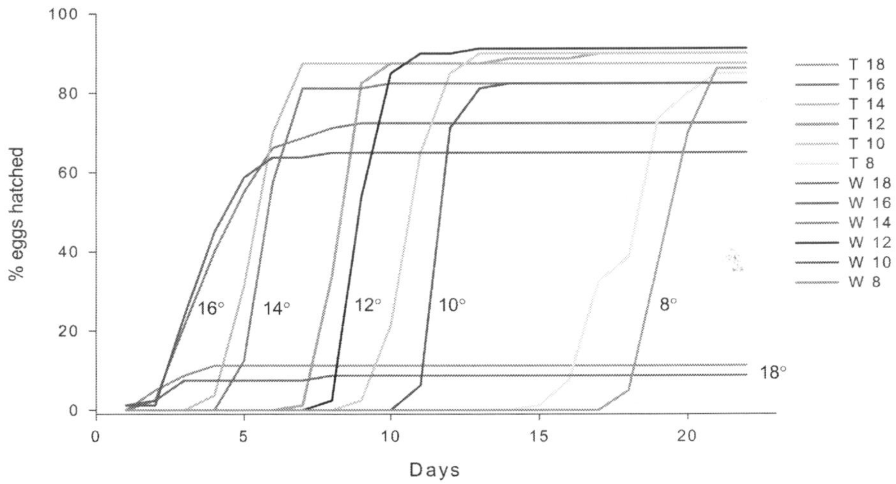

Figure 2: Timing and percentage of eggs hatching.

Wild-type (W) and GH-transgenic (T) coho salmon eggs hatching at different temperatures (8–18°C).

https://doi.org/10.1371/journal.pone.0009980.g002

Table 1: Temperature effects on hatching and survival

https://doi.org/10.1371/journal.pone.0009980.t001

	Hatch timing				Survival to first feeding			
	W		T		W		T	
°C	χ^2	p	χ^2	p	χ^2	p	χ^2	p
8 vs. 10	23.8	0.000	58.9	0.000	1.1	0.293	0.5	0.483
10 vs. 12	74.2	0.000	37.8	0.000	4.1	0.043	0.1	0.799
12 vs. 14	1.7	0.195	47.2	0.000	5.7	0.017	0.2	0.618
14 vs. 16	5.3	0.021	0.2	0.645	23.7	0.000	36.9	0.000
16 vs. 18	52.2	0.000	56.8	0.000	55.4	0.000	58.2	0.000

The table shows statistical output from comparison of temperature effects on the probability and timing of hatching (left side, compare also with Fig. 2), and on the survival to the absorption of the yolk sac (right side) of wild-type (W) and transgenic (T) coho salmon (Kaplan-Meier; Log Rank/Mantel-Cox).
doi:10.1371/journal.pone.0009980.t001

No differences in survival of fry (to the total absorption of the yolk sack stage) were found between different genotypes ($p>0.05$ at all cases; Kaplan-Meier; pair-ways comparison, Log Rank/Mantel-Cox). However we did observe differences in survival of alevin at different temperatures (right side of Table 1) with the maximum survival at 12 degrees (over 90% alive fry in both genotypes) and noticeably decreased survival values for fry at 16° (around 40% alive fry in both genotypes) and 18°C (2.5% live fry in both genotypes). The number of days from the date of hatching until the time when 50% of the group had absorbed their yolk sacks decreased with increasing temperature and differed between genotypes (from 35 days at 8°C to 25 days at 16°C in wild fish and from 28 days at 8° to 22 days at 16°C in transgenics). The number of day-degrees between hatching and buttoning up (closure of the abdomen) was higher at higher temperatures but lower in transgenics than in the wild fish (280 in wild alevin and 224 in transgenic alevin at 8°C; 400 in wild alevin and 352 in transgenic alevin at 16°C). There was a significant interaction effect between temperature and genotype on the weights of alevin at the time when 50% of the group had absorbed their yolk sacks ($F_{4.601}=5.1$; $p<0.001$). There were no differences at 8 and 18 degrees, but transgenic alevin were found to be lighter than wild-type fish at intermediate temperatures (Fig. 3).

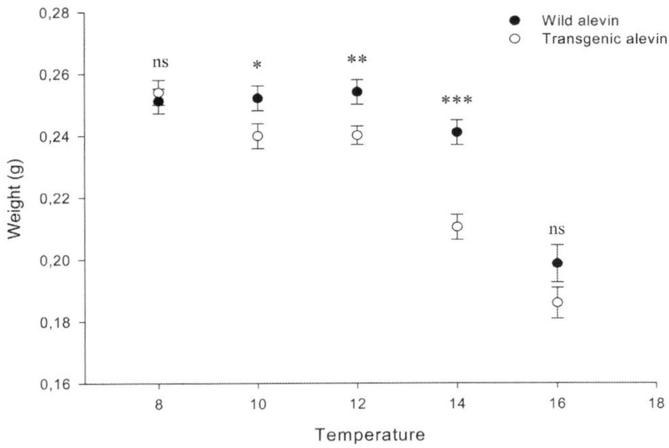

Figure 3: Average weights of alevin.

Wild-type and GH-transgenic coho salmon alevin reared at different temperatures at the time when half of the group had completely absorbed their yolk sacks.

https://doi.org/10.1371/journal.pone.0009980.g003

Growth

Based on the size of fish at 120 days of growth post first feeding, there was a clear interaction between genotype and temperature, with transgenic fish growth being relatively more stimulated as temperature increased (Fig. 4). Since variances could not be made homogenous when analyzing the weight data from fry/juveniles (due to the large difference in growth between transgenic and wild-type fish), analyses of differences between temperatures were done separately for the two genotypes. For both transgenic ($F_{5, 269}=91.5$, $p<0.001$) and wild-type ($F_{5, 265}=91.5$, $p<0.001$) fish, temperature affected growth in a positive direction. In both genotypes, fish at 8°C grew slowest with an increasing temperature having a relatively larger effect on transgenic fish than on wild-type fish. The coefficient of variation for final weights was higher in transgenic than in wild-type at all temperatures (Table 2).

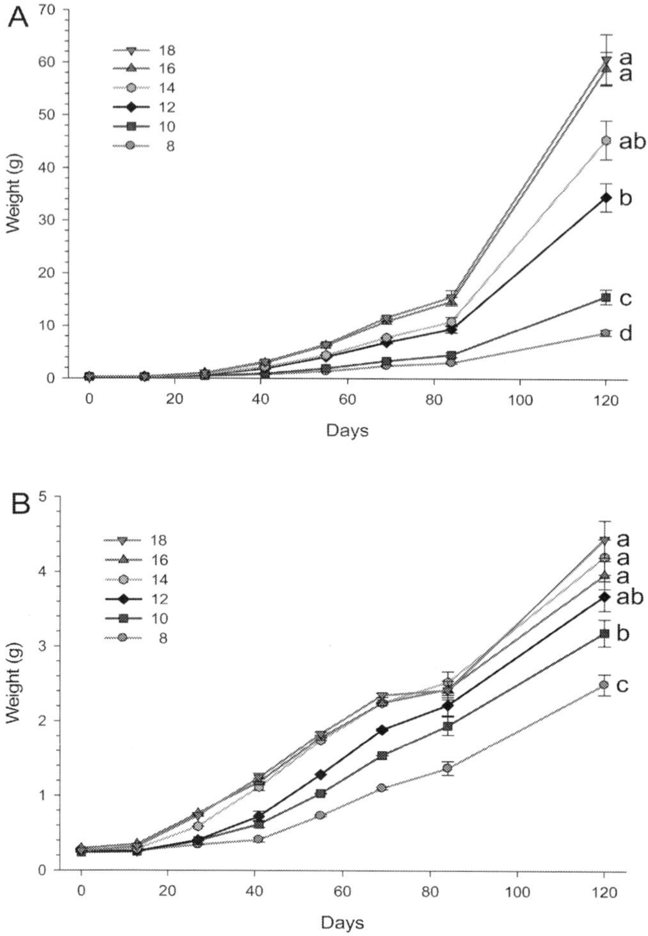

Figure 4: Growth of transgenic (A) and wild-type (B) coho salmon juveniles.

Juveniles were reared from first-feeding until an age of 120 days at temperature from 8 to 18°C. Different letters denote significant differences at the $p<0.05$ level (Tukey›s post hoc test) on ln-transformed values after 120 days (figures based on raw values). Note the more than 10-fold larger Y-axis scale for the transgenic fish.

https://doi.org/10.1371/journal.pone.0009980.g004

Table 2: Coefficient of variation (Cv) and growth coefficients (Gc)

https://doi.org/10.1371/journal.pone.0009980.t002

	wild-type		transgenic	
T°C	Cv	Gc	Cv	Gc
8	0.38	0.77	0.47	2.96
10	0.37	0.82	0.57	4.16
12	0.31	0.79	0.46	7.94
14	0.40	0.78	0.54	8.96
16	0.34	0.64	0.39	10.18
18	0.36	0.65	0.54	9.34

The table shows the Cv of weight (Cv = standard deviation/average weight) at the end of the growth trial and Gc [29] for wild-type and transgenic coho salmon at different temperatures.

doi:10.1371/journal.pone.0009980.t002

The percentage of juvenile coho salmon surviving to the day 120 of the experiment at 8° 10°, 12°, 14°, 16° and 18°C for wild-type fish were 78%, 73%, 55%, 85%, 88% and 72% respectively, and for transgenic fish 86%, 72%, 56%, 75%, 88% and 80% respectively. Tank effects are not suspected as effects were seen in all four tanks randomly distributed in the experimental design). Survival of the fish did not differ between genotypes (binary logistic regression; Wald χ^2=0.12, p=0.73) but did vary with temperature (Wald χ^2=20.9, p<0.001) with no significant interaction (Wald χ^2=4.3, p=0.50). Because 16°C fish had the highest survival, this group was used as reference category to which survival of the others groups was compared (simple contrast coding in SPSS). This revealed that fish at 10°C (Wald χ^2=4.2, p=0.042), 12°C (χ^2=14.5, p<0.001) and 18°C (χ^2=4.9, p=0.026) had lower survival than at 16°C with no difference between 16°C and 8°C or 14°C (both p>0.14).

In general the growth coefficient (Gc) values of the wild-type fish were found to be lower than predicted by the model of Iwama and Tautz [29], and further decreasing at higher temperatures. In contrast, the situation was the opposite in the transgenic fish (Table 2). Slopes estimated by the function $W = a \times e^{b \times time}$ are presented in Table 3 and Fig. 5. Transgenic salmon showed increasing growth rate slopes with temperature up to

14°C (Spearman r =0.81, P=0.05), whereas wild-type fish showed a decreasing response to temperature (Spearman r =−0.88, P=0.020).

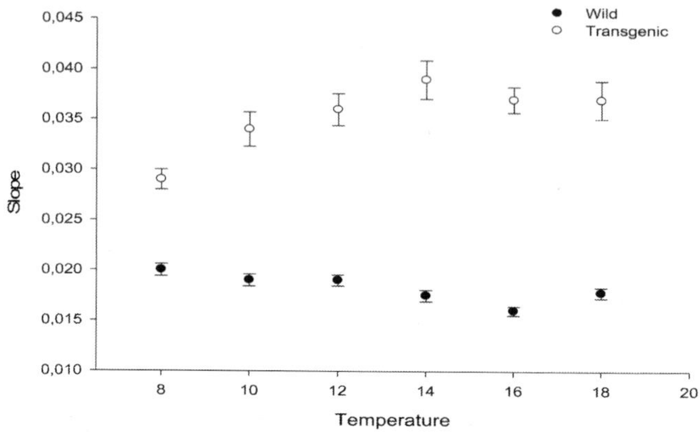

Figure 5: The slope (*b*) values of fitting the growth data to a classic exponential growth function: $W = a \times e^{b \times time}$.

Error bars represent 95% confidence intervals.

https://doi.org/10.1371/journal.pone.0009980.g005

Table 3: Estimates of the growth function $W = a \times e^{b \times time}$

https://doi.org/10.1371/journal.pone.0009980.t003

Temp	Parameters	Transgenic			Wild-type		
		estimate	SE	R^2	estimate	SE	R^2
8	a	0.260	0.032	0.76	0.240	0.015	0.76
	b	0.029	0.001		0.020	0.001	
10	a	0.260	0.050	0.69	0.320	0.020	0.76
	b	0.034	0.002		0.019	0.001	
12	a	0.480	0.087	0.79	0.370	0.021	0.81
	b	0.036	0.002		0.019	0.001	
14	a	0.430	0.099	0.73	0.530	0.031	0.74
	b	0.039	0.002		0.018	0.001	
16	a	0.670	0.100	0.83	0.590	0.027	0.77
	b	0.037	0.001		0.016	0.000	
18	a	0.720	0.160	0.72	0.540	0.029	0.77
	b	0.037	0.002		0.018	0.001	

The table illustrates the estimates of intercept (*a*) and slope (*b*) and the estimate of the least-square fit to the function (R^2). SE is based on the number of individuals.
doi:10.1371/journal.pone.0009980.t003

Morphometrics

Body shapes of fish were influenced by both centroid size ($F_{18, 150}=6.5$, $p<0.001$) and interactions between genotype and temperature ($F_{90, 770}=6.5$, p<0.001). Seven of 18 partial warps were significantly (p<0.05) affected by centroid size, nine were significantly different for genotype and nine for temperature, and six were significant for the interaction between genotype and temperature.

The first discriminant function explained 72.6% and the second function explained 12.6% of the variation in shape. Means of the populations in the discriminant space were significantly different between all groups except the wild fish at 12° and 10°C and between wild fish 14° and 16°C, with no overlap between the genotypes. None of the wild fish at neighboring temperatures were significantly different after multiple test corrections, and transgenic fish from the 14–18°C groups were not significantly different. Separation of the groups was still high, with 97.8% of the fish being classified to the correct category (Fig. 6). Three wild fish were misclassified: one wild fish at 8° classified as a wild fish 10°C, one wild fish 16° as a wild fish 14°C, and one wild fish 18° as a wild fish 14°C. One 10° transgenic fish was misclassified as a 16° transgenic.

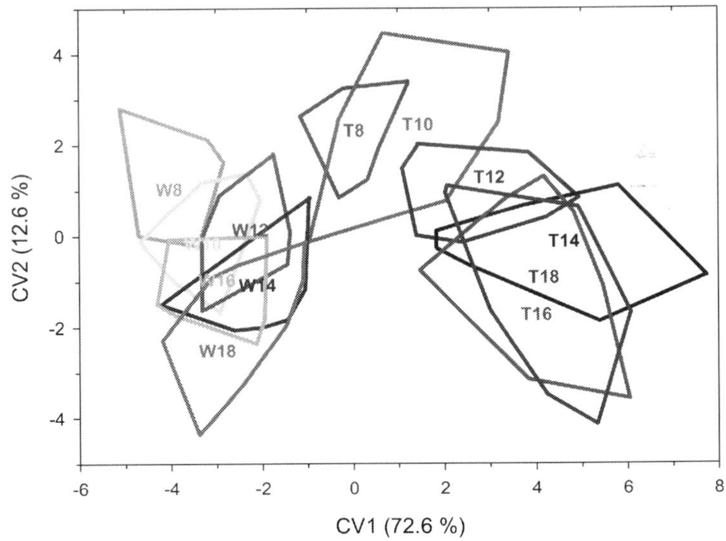

Figure 6: Canonic variate plot (CV).

CV of transgenic (T) and wild-type (W) coho salmon body morphology reared at 8–18°C. Means of the populations in the discriminant space were significant between genotypes and groups except in wild-type fish at neighboring temperatures. and transgenic fish between 14–18°C.

https://doi.org/10.1371/journal.pone.0009980.g006

Visualization plots of the first two relative warps (RW) explaining together 50.2% of the variation in shape illustrates the clear difference between wild-type and transgenic genotypes (Fig. 7). Associated with negative values on the first RW were transgenic fish with relatively deeper bodies and smaller heads, especially the distance between the eye and the tip of the nose. Hence, this RW explained mostly deformations (change in landmark location) in vertical space. In wild-type fish, slender bodies are evident and relatively larger heads. Shape effects associated with RW2 were less clear but had a tendency to associate positive values with increasing temperature. The deformation grid suggests that most deformation for this RW was in horizontal space with positive values being associated with longer ventral and shorter dorsal arcs, with the opposite for negative values.

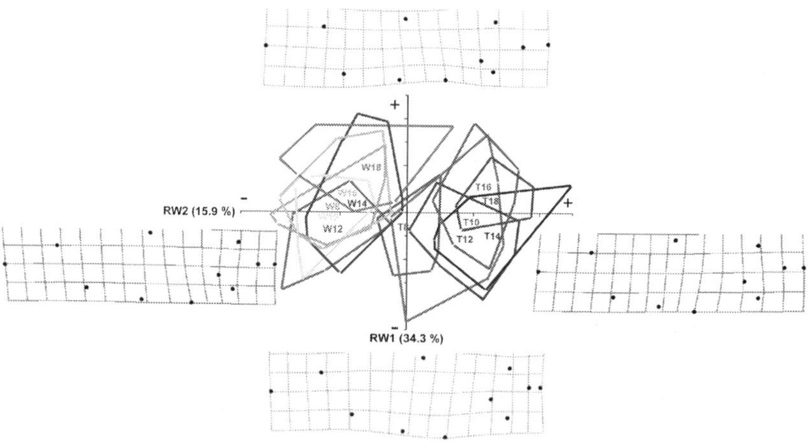

Figure 7: Relative warps.

The plot shows the first two relative warps with corresponding deformation grid from the extreme of the relative warps axes. Derived from morphometric analyses of wild-type (W) and transgenic fish (T) reared at 8–18°C.

https://doi.org/10.1371/journal.pone.0009980.g007

DISCUSSION

Defining the effects of temperature on biological functions is important to understand the contribution of climatic factors to the fitness and the ecological interactions of organisms at both individual and population levels [6]. Hormones play an essential role in the control of these mechanisms since many of them act on or are produced by the central nervous system and in that way link environmental stimuli to behaviours and physiology [36]. The present study shows a strong effect of temperature on developmental rate, juvenile growth and morphology that differed between wild-type and GH-transgenic coho salmon.

A positive correlation between growth rate and temperature is well known for most poikilotherm fish species, including salmonids [1]. Salmonid fishes show consistently increasing growth rates in temperatures ranging from 0°C to approximately 15°C [1], [37]. Within this interval, growth acceleration is achieved through higher metabolism supported by increased food consumption and conversion efficiency [6]. At very high temperatures, growth enhancement ultimately cannot be maintained because of rapidly ascending metabolic costs [1]. However, further increase in growth is still possible within a few degrees over the optimal, if demands for maintaining growth rate can be met by increased rations and sufficient oxygen supply. At very high temperatures (more than 20°C) the feeding rate of most salmonids declines sharply, and is completely inhibited at temperatures several degrees below the incipient lethal level. In this state, the non-optimal temperature limits the oxygen supply in the organism leading to hypoxemia and rapidly falling performance in fish [5].

Consistent with previous studies [37], we observed reduced egg and alevin survival rate at the highest study temperatures. At 18°C the percentage of eggs surviving to hatching averaged 9% for the two genotypes, but rapidly increased to 70% at 16°C and 90% at 12°C. Post-hatch alevin survival of both genotypes was highest at 12°C (ca 90% of hatchlings), whereas at 16°C less than 40% of alevin survived to complete absorption of the yolk sack.

GH and IGF-I are major regulators of somatic growth, and according to several studies, there is a close positive relationship between the levels of these hormones in fish and the ambient water temperature [38], [39], [40], [41], [42], [43], [44]. Both of these genes have been shown to be active in salmonid embryos, even before the development of the pituitary gland [43], [45], [46]. Yet, variation of temperature has been shown not to change (aside from timing arising from different rates of development) the levels of GH and IGF-I protein and transcript in salmonid embryos, implying that these hormones play a lesser role in mediating temperature effects during embryogenesis [43], [45], [46]. High produion of GH in salmon embryos in the present study, however, did, at lower temperatures (8° to 14°C), shorten the development time of eggs compared to wild-type [15], [20]. Thus, even though GH may not play a primary role in the regulation of temperature effect in salmon embryos, as evident in the present study, an over-expression of GH still promotes development and decrease the incubation time in transgenic coho embryos compared to the wild-type.

According to Gabillard et al. [39] the thermal influence on embryonic development of salmonids is best explained by the actions of another growth factor – IGF-II, as the amount of mRNA for this hormone does increase with temperature. IGF-II is known in mammals to act primarily as a mediator of growth prenatally [47]. The strain of GH transgenic salmon used in the present study does not display elevated levels of IGF-II mRNA at later stages of development (i.e. fingerling) [48], but it is not known whether this gene is activated by GH transgenesis at earlier stages. It is possible that the earlier hatching at higher temperatures (16° to 18°C) in both genotypes may have been mediated by temperature-mediated IGF-II action which was able to dominate over effects of GH. At later stages, GH transgenesis and temperature are both able to stimulate growth rate.

The temporal aspect of emergence of fry is a critical element in salmon life histories, and fry that emerge at appropriate times will have adaptive advantages in initial feeding and predator avoidance compared to the fry emerging too early or too late. Consequently, natural selection should time spawning events and egg development rates to correspond with the optimum time for fry emergence in each particular stream environment [49]. During the period of emergence the usage rate of yolk stores sustaining growth until the fry emerge depend strongly on environmental temperature [49], [50]. Fry that emerge at appropriate time will therefore have adaptive advantages in initial feeding, predator avoidance and in the temporal integration with other life history needs compared to the fry emerging too early or too late in the cycle.

Thermal conditions in nature fluctuate between years, creating slightly different temporal optima for each year in the spawning and incubation time. Fry with high production of GH will most likely hatch and emerge earlier than the fry with the lower expression of GH at all physiologically adaptive temperatures. This shift towards earlier hatching and emergence of eggs and fry of the GH-enhanced salmon may cause a fitness disadvantage in the life-histories of these fish at temperatures that are experienced by conspecifics under natural conditions. However, it is possible that being first to emerge could also confer an advantage for fry with high GH production, if they are able to establish territories before the conspecifics with lower GH production emerge [51]. The present data showed a greater decrease in body weight in transgenic alevin with increasing temperatures than in the wild-type fish (caused by a general increased usage rate of the yolk sack). Consequently, increased GH-production may lead to greater sensitivity to environmental conditions and an enhanced necessity to find sufficient food supplies compared to wild genotype [15]. Indeed, populations of GH transgenic fry are less able to withstand periods of limited food availability than are wild type populations [52].

Although the majority of the weight gain of anadromous salmonids occurs in the ocean environment, growth rate of juvenile fish in their home streams is also of great importance, especially for salmonids that must acquire minimum sizes before they can smolt and migrate to the sea. The size achieved by juvenile coho salmon at the end of their first summer has a strong effect on their later success of over-wintering and smolting. Larger size has also been shown to result in competitive advantages in processes of feeding and establishing territories [6]. As even minor temperature shifts strongly influence the growth and development in salmonids and cause essential alterations in life history patterns, any long-term non-adaptive variation in response to thermal conditions is likely to have important consequence on individual physiology and fitness [49].

GH secretion can be influenced by temperature [12], however, Danzmann et al. [53] found a reduced capacity of GH to influence growth of juvenile domesticated rainbow trout (*Oncorhyncus mykiss*) reared at high temperatures. In contrast, the present data showed a clear positive effect of temperature on growth rates of transgenic juvenile coho salmon, indicating that the over-production of GH does promote growth in this species even at high rearing temperatures. These two studies used different modes of administration of GH (injection, vs. endogenous overproduction in transgenics) which may result in differential rates of turnover of active GH protein that could influence growth stimulation effects. The cause for the low survival of coho salmon juveniles in this experiment at 12° is not known, however, it is possible that a pathogen has its peak of virulence at this temperature; however, this is only a speculation.

It is well-known that metabolic costs and requirements of oxygen of fish increase at higher temperatures, making increased growth rates difficult to maintain [1], [4]. The present study suggests that the increased growth rates of fish with elevated GH levels were supported at temperatures several degrees higher than in the wild-type fish. From the mathematical estimations of growth rates (Fig. 6) we observe an increase in the values of growth slopes of GH-enhanced salmon up to 14° C, whereas slopes were slightly negative over the entire range of temperatures in wild-type fish. In other words, the relationship between growth rate and relative size of fish is negatively affected by temperature in wild fish, but increases (up to 14° C) in fish with elevated GH-production. Thus, over-production of GH in coho salmon allows them to meet the increasing demands of growth by enhanced nutritional intake to a larger degree than occurs in wild-type, suggesting that the optimal thermal conditions for GH-enhanced coho salmon might be higher than for the wild-type fish.

Iwama and Tautz [29] developed a simple model for predicting the growth of salmonids at different temperatures in intensive aquaculture conditions. Their model includes assumptions (such as growth increasing steadily with increasing temperature) that may not be true under all circumstances, and thus should be applied with this knowledge in mind [54]. However, their model is still useful to compare growth slopes of groups of salmonids (for instance between strains with different growth rates, or populations in different environmental conditions). In the present experiment we calculated growth coefficients (Gc), comparing the actual growth rate represented by our data with the theoretical growth rate of the model. Both genotypes showed best agreement with the model at lower temperatures but reacted to increased temperatures in opposite ways: wild-type fish showed a lower than predicted growth rate at higher temperatures, whereas the Gc values for GH-transgenic fish were up to 10 times higher at high temperatures than predicted by the model. These relative differences in Gc between genotypes at different temperatures suggest a greater response to temperature in fish with elevated GH levels than predicted by the model. When comparing our results to the Gc values for various stocks of salmonids (see Table 1 in Iwama [28]) we find that the Gc values of the wild-type fish in general agree with data from other salmonids, whereas the Gc for GH-enhanced fish differs greatly (as predicted from their known accelerated growth).

Different rearing temperatures not only affected the growth rate but also the body morphology of juvenile wild-type and GH-enhanced coho salmon. While the change in body shape in wild-type fish seemed to be gradual over the temperatures, in GH-enhanced salmon the effect was sharper, giving rise to a group of fish (reared at 14°–18°C) characterized by very large bodies and relatively small heads. Earlier studies analyzing the morphological effects of GH-transgenesis in fish have demonstrated noticeable differences in the shape and development of the head and in most cases also a change in the head/body size ratio compared to the wild fish [55], [56], [57], [58]. In accordance with these morphological changes, fish with increased GH-production in this study in general showed relatively deeper bodies and smaller heads with a reduced distance between the eye and the tip of the nose, whereas wild-type fish were characterized by slender bodies and relatively larger heads. However a tendency to longer ventral and shorter dorsal arcs was noticed at higher temperatures in both genotypes.

Changes in body shape could affect the swimming capacity of fish and change their ability to escape predators [59]. A lower swimming speed of fast-growing GH-transgenic coho salmon of the same size as wild-type has been observed [60]. A genotype effect on body shape was evident in the present study which could influence swimming capacity in transgenic fish compared to wild-type, but it could also reduce susceptibility to predators, particularly when these are gape-limited [61]. Thus it is conceivable that GH-levels in fish in natural surroundings would be heavily exposed to balancing natural selection modifying the rates of somatic growth. The direction of the selection will likely depend on multiple factors such as abundance and type of predators and food availability.

The present study has observed effects of temperature on growth, survival and body morphology in GH transgenic and wild type coho salmon. Importantly, the two genotypes respond differently to temperature, indicating genotype by environment interactions are influencing phenotypic development between these strains.

AUTHOR CONTRIBUTIONS

Conceived and designed the experiments: ML RD. Performed the experiments: ML. Analyzed the data: ML LFS MB. Contributed reagents/materials/analysis tools: RD. Wrote the paper: ML LFS MB RD.

REFERENCES

1. Brett JR (1979) Environmental factors and growth. In: Hoar WS, Randall DJ, Brett JR, editors. Fish Physiology. London: Academic Press, Inc. pp. 599–675. JR Brett1979Environmental factors and growth. WS HoarDJ RandallJR BrettFish PhysiologyLondonAcademic Press, Inc599675
2. Welsh HH, Hodgson GR, Harvey BC, Roche MF (2001) Distribution of Juvenile Coho Salmon in Relation to Water Temperatures in Tributaries of the Mattole River, California. North American Journal of Fisheries Management 21: 464–470. HH WelshGR HodgsonBC HarveyMF Roche2001Distribution of Juvenile Coho Salmon in Relation to Water Temperatures in Tributaries of the Mattole River, California. North American Journal of Fisheries Management21464470
3. Hazel JR, Prosser CL (1974) Molecular mechanisms of temperature compensation in poikilotherms. Am Physiological Soc. pp. 620–677. JR HazelCL Prosser1974Molecular mechanisms of temperature compensation in poikilotherms. Am Physiological Soc620677
4. Pörtner HO (2002) Climate variations and the physiological basis of temperature dependent biogeography: systemic to molecular hierarchy of thermal tolerance in animals. Comparative Biochemistry and Physiology - Part A: Molecular & Integrative

Physiology 132: 739–761. HO Pörtner2002Climate variations and the physiological basis of temperature dependent biogeography: systemic to molecular hierarchy of thermal tolerance in animals. Comparative Biochemistry and Physiology - Part A: Molecular & Integrative Physiology132739761

5. Pörtner HO, Farrell AP (2008) Physiology and Climate Change. Science 322: 690–692. HO PörtnerAP Farrell2008Physiology and Climate Change. Science322690692
6. Sullivan K, Martin DJ, Carwell RE, Toll JE, Duke S (2000) An analysis of the effects of the temperature on salmonids of the Pacific Northwest with implications for selecting temperature criteria. Portland, Oregon: Sustainable Ecosystems Institute. K. SullivanDJ MartinRE CarwellJE TollS. Duke2000An analysis of the effects of the temperature on salmonids of the Pacific Northwest with implications for selecting temperature criteria. Portland, OregonSustainable Ecosystems Institute
7. McCormick SD, Moriyama S, Bjornsson BT (2000) Low temperature limits photoperiod control of smolting in Atlantic salmon through endocrine mechanisms. American Journal of Physiology 278: R1352–R1361. SD McCormickS. MoriyamaBT Bjornsson2000Low temperature limits photoperiod control of smolting in Atlantic salmon through endocrine mechanisms. American Journal of Physiology278R1352R1361
8. Ball GF, Balthazart J (2008) Review. Individual variation and the endocrine regulation of behaviour and physiology in birds: a cellular/molecular perspective. Philosophical Transactions of the Royal Society B: Biological Sciences 363: 1699–1710. GF BallJ. Balthazart2008Review. Individual variation and the endocrine regulation of behaviour and physiology in birds: a cellular/molecular perspective. Philosophical Transactions of the Royal Society B: Biological Sciences36316991710
9. Williams TD (2008) Individual variation in endocrine systems: moving beyond the 'tyranny of the Golden Mean'. Phil Trans R Soc B 363: 1687–1698. TD Williams2008Individual variation in endocrine systems: moving beyond the 'tyranny of the Golden Mean'. Phil Trans R Soc B36316871698

10. Adkins-Regan E (2008) Review. Do hormonal control systems produce evolutionary inertia? Philosophical Transactions of the Royal Society B: Biological Sciences 363: 1599–1609. E. Adkins-Regan2008Review. Do hormonal control systems produce evolutionary inertia?Philosophical Transactions of the Royal Society B: Biological Sciences36315991609
11. Björnsson BT, Johansson V, Benedet S, Einarsdottir IE, Hildahl J, et al. (2002) Growth hormone endocrinology of salmonids: regulatory mechanisms and mode of action. Fish Physiology and Biochemistry 27: 227–242. BT BjörnssonV. JohanssonS. BenedetIE EinarsdottirJ. Hildahl2002Growth hormone endocrinology of salmonids: regulatory mechanisms and mode of action. Fish Physiology and Biochemistry27227242
12. Björnsson BT (1997) The biology of salmon growth hormone: From daylight to dominance. Fish Physiology and Biochemistry 17: 9–24. BT Björnsson1997The biology of salmon growth hormone: From daylight to dominance. Fish Physiology and Biochemistry17924
13. Jönsson E, Johansson V, Björnsson BT, Winberg S (2003) Central nervous system actions of growth hormone on brain monoamine levels and behavior of juvenile rainbow trout. Hormones and Behavior 43: 367–374. E. JönssonV. JohanssonBT BjörnssonS. Winberg2003Central nervous system actions of growth hormone on brain monoamine levels and behavior of juvenile rainbow trout. Hormones and Behavior43367374
14. Johansson V, Winberg S, Bjornsson BT (2005) Growth hormone-induced stimulation of swimming and feeding behaviour of rainbow trout is abolished by the D1 dopamine antagonist SCH23390. General and Comparative Endocrinology 141: 58–65. V. JohanssonS. WinbergBT Bjornsson2005Growth hormone-induced stimulation of swimming and feeding behaviour of rainbow trout is abolished by the D1 dopamine antagonist SCH23390. General and Comparative Endocrinology1415865
15. Sundström LF, Lõhmus M, Devlin RH (2005) Selection on increased intrinsic growth rates in coho salmon *Oncorhynchus kisutch*. Evolution 59: 1560–1569. LF SundströmM. LõhmusRH Devlin2005Selection on increased intrinsic growth rates in coho salmon *Oncorhynchus kisutch*. Evolution5915601569

16. Tatar M (2000) Transgenic organisms in evolutionary ecology. Trends in Ecology & Evolution 15: 207–211. M. Tatar2000Transgenic organisms in evolutionary ecology. Trends in Ecology & Evolution15207211
17. Devlin RH, Yesaki TY, Biagi CA, Donaldson EM, Swanson P, et al. (1994) Extraordinary salmon growth. Nature 371: 209–210. RH DevlinTY YesakiCA BiagiEM DonaldsonP. Swanson1994Extraordinary salmon growth. Nature371209210
18. Mori T, Devlin RH (1999) Transgene and host growth hormone gene expression in pituitary and nonpituitary tissues of normal and growth hormone transgenic salmon. Molecular and Cellular Endocrinology 149: 129–139. T. MoriRH Devlin1999Transgene and host growth hormone gene expression in pituitary and nonpituitary tissues of normal and growth hormone transgenic salmon. Molecular and Cellular Endocrinology149129139
19. Raven PA, Devlin RH, Higgs DA (2006) Influence of dietary digestible energy content on growth, protein and energy utilization and body composition of growth-hormone transgenic and non-transgenic coho salmon (*Oncorhynchus kisutch*). Aquaculture 254: 730–747. PA RavenRH DevlinDA Higgs2006Influence of dietary digestible energy content on growth, protein and energy utilization and body composition of growth-hormone transgenic and non-transgenic coho salmon (*Oncorhynchus kisutch*). Aquaculture254730747
20. Devlin RH, Biagi CA, Yesaki TY (2004) Growth, viability and genetic characteristics of GH transgenic coho salmon strains. Aquaculture 236: 607–632. RH DevlinCA BiagiTY Yesaki2004Growth, viability and genetic characteristics of GH transgenic coho salmon strains. Aquaculture236607632
21. Oakes JD, Higgs DA, Eales JG, Devlin RH (2007) Influence of ration level on the growth performance and body composition of non-transgenic and growth-hormone-transgenic coho salmon (*Oncorhynchus kisutch*). Aquaculture 265: 309–324. JD OakesDA HiggsJG EalesRH Devlin2007Influence of ration level on the growth performance and body composition of non-transgenic and growth-hormone-transgenic coho salmon (*Oncorhynchus kisutch*). Aquaculture265309324

22. Devlin RH, Johnsson JI, Smailus DE, Biagi CA, Jonsson E, et al. (1999) Increased ability to compete for food by growth hormone-transgenic coho salmon *Oncorhynchus kisutch* (Walbaum). Aquaculture Research 30: 479–482. RH DevlinJI JohnssonDE SmailusCA BiagiE. Jonsson1999Increased ability to compete for food by growth hormone-transgenic coho salmon *Oncorhynchus kisutch* (Walbaum). Aquaculture Research30479482

23. Sundström LF, Lõhmus M, Devlin RH, Johnsson JI, Biagi CA, et al. (2004) Feeding on profitable and unprofitable prey: Comparing behaviour of growth-enhanced transgenic and normal coho salmon (*Oncorhynchus kisutch*). Ethology 110: 381–396. LF SundströmM. LõhmusRH DevlinJI JohnssonCA Biagi2004Feeding on profitable and unprofitable prey: Comparing behaviour of growth-enhanced transgenic and normal coho salmon (*Oncorhynchus kisutch*). Ethology110381396

24. Lõhmus M, Raven PA, Sundström LF, Devlin RH (2008) Disruption of seasonality in growth hormone-transgenic coho salmon (*Oncorhynchus kisutch*) and the role of cholecystokinin in seasonal feeding behavior. Hormones and Behavior 54: 506–513. M. LõhmusPA RavenLF SundströmRH Devlin2008Disruption of seasonality in growth hormone-transgenic coho salmon (*Oncorhynchus kisutch*) and the role of cholecystokinin in seasonal feeding behavior. Hormones and Behavior54506513

25. Sundström LF, Devlin RH, Johnsson JI, Biagi CA (2003) Vertical position reflects increased feeding motivation in growth hormone transgenic coho salmon (*Oncorhynchus kisutch*). Ethology 109: 701–712. LF SundströmRH DevlinJI JohnssonCA Biagi2003Vertical position reflects increased feeding motivation in growth hormone transgenic coho salmon (*Oncorhynchus kisutch*). Ethology109701712

26. Abrahams MV, Sutterlin A (1999) The foraging and antipredator behaviour of growth-enhanced transgenic Atlantic salmon. Animal Behaviour 58: 933–942. MV AbrahamsA. Sutterlin1999The foraging and antipredator behaviour of growth-enhanced transgenic Atlantic salmon. Animal Behaviour58933942

27. Sundström LF, Lõhmus M, Johnsson J, Devlin RH (2004) Growth hormone transgenic salmon pay for growth potential with increased

predation mortality. Proceedings of the Royal Society Biological Sciences Series B 271: S350–S352. LF SundströmM. LõhmusJ. JohnssonRH Devlin2004Growth hormone transgenic salmon pay for growth potential with increased predation mortality. Proceedings of the Royal Society Biological Sciences Series B271S350S352

28. Iwama GK (1996) Growth of Salmon. In: Pennell W, Barton BA, editors. Principles of salmon culture. Amsterdam: Elsevier. pp. 467–515. GK Iwama1996Growth of Salmon. W. PennellBA BartonPrinciples of salmon cultureAmsterdamElsevier467515
29. Iwama GK, Tautz AF (1981) A Simple Growth Model for Salmonids in Hatcheries. Canadian Journal of Fisheries and Aquatic Sciences 38(6):: 649–656: 649–656. GK IwamaAF Tautz1981A Simple Growth Model for Salmonids in Hatcheries. Canadian Journal of Fisheries and Aquatic Sciences38(6):649–656: 649656
30. Slice DE (2007) Geometric Morphometrics. Annual Review of Anthropology 36: 261–281. DE Slice2007Geometric Morphometrics. Annual Review of Anthropology36261281
31. Adams DC, Rohlf FJ, Slice DE (2004) Geometric morphometrics: ten years of progress following the 'Revolution'. Italian Journal of Zoology 71: 5–16. DC AdamsFJ RohlfDE Slice2004Geometric morphometrics: ten years of progress following the 'Revolution'. Italian Journal of Zoology71516
32. Rohlf FJ (2008) FJ Rohlf2008tpsDig, digitize landmarks and outlines, version 2. 12. Department of Ecology and Evolution, State University of New York at Stony Brook, © 2005 by F James Rohlf. tpsDig, digitize landmarks and outlines, version 2. 12. Department of Ecology and Evolution, State University of New York at Stony Brook, © 2005 by F James Rohlf.
33. Rohlf FJ (2005) FJ Rohlf2005tpsSmall, version 1. 15. Department of Ecology and Evolution, State University of New York at Stony Brook, NY 11794-5245, © 2005 by F James Rohlf. tpsSmall, version 1. 15. Department of Ecology and Evolution, State University of New York at Stony Brook, NY 11794-5245, © 2005 by F James Rohlf.
34. Rohlf FJ (2007) FJ Rohlf2007tpsRelw, relative warp analysis, version 1.47. Department of Ecology and Evolution, State University

of New York at Stony Brook, © 2007 by F James Rohlf. tpsRelw, relative warp analysis, version 1. 47. Department of Ecology and Evolution, State University of New York at Stony Brook, © 2007 by F James Rohlf.

35. Tabachnik BG, Fidell LS (2001) Using Multivariate Statistics 4th ed. Needham Heights, MA: Allyn & Bacon. 966 p. BG TabachnikLS Fidell2001Using Multivariate Statistics 4thed. Needham Heights, MAAllyn & Bacon966
36. Wingfield JC, Visser ME, Williams TD (2008) Integration of ecology and endocrinology in avian reproduction: a new synthesis. Phil Trans R Soc B 363: 425–441. JC WingfieldME VisserTD Williams2008Integration of ecology and endocrinology in avian reproduction: a new synthesis. Phil Trans R Soc B363425441
37. Richter A, Kolmes SA (2005) Maximum temperature limits for chinook, coho, and chum salmon, and steelhead trout in the Pacific Northwest. Reviews in Fisheries Science 13: 23–49. A. RichterSA Kolmes2005Maximum temperature limits for chinook, coho, and chum salmon, and steelhead trout in the Pacific Northwest. Reviews in Fisheries Science132349
38. Mingarro M, Vega RdCS, Astola A, Pendon C, Martinez VM, et al. (2002) Endocrine mediators of seasonal growth in gilthead sea bream (Sparus aurata): The growth hormone and somatolactin paradigm. General and Comparative Endocrinology 128: 102–111. M. MingarroRdCS VegaA. AstolaC. PendonVM Martinez2002Endocrine mediators of seasonal growth in gilthead sea bream (Sparus aurata): The growth hormone and somatolactin paradigm. General and Comparative Endocrinology128102111
39. Gabillard JC, Weil C, Rescan PY, Navarro I, Gutierrez J, et al. (2005) Does the GH/IGF system mediate the effect of water temperature on fish growth? A review. Cybium 29: 107–117. JC GabillardC. WeilPY RescanI. NavarroJ. Gutierrez2005Does the GH/IGF system mediate the effect of water temperature on fish growth? A review. Cybium29107117
40. Gabillard JC, Weil C, Rescan PY, Navarro I, Gutierrez J, et al. (2003) Effects of environmental temperature on IGF1, IGF2, and IGF type I receptor expression in rainbow trout (Oncorhynchus

mykiss). General and Comparative Endocrinology 133: 233–242. JC GabillardC. WeilPY RescanI. NavarroJ. Gutierrez2003Effects of environmental temperature on IGF1, IGF2, and IGF type I receptor expression in rainbow trout (Oncorhynchus mykiss). General and Comparative Endocrinology133233242

41. Gabillard JC, Weil C, Rescan PY, Navarro I, Gutierrez J, et al. (2003) Environmental temperature increases plasma GH levels independently of nutritional status in rainbow trout (Oncorhynchus mykiss). General and Comparative Endocrinology 133: 17–26. JC GabillardC.WeilPYRescanI.NavarroJ.Gutierrez2003Environmental temperature increases plasma GH levels independently of nutritional status in rainbow trout (Oncorhynchus mykiss). General and Comparative Endocrinology1331726
42. Gabillard JC, Yao K, Vandeputte M, Gutierrez J, Le Bail PY (2006) Differential expression of two GH receptor mRNAs following temperature change in rainbow trout (Oncorhynchus mykiss). Journal of Endocrinology 190: 29–37. JC GabillardK. YaoM. VandeputteJ. GutierrezPY Le Bail2006Differential expression of two GH receptor mRNAs following temperature change in rainbow trout (Oncorhynchus mykiss). Journal of Endocrinology1902937
43. Gabillard JC, Rescan PY, Fauconneau B, Weil C, Le Bail PY (2003) Effect of temperature on gene expression of the GH/IGF system during embryonic development in rainbow trout (Oncorhynchus mykiss). Journal of Experimental Zoology 298A: 134–142. JC GabillardPY RescanB. FauconneauC. WeilPY Le Bail2003Effect of temperature on gene expression of the GH/IGF system during embryonic development in rainbow trout (Oncorhynchus mykiss). Journal of Experimental Zoology298A134142
44. Taylor JF, Porter MJR, Randall CF, Bromage NR (2003) The interactions of photoperiod and insulin-like growth factor-I (IGF-I) in the control of rainbow trout reproduction Fish Physiology and Biochemistry. 28. : 449–450. JF TaylorMJR PorterCF RandallNR Bromage2003The interactions of photoperiod and insulin-like growth factor-I (IGF-I) in the control of rainbow trout reproduction Fish Physiology and Biochemistry28449450
45. Li M, Raine JC, Leatherland JF (2007) Expression profiles of growth-related genes during the very early development of

rainbow trout embryos reared at two incubation temperatures. General and Comparative Endocrinology 153: 302–310. M. LiJC RaineJF Leatherland2007Expression profiles of growth-related genes during the very early development of rainbow trout embryos reared at two incubation temperatures. General and Comparative Endocrinology153302310

46. Li M, Greenaway J, Raine J, Petrik J, Hahnel A, et al. (2006) Growth hormone and insulin-like growth factor gene expression prior to the development of the pituitary gland in rainbow trout (Oncorhynchus mykiss) embryos reared at two temperatures. Comparative Biochemistry and Physiology, Part A 143: 514–522. M. LiJ. GreenawayJ. RaineJ. PetrikA. Hahnel2006Growth hormone and insulin-like growth factor gene expression prior to the development of the pituitary gland in rainbow trout (Oncorhynchus mykiss) embryos reared at two temperatures. Comparative Biochemistry and Physiology, Part A143514522

47. Constância M, Hemberger M, Hughes J, Dean W, Ferguson-Smith A, et al. (2002) Placental-specific IGF-II is a major modulator of placental and fetal growth. Nature 417: 945–948. M. ConstânciaM. HembergerJ. HughesW. DeanA. Ferguson-Smith2002Placental-specific IGF-II is a major modulator of placental and fetal growth. Nature417945948

48. Devlin RH, Sakhrani D, Tymchuk WE, Rise ML, Goh B (2009) Domestication and growth hormone transgenesis cause similar changes in gene expression in coho salmon (Oncorhynchus kisutch). Proceedings of the National Academy of Sciences 106: 3047–3052. RH DevlinD. SakhraniWE TymchukML RiseB. Goh2009Domestication and growth hormone transgenesis cause similar changes in gene expression in coho salmon (Oncorhynchus kisutch). Proceedings of the National Academy of Sciences10630473052

49. Brannon EL, Powell MS, Quinn TP, Talbot A (2004) Population structure of columbia river basin chinook salmon and steelhead trout. Reviews in Fisheries Science 12: 99–232. EL BrannonMS PowellTP QuinnA. Talbot2004Population structure of columbia river basin chinook salmon and steelhead trout. Reviews in Fisheries Science1299232

50. Rombough P (2005) Developmental Costs and the Partitioning of Metabolic Energy. In: Warburton SJ, Burggren WW, Pelster B, L RC, editors. Comparative Developmental Physiology: Contributions, Tools, and Trends. New York: Oxford University Press. pp. 99–123. P. Rombough2005Developmental Costs and the Partitioning of Metabolic Energy. SJ WarburtonWW BurggrenB. PelsterRC LComparative Developmental Physiology: Contributions, Tools, and TrendsNew YorkOxford University Press99123

51. Brännäs E (1995) First access to territorial space and exposure to strong predation pressure: A conflict in early emerging Atlantic salmon (*Salmo salar* L.) fry. Evolutionary Ecology 9: 411–420. E. Brännäs1995First access to territorial space and exposure to strong predation pressure: A conflict in early emerging Atlantic salmon (*Salmo salar* L.) fry. Evolutionary Ecology9411420

52. Devlin RH, D›Andrade M, Uh M, Biagi CA (2004) Population effects of growth hormone transgenic coho salmon depend on food availability and genotype by environment interactions. Proceedings of the National Academy of Sciences of the United States of America 101: 9303–9308. RH DevlinM. D›AndradeM. UhCA Biagi2004Population effects of growth hormone transgenic coho salmon depend on food availability and genotype by environment interactions. Proceedings of the National Academy of Sciences of the United States of America10193039308

53. Danzmann RG, Van DKGJ, Chen TT, Powers DA (1990) Metabolic Effects of Bovine Growth Hormone and Genetically Engineered Rainbow Trout Growth Hormone in Rainbow Trout Oncorhynchus-Mykiss Reared at a High Temperature. Canadian Journal of Fisheries and Aquatic Sciences 47: 1292–1301. RG DanzmannDKGJ VanTT ChenDA Powers1990Metabolic Effects of Bovine Growth Hormone and Genetically Engineered Rainbow Trout Growth Hormone in Rainbow Trout Oncorhynchus-Mykiss Reared at a High Temperature. Canadian Journal of Fisheries and Aquatic Sciences4712921301

54. Malcolm J (2003) The thermal growth coefficient (TGC) model of fish growth: a cautionary note. Aquaculture Research 34: 581–584. J. Malcolm2003The thermal growth coefficient (TGC) model of fish growth: a cautionary note. Aquaculture Research34581584

55. Devlin RH, Yesaki TY, Donaldson EM, Hew CL (1995) Transmission and phenotypic effects of an antifreeze GH gene construct in coho salmon (*Oncorhynchus kisutch*). Aquaculture 137: 161–169. RH DevlinTY YesakiEM DonaldsonCL Hew1995Transmission and phenotypic effects of an antifreeze GH gene construct in coho salmon (*Oncorhynchus kisutch*). Aquaculture137161169

56. Rahman MA, Mak R, Ayad H, Smith A, Maclean N (1998) Expression of a novel piscine growth hormone gene results in growth enhancement in transgenic tilapia (Oreochromis niloticus). Transgenic Research 7: 357–370. MA RahmanR. MakH. AyadA. SmithN. Maclean1998Expression of a novel piscine growth hormone gene results in growth enhancement in transgenic tilapia (Oreochromis niloticus). Transgenic Research7357370

57. Dunham RA, Chatakondi N, Nichols AJ, Kucuktas H, Chen TT, et al. (2002) Effect of Rainbow Trout Growth Hormone Complementary DNA on Body Shape, Carcass Yield, and Carcass Composition of F1 and F2 Transgenic Common Carp (Cyprinus carpio. Marine Biotechnology 4: 604–611. RA DunhamN. ChatakondiAJ NicholsH. KucuktasTT Chen2002Effect of Rainbow Trout Growth Hormone Complementary DNA on Body Shape, Carcass Yield, and Carcass Composition of F1 and F2 Transgenic Common Carp (Cyprinus carpio. Marine Biotechnology4604611

58. Ostenfeld TH, McLean E, Devlin RH (1998) Transgenesis changes body and head shape in Pacific salmon. Journal of Fish Biology 52: 850–854. TH OstenfeldE. McLeanRH Devlin1998Transgenesis changes body and head shape in Pacific salmon. Journal of Fish Biology52850854

59. Brönmark C, Miner JG (1992) Predator-induced phenotypical change in body morphology in crucian carp. Science 258: 1348. C. BrönmarkJG Miner1992Predator-induced phenotypical change in body morphology in crucian carp. Science2581348

60. Farrell AP, Bennett W, Devlin RH (1997) Growth-enhanced transgenic salmon can be inferior swimmers. Canadian Journal of Zoology 75: 335–337. AP FarrellW. BennettRH Devlin1997Growth-enhanced transgenic salmon can be inferior swimmers. Canadian Journal of Zoology75335337

61. Domenici P, Turesson H, Brodersen J, Brönmark C (2008) Predator-induced morphology enhances escape locomotion in crucian carp. Proceedings of the Royal Society B: Biological Sciences 275: 195–201. P. DomeniciH. TuressonJ. BrodersenC. Brönmark2008Predator-induced morphology enhances escape locomotion in crucian carp. Proceedings of the Royal Society B: Biological Sciences275195201

CHAPTER 9

FISH NUTRITION AND CURRENT ISSUES IN AQUACULTURE: THE BALANCE IN PROVIDING SAFE AND NUTRITIOUS SEAFOOD, IN AN ENVIRONMENTALLY SUSTAINABLE MANNER

Stefanie M Hixson

Department of Ocean Sciences, Memorial University of Newfoundland, St. John's, Newfoundland and Labrador, Canada

ABSTRACT

Global aquaculture production has increased in recent years and it is predicted that aquaculture will provide the most reliable supply of seafood in the future. However, there are many controversial issues in aquaculture regarding food safety, nutrition, and sustainability; many of which are directly related to the nutrition and feeds for farmed fish. These nutrition-related issues must be considered in order to achieve balance in safe and nutritious food production and sustainability in aquaculture. This review highlights recent studies and discusses new and innovative aspects in fish nutrition. Some issues in the area of fish nutrition require consideration and improvement, such as: feed and nutrient efficiency, overfeeding and waste, fish meal and fish oil replacements, fish health, biotechnology, and human health concerns. The findings reviewed in this manuscript demonstrate promise toward improvement of the aquaculture industry through nutrition. This review is an update in fish nutrition research, and provides insight on the progression and evolution of this field in order to meet the needs of the industry with the purpose to achieve a balance in seafood production and environmental sustainability. The outcome of this review encourages the use of biotechnology as a tool to meet seafood production and environmental sustainability, in order to ensure global food security in the future and to improve our resource use.

Keywords: Aquaculture; Biotechnology; Feeds; Health; Nutrition; Sustainability

INTRODUCTION

Aquaculture requires optimisation of nutrition to efficiently raise fish for the purpose of food production. Fish nutrition is the study of nutrients and energy sources essential for fish health, growth and reproduction. Global consumption of seafood is increasing, while the amount of captured fish is declining, therefore it is predicted that aquaculture will provide the most reliable supply of seafood in the coming years. With the world's rapidly expanding population, it is important to provide safe and nutritious fish; however there are many issues related to fish nutrition that need to be considered in order to achieve balance in food production and sustainability. Sustainability of the aquaculture industry is an environmental, economic and social concern; however, this review focused specifically on environmental sustainability in regards to fish nutrition. Certain issues related to fish nutrition have become controversial because they impact the environment and/or affect the final product for consumption. Some of these issues include: feed and nutrient efficiency, overfeeding and waste, unsustainable feed ingredients, fish health issues, biotechnology and human health concerns. Ultimately, each of these issues can affect the final product for human consumption, either nutritionally, environmentally or economically. Achieving a balance between efficient and safe food production with environmental sustainability will be a challenge for the industry. The following is a review of several issues in farmed fish nutrition, relating to providing quality food products while maintaining environmental sustainability, emphasizing limitations of this balance and strategies for improvement.

Fish Nutrition and Nutrient Efficiency in Aquaculture

The essential nutrients for fish are amino acids, fatty acids, vitamins, minerals and energy-yielding macronutrients (protein, lipid and carbohydrate). Diets for fish must supply all essential nutrients and energy required to meet the physiological needs of growing animals. Guidelines for nutrient adequacy for some farmed fish species suggest the minimum nutrient requirement to promote growth and prevent signs of nutrient deficiency [1]. Protein is required in the diet to obtain amino acids, which are utilized to synthesize new proteins or maintain existing proteins in tissues while excess protein is converted to energy. Lipids supply essential fatty acids and energy in the diet. The requirement of essential fatty acids can only be met by supplying Long Chain (LC) Polyunsaturated Fatty Acids (PUFA) in the diet, specifically α-Linolenic Acid (LNA, 18:3ω3) and Linoleic Acid (LA, 18:2ω6), with varying requirements for eicosapentaenoic acid (EPA, 20:5ω3) and docosahexaenoic acid (DHA, 22:6ω3) depending on species. Dietary lipids are also important structural components of membranes, and act as precursors of steroid hormones and prostaglandins in fish. Dietary carbohydrates can be a source of energy for fish; however their ability to utilize dietary carbohydrate for energy varies depending on the species and their natural diet. Therefore, depending on species, protein and lipid are the main source of energy for fish. Feeds in aquaculture are formulated with a balance of nutrients in order to meet specific nutrient requirements for different species, life stages and other purposes.

The digestibility of nutrients in the feed can affect aquaculture production efficiency and impact the environment. The bioavailability or digestibility of the diet is the proportion of nutrients in the feed that is digested and absorbed by the fish. Data on the digestibility and available digestible energy of feed ingredients in fish diets are essential for optimization of feed formulations [1]. Feeds that are poorly digested result in limited growth and feces with high nutrient content, which pollutes the environment. Therefore, the digestibility of nutrients and the potential for nutrient retention and waste must be considered for efficient and sustainable animal production when considering feed formulations.

Growing fish accrete new tissues and some of the energy supplied in the diet is stored as protein, lipid and some glycogen. Protein deposition depends on the balance of available amino acids in protein and the digestible protein-to-digestible energy ratio. Excess energy intake and low protein levels result in the deposition of lipid as recovered energy, which does not equate to faster growth and is an inefficient use of nutrients. Fish have the ability to utilize lipids for energy, saving protein for deposition and growth [2]; therefore inclusion of lipids in diets for fish is important for both growth and energy purposes. Most commercial feeds today are formulated to increase growth performance by exploiting the protein-sparing effect of high energy lipid, allowing as much of the dietary protein as possible to be converted into muscle protein. As a result, the production efficiency of farmed salmon has significantly improved over time [3]. Today, the use of more highly digestible nutrient-density extruded feeds (46 to 50% protein, 20 to 24% fat) allows commercial farmers to achieve a Feed Conversion Ratio (FCR) of about 0.9-1.2 for rainbow trout (*Oncorhynchus mykiss*) grown to market size [2]. The improved FCR over time has been due to increased digestible nutrient and energy content of the feeds, as well as feed extrusion which has resulted in the production of feeds with higher lipid levels, improved starch gelatinization (increases digestible energy content and utilization), and improved pellet characteristics (durability, buoyancy, etc.) [2,3]. Further research to improve feed ingredients and feed efficiency in the future will help improve sustainability in aquaculture. For example, genetically modified (GM) salmon have better FCR than non-GM salmon (see Biotechnology section).

Feeding and the Environment

Feed is the main source of waste and is responsible for most of the environmental impact of aquaculture [4] feed composition and the FCR affect the amount of waste produced, as well as its physical and chemical composition. Excess feed results in immediate eutrophication of the surrounding environment; while consumed feed is yields products of metabolic processes, such as ammonia, phosphorus and carbon dioxide [5]. The quantity and quality of the waste excreted by fish depend on intake, digestion and metabolism of dietary compounds [6]. Excess feed waste has a much greater capacity than fecal material to impact the environment, in terms of energy content and degradation rate. The particulate organic matter sinks and disperses, which results in environmental toxicity and anoxia [7]. The degree of impact from effluent wastes depends on feed quality, digestion and metabolism of the diet, species, culture method and the nature of the surrounding environment in terms of physics, chemistry and biology [8]. This has a direct impact on marine benthic habitats, with effects such as reducing sediments, hypoxia in the water overlying the sediment, increased sulphate reduction and changes in benthic fauna assemblages in terms of species number, diversity, abundance and biomass [7].

Feeding methods and technologies have advanced in recent years to minimize and eliminate waste, but the issue remains. Improved digestibility, feed utilization and feeding practices should implemented by the feed and production industry to reduce losses from waste. One example is the use of devices to monitor feeding activity from below the surface in order to feed to satiation without overfeeding and consequent feed wastage, a technology that has been used in recent years by several commercial farms [2]. Nutritional strategies to reduce waste include improvements in feed formulations without affecting growth and production efficiency, inclusion of feed ingredients with high phosphorus bioavailability, use of feed additives to improve the apparent digestibility of phosphorus, and processing-refining of ingredients [9]. These efforts have resulted in a significant reduction of waste outputs (per unit of fish produced) by fish culture operations in Canada over the past four decades [6]. However, feeds that are fully digested by the fish cannot totally resolve the impact of fecal waste because the scope of digestion in fish is limited and there will always be a fraction of undigested feed [10]. Removing the solid waste before it is discharged can be a solution for reducing the environmental impact of wastewater [9]. Eco-certification may be a tool to set standards with criteria aimed at reducing eutrophication through the level of inclusion; e.g., specific allowed amounts of nitrogen and phosphorus release from the system and a set limit for solids in the effluent water [11]. Integrated multi-trophic aquaculture (IMTA) of fish in combination with invertebrates and plants can help reduce environmental impacts and maximize the usage of food input. IMTA shows the most promise in terms of new and innovative systems for waste mitigation and production efficiency. The IMTA system reduces organic waste by mimicking trophic relationships found in nature; the waste from one organism is food for another, resulting in decreased organic particle concentrations with increased distance from farm sites as they are consumed by other farmed organisms [12]. The carbon, nitrogen, and phosphorus compositions of feed, fish and faeces were studied at an Atlantic salmon (*Salmo salar*) farm to estimate the release rates of wastes from salmon cages and the qualities of particulate wastes as food resources for integrated multi-trophic aquaculture. The study found that both salmon feed and faeces were adequate food for blue mussels (*Mytilus edulis*) and sea cucumbers (species unknown), and the nutrient content may meet their nutritional requirements, including DHA and EPA contents of feces, which were comparable to those of some phytoplankton species [13]. Other studies have also reported successful incorporation of nutrients from salmon waste feed into the tissue of blue mussels [14,15]. Choice of the extractive species and distance from the feeding location is an important consideration. For example, in a study on a commercial salmon farm in British Columbia, Canada, mussels had significantly higher amounts of DHA compared with other molluscs (chitons, clam, limpets, periwinkles and whelks), which indicates their potential in IMTA. The levels of DHA in mussels showed a significant breakpoint at 339 m from the farm, which suggests that distance should be considered to optimize certain nutrients [16]. However, organic fish waste captured by mussels is limited by the time available to intercept solid wastes contained in the horizontal particle flux, the velocity of the current, available IMTA farm space, and any negative feedback effects on fish culture from flow reduction caused by mussel culture [17]. New research is focused on expanding novel fish and extractive species. A study recently demonstrated that green sea urchins (*Strongylocentrotus droebachiensis*) actively ingest and absorb organic material from the waste produced by sablefish (Anoplopoma fimbria) culture. Further research was recommended to determine the effect of the sablefish waste diet on green sea urchin survivorship, growth, and gonad quality for urchin production [18]. Multi-species production in an IMTA system must be optimized and better utilized in the future to increase productivity and improve sustainability.

Sustainable Feeds and Ingredients

According to the Food and Agriculture Organization of the United Nations (FAO), at least half of the world's recognized fish stocks are fully exploited and 32% are overexploited or depleted [19]. With the production of farmed fish surpassing that of wild fish, aquaculture's share of global Fish Meal (FM) and Fish Oil (FO) consumption has more than doubled over the past decade [20]. For the past 25 years, annual FM and FO production has not increased beyond 1.5 million tons per year; therefore the aquaculture industry cannot continue to rely on finite stocks of marine pelagic fish as a supply of FM and FO [21]. The paradox of FM and FO use in aquaculture has sparked major research into alternative feed ingredients to replace FM and FO over the past two decades. The focus of this research has mainly been on terrestrial plant meals, concentrates and oils; and many of these studies have led to an impressive reduction in the average inclusion of FM and FO in commercial feeds [22]. The FAO predicts reductions in FM over the next decade for different fish diets: from 26% to 12 % for marine fish, from 22% to 12% for salmon and from 3% to 1% for tilapia [3]. The FAO also predicts that the sustainability of the aquaculture industry will likely depend on using terrestrial plant oils for aquaculture feeds rather than solely depending on FO as a lipid source [19]. Other FM and FO alternatives include meat and bone meal, livestock tallow or fat, blood meal and poultry by-product meals; as well as fishery byproducts and zooplankton meals and oils [23]. Considering the use of alternative protein and lipid sources in commercial feeds [24], recent calculations using the Marine Nutrient Dependency Ratio (MNDR) estimate marine nutrient input to marine nutrient output ratios that are less than one; indicating that farmed salmon can be net producers of FM and FO [24]; and actually increases the supply of fish for human consumption by 7-8 million tonnes per year [25]. In fact, globally fed aquaculture is producing more than three times as much fish as it uses in the feed; therefore reducing the amount of FM and FO in the feeds actually increases the effective global supply of fish for human consumption. However, it is a challenge to find sustainable feed ingredients that meet the nutritional requirements of many farmed fish species. It is likely that some production of FM and FO will remain necessary due to nutritional and industry growth constraints.

Terrestrial plant meals and oils will probably continue to be the main choice when replacing FM and FO in aquaculture diets; however, they are fairly limited in their ability to fully replace FM and FO in diets for fish in terms of meeting nutritional requirements. As a protein source, plant meals are often deficient in certain essential amino acids, and contain high amounts of fibre, carbohydrate, and certain antinutritional factors, which cause adverse effects on feed intake, digestion, absorption, leading to reduced growth [26]. A meta-analysis by Collins et al. [27] found differences in the effect of plant ingredients on the growth performance of salmonid fish and that increasing the inclusion level of pea meal, pea protein concentrate, soy protein concentrate, canola/rapeseed meal or canola/rapeseed concentrate decreased the specific growth rate. The reduced growth performance can likely be explained by reduced protein, amino acid, lipid and energy digestibility, if not compensated by increased feed intake [27]. Some improvements in the nutritional quality can be achieved by chemical and mechanical processing, which can eliminate or reduce the concentration of certain anti-nutrients, as well as concentrate protein content [22]. In fact, in 25 years of research, fish growth performance in response to plant protein ingredients has improved due to changes in the chemical and physical properties of the ingredients due to advances in plant breeding, such as reducing the glucosinolate content in crops like canola [27]. Several studies have shown that replacing FO with various terrestrial plant oils in diets for different fish species does not compromise health, growth or feed consumption and are that they readily catabolised by fish as an energy source for growth, as reviewed by Turchini et al. [21]. However the fatty acid profile of terrestrial plant oils is not equivalent to that of FO, so research in this area has been extensive in recent years.

Terrestrial Plant Oils in Aquaculture Feeds

Terrestrial plant oils will likely be the main choice when replacing FO in aquaculture diets; however they are fairly limited in their ability to fully replace FO in diets for fish. Most plant oils are relatively poor sources of ω3 PUFA in comparison to marine FO, and completely lack LC ω3 PUFA. Rather, they are rich sources of ω6 and ω9 fatty acids, mainly LA and 18:1ω9, with the exception of some oilseeds. Although considered an excellent energy source, feeding terrestrial plant oils inevitably results in lower levels of DHA and EPA in tissues of fish fed plant oils [21,28-32], which is detrimental to fish health and compromises the health benefits for humans that consume these fish. Extensive replacement of FO with terrestrial plant oils, particularly those high in LA, cause a high incidence of cardiovascular disorders in fish [33] and also has been suggested to be detrimental to human health after consumption of fish fed soybean oil [34], which is discussed in fish nutrition and human health section in this review. As a result, the immediate thought is that plant oils best suited as a substitute for FO should contain high levels of ω3 PUFA (LNA) and lower amounts of LA, in order to increase the ω3/ω6 ratio. However, a study by Francis et al. [35] contradicts this idea because it was found that ω6 PUFA (sunflower oil diet) appeared to 'spare' the catabolism of ω3 LC PUFA and, as such, resulted in the highest retention of these fatty acids by rainbow trout. These results suggest new nutritional approaches to maximise the maintenance of the qualitative benefits of fish oils when they are used in feeds for aquaculture species [35]. Other recent studies have described that diets that contain high levels of LNA are relatively wasteful, because the fish did not extensively utilize the LNA toward ω3 PUFA biosynthesis, but rather catabolized the ω3 PUFA for energy. A similar observation is true for some fatty acid classes, particularly saturated fatty acids and monounsaturated fatty acids, which also appear to enhance the retention of ω3 LC PUFA in the fillets of some fish species. These observations have been found in Atlantic cod (Gadus morhua) [31], sunshine bass (Morone chrysops x Morone saxatilis) [36], Murray cod (Maccullochella peelii peelii) [37] and rainbow trout [35]. This approach has been given the more specific definition of the 'ω3 LC PUFA sparing effect' as found in Atlantic salmon [36]. Consequently, it could be argued that plant oils high in LNA do not have any nutritional advantage over other commercially available oils with lower levels of ω3 PUFA, provided that DHA and EPA are spared through catabolism of other abundant fatty acids. These results also indicate that feeding a diet based on FO only is an inefficient practice anyway, because substantial amounts of the nutritionally valuable ω3 LC PUFA are oxidized for energy, particularly EPA [38]. Therefore, the use of plant oils in fish diets conserves LC ω3 PUFA in FO for critical physiological functions only. This has been the focus of studies in this area recently, specifically on the dynamics of dietary DHA/EPA/ARA and its effect on fish performance, tissue concentration and immunity (Table 1).

Table 1: Replacement of fish oil with terrestrial plant oils in diets for Atlantic salmon (*Salmo salar*), with particular focus on dietary DHA/EPA/ARA dynamics and the effect on tissue fatty acid composition and/or immune function

Lipid source	Study duration (d)	Diet total lipid (%)	ω3/ω6	EPA/ARA	DHA/EPA	SGR[1]	Reference
Tuna oil/rape-seed oil	75	23.5	1.3	5	2.8	1	Codabaccus et al. [38]
Fish oil/chick-en fat	75	23.5	2	16	0.5	1	Codabaccus et al. [38]
Linseed oil	112	20	3.6	4.9	1.8	2	Francis et al. [35]
Olive oil	112	20	0.8	5.2	2.1	2	Francis et al. [35]
Palm oil	112	20	0.8	5	2	2	Francis et al. [35]
Sunflower oil	112	20	0.2	5	2.2	2	Francis et al. [35]
Fish oil/vegeta-ble oil + EPA	112	18	2.5	18	0.7	-	Martinez-Ru-bio et al. [49]
Fish oil + EPA	112	18	3.6	18	0.9	-	Martinez-Ru-bio et al. [49]
Fish oil/rape-seed oil: high fat	84	34	1.4	22	1.2	1.0	Martinez-Ru-bio et al. [50]
Fish oil/rape-seed oil: low fat	84	20	1.5	26	1.3	0.9	Martinez-Ru-bio et al. [50]
Camelina oil	112	19	1.2	15	1	0.9	Hixson et al. [32]
Olive oil + DHA	62	20	1.3	4	19	1.2	Glencross et al. [70]
Olive + DHA + ARA	62	18	0.5	0.1	6.8	1.2	Glencross et al. [70]
Rapeseed oil	109	27	0.8	16	1.7	0.9	Thomassen et al. [57]
Rapeseed oil + EPA	109	28	1.9	158	0.2	0.9	Thomassen et al. [57]

Rapeseed oil + EPA + DHA	109	28	1.9	114	0.6	0.9	Thomassen et al. 57]
Rapeseed oil	95	26	0.65	18	1.3	1	Hatlen et al. [81]
Rapeseed oil + GM yeast (high EPA source)	95	23	0.8	33	0.3	0.9	Hatlen et al. [81]
Soybean oil + GM yeast (high EPA source)	112	31	0.4	91	z0.1	0.66	Berge et al. [45]

[1]Specific growth rate, % day^{-1}

The ability of fish to synthesize LC PUFA may allow for plant oils to fully replace FO in aquaculture feeds without lowering levels of key fatty acids in the flesh such as DHA and EPA that are significant for fish and human health. The saturated fatty acids 16:0 and 18:0 can be biosynthesized by all known organisms, including fish. Desaturases and elongases are the critical enzymes in the pathways for the biosynthesis of the LC PUFA from the shorter-chain fatty acids to longer, more unsaturated chains. Fish can desaturate 16:0 and 18:0 to yield 16:1ω7 and 18:1ω9 by Δ9 desaturase. However, all vertebrates lack Δ12 and Δ15 desaturases, which are necessary to form LNA and LA, so these fatty acids are considered essential. Subsequently, LNA and LA can be desaturated and elongated to form the physiologically essential EPA, DHA and ARA. However, the degree to which an animal can synthesize these fatty acids from LNA and LA depends on the activities of the elongase and desaturase enzymes (Δ6 and Δ5) in their tissues. Fish species differ in the extent to which they can tolerate diets without FO, and this trait appears to be evolutionarily related to the fatty acid profile of the natural diet. Consequently, carnivorous marine fish have lost much of the capacity to synthesize these fatty acids during evolution since they remained in an environment where such a conversion is not necessary. Freshwater fish have a greater ability to biosynthesize EPA and DHA from LNA, since the natural prey of many freshwater fish is not rich in EPA and DHA, but rather LNA and LA. The fatty acid desaturation and elongation pathway has been extensively studied in fish at both the molecular and enzymatic level, with fatty acyl elongase (ELOVL) and fatty acyl desaturase (FAD) identified and functionally characterized in several marine and freshwater species. The dietary fatty acid profile is influential to the expression of ELOVL and FAD genes. The tissue fatty acid profile has been found to be significantly correlated with FAD and

ELOVL gene expression in Atlantic cod fed diets without FO; while FAD and ELOVL expression were significantly correlated with each other [38]. This is evidence that the regulation of these genes is signalled by a change in the fatty acid profile of the tissue. However, up-regulation of these genes is not necessarily reflected phenotypically because levels of DHA and EPA in fish fed plant oil diets are significantly lower than DHA and EPA levels in fish fed a FO diet. However, these PUFA could be present in even lower amounts if ELOVL and FAD were not facilitating fatty acid biosynthesis. Often this effort cannot fully compensate for low levels of ω3 LC PUFA intake from plant oil diets; although results show that it is this low dietary level that may trigger the up-regulation of genes involved in their synthesis.

Using molecular tools to identify ELOVL and FAD gene expression has often been the centre of research studies on FO replacements. However, there are other tools that can be used to quantify LC PUFA biosynthesis and can verify phenotypically the results found at the gene expression level. Mathematically, the level of LC PUFA synthesis can be quantified using the fatty acid mass balance equation. The fatty acid mass balance method was developed for fish by Turchini et al. [40] to quantify ω3 fatty acid synthesis in fish to determine the level of elongation and desaturation that occurred over the course of a feeding experiment. The method involves computation of the fatty acid intake, accumulation, and appearance or disappearance of the selected fatty acids in the ω3 pathway and computes the percentage of synthesized ω3 LC PUFA from dietary intake of LNA. An application of this method revealed that Atlantic cod synthesized 6% of their own LC ω3 PUFA [31], and 12% for rainbow trout [41] after fed a plant oil-based diet. Another interesting method to quantify LC PUFA biosynthesis is the use of compound specific stable isotope analysis. Fatty acid isotopic signatures are frequently used in food web studies to determine the transfer of fatty acids from prey to predator based on their 13C/12C ratio [42]. Fatty acids from terrestrial plant oils have distinctly different isotopic signatures than the same fatty acids in marine sources like fish oil due to differences in the source of carbon (terrestrial carbon in the form of CO_2 gas vs. marine carbon as carbonate) [43]. Using CSIA and a mixing model calculation [42], the proportion of synthesized LC PUFA (i.e., DHA) from LNA can be determined. Using this method it has been found that rainbow trout can synthesize up to 27% of the DHA in the muscle tissue from dietary camelina oil [31]. Quantifying fatty acid biosynthesis in fish using quick and efficient methods will become even more important in subsequent years; particularly if breeding programs are designed to select fish that have superior ELOVL and FAD expression and express these traits phenotypically.

Future research in this area will be dedicated to exploring new plant resources that have high levels of ω3 fatty acids, particularly plant sources that contain substantial levels of EPA and DHA. Single-celled microalgae and yeast can produce their own EPA and DHA and are renewable resources [44,45] ; however, high production costs make commercialization very limiting. Research in this area is needed to optimize time and cost of production. However, the most promising and upcoming FO replacements are genetically modified LC ω3 PUFA enriched crop production [46,47] and is further discussed in the Biotechnology section of this review. The use of plant ingredients in aquaculture inevitable; however it also raises questions regarding sustainability of crop production for aquaculture feeds. Measurements to quantify the amounts of land, water, nutrients and energy required for crops per unit of fish production should be calculated to assess environmental impact and sustainable development, and also could be compared to other types of animal production. Nevertheless, the conservation benefit of substituting plant meal and oil for FM and FO is obvious.

Fish Nutrition and Fish Health

It could be argued that the activity of intensive fish farming is, by definition, a stressful procedure, but such a consideration has to be balanced against the need to produce food for human consumption. Therefore, fish health is of upmost importance, in terms of an ethical and production point of view. Proper nutrition plays a critical role in maintaining normal growth and health of aquatic organisms. Good nutrition can help mitigate the effects of stress, decrease the susceptibility to disease, and boost the immune system; therefore it is important to consider their diet in terms of stress and disease prevention, while the diet may also serve as a primary method of administering medications, immuno-stimulants and functional ingredients for fish.

It is well established that all essential nutrients are required in sufficient quantity to sustain normal health. A deficiency in any required nutrient can adversely affect health by impairing metabolic functions and increasing susceptibility to disease. Lipids, for example, are directly involved in aspects of the immune response. LC PUFAs are precursors and modulators for eicosanoid production, including prostaglandins and thromboxane via a synthesis pathway involving cyclooxygenase and lipoxygenase enzyme activity. Arachadonic acid (ARA; 20:4ω6) and EPA compete for the same lipoxygenase or cyclooxygenase, therefore the dietary ω3/ω6 ratio determines the ratio of high activity (inflammation) to low activity eicosanoids (anti-inflammatory). Given the recent focus on FO replacements, studies have compared the immune and inflammatory response of fish fed different plant oil diets after infection by a virus or pathogen. For example, barramundi (Lates calcarifer) fed a diet with either FO or echium oil had a longer lasting and enduring response in eicosanoid concentrations following a week after bacterial infection, compared with those fed on rapeseed oil. Echium oil has a comparatively higher ω3/ω6 ratio than rapeseed oil, and proved more effective in compensating for stress [48]. Another study tested functional feeds for Atlantic salmon infected with heart and skeletal muscle inflammation, an emerging viral disease caused by a novel salmon reovirus [49]. The functional feeds were formulated with reduced digestible energy levels achieved through lower lipid content, and specifically increased the EPA level and increased ω3/ ω6 PUFA and EPA/ARA levels. The LC PUFA biosynthesis activity in salmon fed the standard diet was low compared to those fed the functional feed, limiting the availability of anti-inflammatory LC ω3 PUFA, which highlighted the crucial role of the diet in maintaining the availability of ω3 PUFA in membrane phospholipids. Both EPA and ARA reduced considerably after 16 weeks with infection; these losses were mitigated by the functional feed in comparison to those fed the standard feed, which likely contributed to the better outcome of fish fed the functional feed. In addition, reduced dietary lipid content altered the expression of key genes (Δ5 and Δ6 FAD) in fish fed the functional feeds, potentially increasing their capacity for endogenous production and availability of anti-inflammatory EPA [49,50]. These results highlight the benefits of using diets with reduced dietary lipid and increased EPA levels on disease outcome. Factors that modulate and dampen the inflammatory process might be the key to mitigating clinical symptoms and improving performance, notably the ω3/ω6 PUFA and EPA/ARA levels. Clinical nutrition through functional feeding has been shown to be an effective complementary therapy for emerging salmon viral diseases associated with long-term inflammation [49]. Therefore, future dietary formulations containing plant oil should consider the ω3/ω6 ratio as it strongly affects eicosanoid production and can help deal with stress and disease; which has the focus of several recent studies related with fish oil replacements (Table 1).

Achieving good health through nutrition is relatively simple by providing essential nutrients in the right amount to cater to the specific requirements of different species and life stages. However, disease caused by infections and microorganisms is known to be one of the major constraints in the aquaculture industry, and fish may be susceptible to such diseases, regardless of nutritional status. Feeds containing antibiotics are commonly used to treat diseases. However, sick fish generally do not feed vigorously and therefore may not consume enough medicated feed to ensure proper dosage. The use of antibiotics in aquaculture is also problematic because of the potential to develop antibiotic resistant bacteria, the presence of antibiotic residues in seafood, destruction of microbial populations in the environment and suppression of the fish immune system [51,52]. Therefore, research has focused on functional constituents in feeds to improve growth, feed efficiency, health status, stress tolerance and resistance to disease to reduce antibiotic utilization in farms.

The concept of functional feeds is an emerging paradigm in the aquaculture industry, which are diets supplemented with feed additives to improve health and disease resistance with minimal negative effects on the fish, consumers and the environment [53]. Dietary supplementation of different non-digestible feed additives, such as immunostimulants, probiotics and prebiotics have been found to be beneficial for improving immune status, feed efficiency and growth performance in various different fish species. Feeding animals with immunostimulants prior to an infection or in situations known to result in stress will elevate defences and thus provide protection against otherwise potentially severe or lethal conditions [52]. Probiotics are live microbial organisms, non-pathogenic and non-toxic, which beneficially affect intestinal microbial products, reduce gut pH and release chemicals with bactericidal effects on other microbial populations, thus improving the immune response. The improvement in innate immunity in the fish helps against pathogenic bacteria, as well as against environmental stressors [54]. One of the major bottlenecks in the industry is high mortality during larval first feeding. Probiotics may be administered to live feed which act as live carriers of the bacteria and subsequently fed to fish larvae. This method has proven to be successful in European sea bass larvae (Dicentrarchus labrax) fed artemia cultured with bacterial enrichment, which affected the larval immune response and reduced mortality rates [55] Evidence of the beneficial effects of probiotics gave rise to the concept of prebiotics, which are indigestible oligosaccharides and dietary fibre that selectively stimulate growth and/or activate the metabolism of health-promoting bacteria in the gut, and depress the proliferation of harmful microbes, thus improving intestinal balance in the fish [51]. Beta-glucan for example, can act as both an immunostimulant and a prebiotic. It is a polysaccharide with glucose as its only structural component and forms the major constituents of the cell wall of some plants, fungi, bacteria, mushrooms, yeast, and seaweeds. Beta glucan appears to be the most promising feed additive to enhance growth, survival and immunity for a several different of farmed fish species, supplemented at levels between 0.1 to 1.0% (Table 2). Functional feed additives, such as beta glucan, can improve immunity, feed efficiency and growth performance of farmed fish [2] with minimal impact on the environment and the consumer; therefore commercial feeds may standardize its use in feeds.

Table 2: Various forms of beta-glucan in functional feeds found to improve fish health and immunity for different farmed fish species

Compound	Species	Inclusion	Result	Reference
Immunogen®	Rainbow trout (Oncorhynchus mykiss)	0.2%	Up-regulated immune related genes (lysozyme, TNFα), lower heat shock protein (Hsp70) gene expression, stimulated humeral immune response, elevated disease resistance	Ahmadi et al. [80]
Brown algae beta glucan Laminaria digitata	Rainbow trout	0.2%	Increase in phagocytic activity in head kidney macrophages and a significant increase in the production of TNFα and IL-8 in gill tissue	Morales-Lange et al. [89]
Yeast beta glucan Saccharomyces cervisiae		0.2-0.4%	Increased compliment and immunoglobulin	Rozita et al [91]
Yeast beta glucan (Aquate®)		0.2%	Successfully induced inflammatory gene expression (interleukin-1β) in the head kidneys of fish infected with Lepeophtheirus salmonis; lice burdens were lower on fish fed Aquate® (19%)	Poley et al [90]
Yeast-derived beta glucan		0.1%	Up-regulation of IL-1β after challenged with Vibrio anguillarum	Lokesh et al. [88]
Beta glucan		0.1%	Significantly increased growth, antioxidant activity, and digestive enzyme activity after exposure to lipopolysaccharides	Guzman-Villanueva et al. [89]
Oyster mushroom beta glucan Pleurotus ostreatus		0.5-2.0%	Significant shifts in white blood cell, lymphocyte, monocyte, neutrophil granulocytes-segments, as well as plasma concentrations of glucose, albumins, cholesterol, antrium and chlorides	Dobsikova et al. [84]
Macro Gard®		1%	Differentiated the regulation of mRNA expression of claudin genes and prevented an intestinal inflammatory response post Aeromonas hydrophil intubation	Syakuri et al. [93]
Macro Gard®		1-2%	Fish fed diets containing 1% and 2% MacroGard® showed significant improvements in weight gain, specific growth rate and feed conversion ratio compared to fish fed both the control and the 0.1% MacroGard® containing diet	Kuhlwein et al. [87]
Mushroom beta glucan Ganoderma lucidum and Coriolus versicolor		0.2%	Enhanced the lysozyme activity, alternative complement activity, phagocytic activity and respiration burst; significantly enhanced the protection of grouper against Vibrio alginolyticus up to 16% compared to control group	Chang et al. [82]
Macrogard®		0.1%	Significantly reduced mortalities after challenging with Aeromonas hydrophila with significant improvement in red blood corpuscles, white blood corpuscles, pack cell volume, haemoglobin concentration, erythrocyte sedimentation rate and serum protein content in treated groups over the control	Talpur et al. [94]

Biotechnology to Improve Fish Nutrition

Bridging the gap between problems and solutions in aquaculture will require insight and creative applications. Biotechnology research and applications can help the aquaculture industry develop sustainably, and efficiently produce good quality food for a growing human population. There are many new applications of biotechnology in aquaculture; however this review will focus on how biotechnology can improve fish nutrition for the purpose of safe, nutritious and sustainable food production. In particular, genetic modification will have the most significant impact on fish nutrition in aquaculture.

Producing environmentally sustainable and nutritionally sound feeds for farmed fish has proved to be a challenge. The application of biotechnology to produce feed ingredients that are specifically enhanced for aquaculture feeds can help alleviate pressure on wild fisheries that produce FM and FO. Research on GM crops that are enhanced specifically for fish feeds is in the forefront of both plant and aquaculture research. The limiting factor in using plant oils is the lack of LC ω3 PUFA, namely DHA and EPA. GM plants that can produce their own DHA and EPA, and that can be grown sustainably, is certainly well sought after by the aquaculture industry. Development of such a crop is in the research and development phase, but is not yet available commercially. A study by

Ruiz-Lopez et al. [56] demonstrated fish oil-like levels of DHA and EPA produced by the transgenic oilseed camelina, while avoiding accumulation of undesirable fatty acids. Atlantic salmon fed GM rapeseed oil and yeast containing high levels of DHA and EPA showed high retention of LC ω3 PUFA in the flesh [45,57]. The benefits of producing LC ω3 PUFA in plants are clear: a sustainable and non-contaminated source of important fatty acids essential to human nutrition. GM crops may also have other enhanced features that can directly benefit aquaculture production, such as improving feed efficiency, performance, product quality and health; and can indirectly benefit production, such as pest and disease resistance, and increased yield [58].

A controversial and interesting topic of discussion is GM salmon. Aqua Bounty Technologies was the first company to submit an application to the Food and Drug Administration (FDA) for a GM animal for food consumption in 1995 and has been in the regulation system to date. In order to acquire FDA approval, Aqua Bounty has had to demonstrate the food's safety, and gauge the environmental risk of the sterile fish escaping its tanks and successfully breeding with wild salmon. By contrast, the FDA approved the first GM crop for human consumption (the Flavr Savr tomato) after just three years of regulatory consideration [59]. Research has shown that growth enhanced transgenic Atlantic and coho salmon (Oncorhynchus kisutch), have the capacity for increased daily feed intake, increased growth rates and improved feed utilization due to an enhanced capacity for protein biosynthesis over non-transgenic fish [60,61,62]. Transgenic coho salmon demonstrate an enhanced ability to utilize dietary carbohydrates (in addition to lipid) as a major source of non-protein energy, which is not typical for non-transgenic salmon [63]. The utilization of carbohydrate allows for greater amount of dietary protein to be spared for rapid tissue growth [62]. The altered gene expression and different metabolic enzyme capacity in transgenic salmon allows for a shift towards the use of fatty acids and glucose for energy-production purposes, conserving amino acids for protein biosynthesis [64]. In a study by Tibbetts et al. [62], transgenic Atlantic salmon consumed a significantly higher amount of feed on a daily basis, but also had better FCRs, higher nitrogen retention efficiency and achieved target weight gain in a shorter period (40%) than non-transgenic fish. The transgenic salmon in this study demonstrated a higher cellular capacity to direct dietary non-protein energy towards satisfying their daily metabolic energy requirements, allowing for a higher proportion of dietary amino acids to be directed towards protein biosynthesis rather than catabolised as a dietary energy source. Dietary protein represents the highest proportion of salmon feeds and also the major source of nitrogenous pollution from salmon farming; therefore transgenic salmon could improve production efficiency through better feed utilization, which could result in more sustainable salmon production. Furthermore, transgenic salmon were found to have a higher expression of Δ5 and Δ6 FAD, which might help compensate for a potentially higher demand for EPA and DHA resulting from rapid growth [64]. This also has positive implications for the use of alternative FO feeds for transgenic fish in terms of their ability to synthesize EPA and DHA at an increased rate. However, there are environmental concerns related to GM salmon, particularly the potential impacts of escaped GM salmon on wild salmon; although GM salmon produced by Aqua Bounty are sterile, thus eliminating the potential of interbreeding between escaped GM salmon and wild salmon. There are also health and safety concerns that pertain to the immediate and long-term effects on human health after consumption of GM salmon products. Initial tests on the safety of consuming transgenic carp and tilapia show no apparent negative health impacts on mice and humans [65]. However, allergens or toxins may be produced as a result of gene transfer if the transgene codes for a protein that induces expression of a previously inactive toxin (e.g., shellfish protein used to create a transgenic fish), and the expression of bioactive proteins which may continue to possess bioactive properties following consumption (e.g., growth hormone) [65]. Research on the effect of GM fish on human health should be conducted, as well as the effect on product quality.

Fish Nutrition and Human Health

The world will be very different in 2050: the global population will likely increase to nine billion, and about 50% more food will be needed to sustain the quality of human life [66]. Rapid population growth and increases in fish consumption has led to rapid increases in global fish demand. The expansion of aquaculture will fill this gap and relieve pressure on capture fisheries, which have been steadily declining [67]. The benefits of consuming a diet rich in seafood are based on high levels of LC ω3 PUFA and high quality animal protein, essential amino acids, vitamins, and minerals [67]. There are numerous benefits of aquaculture being the main supply of seafood for the world, from a nutritional, economic and social standpoint; however the industry has received much public scrutiny regarding toxins, contaminants, hormones, and antibiotics in farmed fish, which raises questions about public health and safety [68].The composition of the feed affects the final composition of the product; therefore fish nutrition directly impacts consumers and must be considered in terms of safety and human nutrition.

It is obvious that the use of FM and FO must be significantly reduced in aquaculture feeds in order to be environmentally sustainable. However, using sustainable alternative feed ingredients to feed fish could potentially affect human health. For example, terrestrial plant oils are typically high in ω6 fatty acids, specifically LNA, which get stored in significant amounts in flesh of the fish that are consuming these feeds. From a human health standpoint, increases in ω6 fatty acids in salmon fillets are a concern. When mice were fed farmed salmon raised on diets with 100% soybean oil, the excessive dietary LA elevated endocannabinoids in the liver, increased weight gain and counteracted the anti-inflammatory properties of EPA and DHA [69] and also exaggerated insulin resistance and increased accumulation of fat in the liver in rats [34]. The typical "Western" diet is already dangerously high in ω6 fatty acids, which is why it is so important to consume fish that are high in ω3 fatty acids to balance this ratio. However, in addition to the high level of ω6 fatty acids, terrestrial plant oils do not contain EPA and DHA, which causes a lower concentration of these PUFA in farmed seafood products compared to farmed fish that were fed a FObased diet. Most consumers are not fully aware of the lower DHA and EPA content now occurring in many farmed fish species as a result of using alternative feeds in recent years. A study by Nichols et al. [70] investigated the LC ω3 PUFA content in farmed Atlantic salmon and barramundi samples from 2002 compared to samples from 2010-2013. In comparison to the 2002 samples, both species contained decreased levels and contents of LC ω3 PUFA. Atlantic salmon in 2002 contained 2014 mg/100 g, decreasing to 975 mg/100g in 2013. Barramundi in 2002 contained 1970 mg/100 g, decreasing to 790 mg/100 g in 2013 [70]. These changes have resulted from the use of new, sustainable and lower cost ingredients in the feed.

Despite significantly lower LC ω3 PUFA amounts in farmed salmon compared to a decade ago, a human study found that addition of farmed Atlantic salmon to the diet twice per week for 4 weeks at portions of 180 g and 270 g modifies phospholipid fatty acid proportions of ω3 and ω6 in a level associated with decreased risk for cardiovascular disorders [71]. Regardless, there are strategies to increase the ω3 PUFA content in the lipid profile of farmed fish. In the present time, even using plant oils that are high in LNA but do not supply DHA and EPA is still beneficial for humans simply to achieve a higher ω3/ω6 ratio in the diet. Finishing oil diets with higher inclusion of marine oils at the end of the production cycle can elevate DHA and EPA levels immediately before harvest. This may result in grades of farmed seafood products that contain a higher content of LC ω3 PUFA than standard (and lower cost) products grown using a mainly nonmarine oil based diet [70]. In the future it is likely that GM plants with endogenous DHA and EPA will be used in fish feeds. The effect of GM foods on human health is not well known, however, many different types of produce and processed foods available today are non-labelled GM, so many North American consumers have purchased GM foods on a regular basis for at least a decade. However, there have been no studies to date on the long term effect of GM ingredients fed to farmed fish on human health. This will be a critical area of research in the future.

Dietary changes may also affect the product quality of the fish, which may change public perception of farmed fish. Sensory evaluations, including taste, colour and texture are frequently part of studies that evaluate changes after feed ingredients have been altered in the diet, particularly replacements of FM and FO. Although many studies have reported that alternative protein and lipid sources can affect the sensory quality of farmed fish, results in the literature are divided and contradictory [21], therefore it is difficult to pinpoint if certain ingredients clearly affect sensory quality because it can depend on the study. Interestingly, some studies showed that salmon fed plant oils had less rancid and marine characteristics and were preferred over salmon fed FO [72]. Consumer perception and satisfaction are an important aspect of fish nutrition; therefore the use of GM ingredients or the production of GM salmon needs to be evaluated for sensory quality due to possible biochemical changes that affect sensory characteristics.

Contamination of fish tissues with organic and inorganic contaminants has been a pervasive environmental and public health issue. From a human health perspective, concerns regarding relatively high levels of Persistent Organic Pollutants (POPs) (e.g. Polychlorinated Biphenyls, PCBs) and inorganic contaminants (e.g. heavy metals) in farmed salmon have raised questions regarding the relative health risks and benefits of farmed salmon consumption [73]. However, the accumulation of some contaminants is the result of indirect exposure, and can equally affect wild salmon [74]. This review will only focus on POPs accumulated in the flesh as a direct result of the feed. Fish that are fed FM and FO accumulate POPs that are derived from marine feed components, and several studies have reported elevated levels of environmental contaminants, particularly PCBs in feeds and farmed Atlantic salmon flesh [75]. One of the benefits of using plant meal and oil to replace FM and FO is the reduced levels of toxins in the feeds, and subsequently in the final product. Studies have shown that using plant oils to replace FO for most of the production cycle show large reductions in flesh POPs in salmon; thereby providing a low contaminant product for human consumption [75,76]. Interestingly, Atlantic salmon fed a diet that replaced 75% FO with flaxseed oil showed 61% lower levels of POPs in the flesh compared with fish fed diets with 100% FO, without compromising growth performance [77]. Farrell et al. [78] concluded that plant oils in salmon feeds considerably lower contaminant levels in farmed salmon without reducing the human health benefits and compromising production efficiency. Furthermore, rats that were fed farmed salmon raised on 100% soybean oil had significantly lower levels of POPs in their tissues compared to rats fed salmon raised on FO [34]. However, considering high levels of LA and significantly reduced levels of DHA and EPA in salmon after fed a plant oil diet, there is certainly a trade-off between using plant oils and marine oils. The best possible outcome to achieve environmental sustainability and reduce dependence on marine resources, while maximizing the amount of DHA and EPA in farmed seafood would be using GM oilseeds with equivalent levels of DHA and EPA to that of FO.

CONCLUSION

Fish nutrition certainly has an impact on the aquaculture industry. The areas discussed in this review were meant to highlight current issues in aquaculture related specifically to the nutrition of farmed fish and how these issues can impact food safety and quality, as well as environmental sustainability. In a world where natural resources are being consumed at 50 times the replenishment rate, the future belongs to those who can manage elements of population growth, climate change, and food and energy security. Aquaculture has a huge role to play in the future where safe, nutritious, quality food is in demand and environmental sustainability is the key to life [79]. It is difficult to quantify environmental impacts and sustainability in aquaculture since different species and production systems are more sustainable than others because of their feeds and feeding practices. The result is a trade-off of positive and negative outcomes, and both food security and sustainability must be balanced. The future advancements in aquaculture to achieve this balance will inevitably require the use of biotechnology. Scientists should be able to develop and effectively use new technologies, within reason, to benefit of our society. The use of GM plants and animals to increase nutrient utilization and production efficiency, and to reduce environmental impacts, will become necessary in the future. However, food safety is of upmost importance, therefore diligent research and stringent regulations must be required before commercialization of GM products. The balance between nutritious, safe food production and maintaining environmental sustainability will rely on our ability to use new technologies to improve farming practices, and our ability to use as few resources as possible to provide for our society [80-95].

ACKNOWLEDGEMENTS

This review was supported by the Research and Development Corporation of Newfoundland (RDC) and Memorial University of Newfoundland. The author would like to thank Dr. Christopher C. Parrish for review of this paper.

REFERENCES

1. National Research Council (NRC) (2011) Nutritional requirements of fish and shrimp. National Acadamies Press, Washington.
2. Ganguly S, Krushna C, Sarkar S, Chowdhury S (2013) Supplementation of prebiotics in fish feed: a review. Rev

Fish Biol Fisheries 23:195-199.

3. Tacon A, Hasan M, Metian M (2011) Demand and supply of feed ingredients for farmed fish and curstaceans- trends and prospects. FAO fisheries technical paper, vol. 564.
4. d'Orbcastel E, Blancheton J, Aubin J (2009) Towards environmentally sustainable aquaculture: Comparison between two trout farming systems using life cycle assessment. AquacultEng 40:113-119.
5. Schneider O, Amirkolaie A, Vera Cartas J, Eding E, Schrama J, et al. (2004) Digestibility, faeces recovery, and related C, N, P balances of five feed ingredients evaluated as fishmeal alternatives in Oreochromisniloticus L. Aquaculture Research 35:1370-1379.
6. Bureau D, Hua K (2010) Towards effective nutritional management of waste outputs in aquaculture, with particular reference to salmonid aquaculture operations. Aquacult Rev 41:777-792.
7. Piedecausa M, Aguado-Gimenez F, Valverde J, Llorente M, Garcia-Garcia B (2012) Influence of fish food and faecal pellets on short-term oxygen uptake, ammonium flux and acid volatile sulphide accumulation in sediments impacted by fish farming and non-impacted sediments. Aquacult Res 43:66-74.
8. Boyd C, Tucker C, McNevin A, Bostick K, Clay J (2007) Indicators of resource use efficiency and environmental performance in fish and crustacean aquaculture. Rev Fish Sci 15:327-360.
9. Amirkolaie A (2011) Reduction in the environmental impact of waste discharged by fish farms through feed and feeding. Rev Aqua 3:19-26.
10. Cho C, Hynes J, Wood K, Yoshida H (1994) Development of high-nutrient-dense, low-pollution diets and prediction of aquaculture wastes using biological approaches. Aquaculture 124:293-305.
11. Jonell M, Phillips M, Ronnback P, Troell M (2013) Eco-certification of farmed seafood: will it make a difference? AMBIO 42:659-674.
12. Reid G, Liutkus M, Bennett A, Robinson S, MacDonald B, et al. (2010) Absorption efficiency of blue mussels (Mytilusedulis and M. trossulus) feeding on Atlantic salmon (Salmosalar) feed and fecal particulates: implications for integrated multi-trophic aquaculture. Aquaculture 299:165-169.

13. Wang X, Andersen K, Handa A, Jense B, Reitan K, et al. (2013) Chemical composition and release rate of waste discharge from an Atlantic salmon farm with an evaluation of IMTA feasibility. AquacultEnvInterac 4:147-162.
14. Handa A, Min H, Wang X, Broch O, Reitan K, et al. (2012) Incorporation of fish feed and growth of blue mussels (Mytilusedulis) in close proximity to salmon (Salmosalar) aquaculture: Implications for integrated multi-trophic aquaculture in Norwegian coastal waters. Aquaculture. 356:328-341.
15. Redmond K, Magnesen T, Hansen P, Strand O, Sonnich M (2010) Stable isotopes and fatty acids as tracers of the assimilation of salmon fish feed in blue mussels (Mytilusedulis). Aquaculture. 298: 202–210.
16. George E, Parrish C (2013) Invertebrate uptake of lipids in the vicinity of Atlantic salmon (Salmosalar) aquaculture sites in British Columbia. Aquacult Res.
17. Cranford P, Reid G, Robinson S (2013) Open water integrated multi-trophic aquaculture: constraints on the effectiveness of mussels as an organic extractive component. AquacultEnvInterac 4:163-173.
18. Orr L, Curtis D, Cross S, Gurney-Smith H, Shanks A, et al. (2014) Ingestion rate, absorption efficiency, oxygen consumption, and fecal production in green sea urchins (Strongylocentrotusdroebachiensis) fed waste from sablefish (Anoplopoma fimbria) culture. Aquaculture 422:184-192.
19. FAO (Food and Agricultural Organization of the United Nations) (2010) The state of the world fisheries and aquaculture 2010.
20. Naylor R, Hardy R, Bureau D, Chiu A, Elliott M, et al. (2009) Feeding aquaculture in an era of finite resources. PNAS 106:15103-15110.
21. Turchini G, Torstensen B, Ng W (2009) Fish oil replacement in finfish nutrition. Rev Aqua 1:10-57.
22. Olsen R, Hasan M (2012) A limited supply of fishmeal: Impact on future increases in global aquaculture production. Trends Food Sci Tech 27:120-128.
23. Hertrampf J, Piedad-Pascual F (2000) Handbook on ingredients for aquaculture feeds. Kluwer Academic Publishers, Netherlands.

24. Crampton V, Nanton D, Ruohonen K, Skjervold P, El-Mowafi A (2010) Demonstration of salmon farming as a net producer of fish protein and oil. AquacultNutr 16:437-446.

25. Jackson A (2012) Fish meal and fish oil and its role in sustainable aquaculture. International aquafeed 15:18-21.

26. Krogdahl A, Penn M, Thorsen J, Refstie S, Bakke A (2010) Importantantinutrients in plant feedstuffs for aquaculture: an update on recent findings regarding responses in salmonids. Aquacult Res 41:333-344.

27. Collins S, Overland M, Skrede A, Drew M (2013) Effect of plant protein sources on growth rate in salmonids: Meta-analysis of dietary inclusion of soybean, pea and canola/rapeseed meals and protein concentrates. Aquaculture 400:85-100.

28. Bell G, Pratoomyot J, Strachan F, Henderson R, Fontanillas R, et al. (2010) Growth, flesh adiposity and fatty acid composition of Atlantic salmon (Salmosalar) families with contrasting flesh adiposity: Effects of replacement of dietary fish oil with vegetable oils. Aquaculture 306:225-232.

29. Alhazzaa R, Bridle A, Nichols P, Carter C (2011) Replacing dietary fish oil with echium oil enriched barramundi with C18 PUFA rather than long-chain PUFA. Aquaculture 312:162–171.

30. Hixson S, Parrish C, Anderson D (2013) Effect of replacement of fish oil with camelina (Camelinasativa) oil on growth, lipid class and fatty acid composition of farmed juvenile Atlantic cod (Gadusmorhua). Fish PhysiolBiochem 39:1441-1456.

31. Hixson S, Parrish C, Anderson D (2014) Full substitution of fish oil with camelina oil, with partial substitution of fish meal with camelina meal, in diets for farmed Atlantic salmon (Salmosalar) and its effect on tissue lipids and sensory quality. Food Chem 157:51-61.

32. Hixson S, Parrish C (2014) Substitution of fish oil with camelina oil and inclusion of camelina meal in diets of Atlantic cod (Gadusmorhua) and its effect on growth and tissue lipid classes and fatty acids. J AnimSci 92:1055-67.

33. Waagbo R (2008) Reducing production related diseases in farmed fish. In: Lie O (ed) Improving Farmed Fish Quality and Safety. VS

Woodhead Publishing, UK, pp 363–398.

34. Mitdbo L, Ibrahim M, Myrmel L, Aune U, Alvheim A, et al. (2013) Intake of farmed Atlantic salmon fed soybean oil increases insulin resistance and hepatic lipid accumulation in mice. PLOS One 8:1-11.
35. Francis D, Thanuthong T, Senadheera S, Paolucci M, Coccia E, et al. (2014) n-3 LC-PUFA deposition efficiency and appetite-regulating hormones are modulated by the dietary lipid source during rainbow trout grow-out and finishing periods. Fish PhysiolBiochem 40:577–593.
36. Trushenski J (2009) Saturated lipid sources in feeds for sunshine bass: alterations in production performance and tissue fatty acid composition. North Am J Aquacult 71:363-373.
37. Turchini G, Francis D, Senadheera S, Thanuthong T, De Silva S (2011) Fish oil replacement with different vegetable oils in Murray cod: Evidence of an "omega-3 sparing effect" by other dietary fatty acids. Aquaculture 315:250-259.
38. Codabaccus B, Carter C, Bridle A, Nichols P (2012) The "n−3 LC-PUFA sparing effect" of modified dietary n−3 LC-PUFA content and DHA to EPA ratio in Atlantic salmon smolt. Aquaculture 356:135-140.
39. Xu X, Feng C, Hixson S, Johnstone K, Anderson D, et al. (2014) Characterization of fatty acyl elongase (elovl) gene family, and hepatic elovl and delta-6 fatty acyl desaturase (fadsd6) transcript expression and fatty acid responses to diets containing camelina oil in Atlantic cod (Gadusmorhua).
40. Turchini G, Francis D, De Silva S (2007) A whole body, in vivo, fatty acid balance method to quantify PUFA metabolism (desaturation, elongation and beta-oxidation). Lipids 42: 1065-1071.
41. Turchini G, Francis D (2009) Fatty acid metabolism (desaturation, elongation and b-oxidation) in rainbow trout fed fish oil- or linseed oil-based diets. Br J Nutr 102:69-81.
42. Budge S, Wooller M, Springer A, Iverson S, McRoy C, et al. (2008) Tracing carbon flow in an arctic marine food web using fatty acid stable isotope analysis. Oecologia 157: 117-129.
43. O'Leary M (1988) Carbon isotopes in photosynthesis. Bioscience

38:328-336.

44. Bendiksen E, Johnsen C, Olsen H, Jobling M (2011) Sustainable aquafeeds: Progress towards reduced reliance upon marine ingredients in diets for farmed Atlantic salmon (Salmosalar). Aquaculture 314:132-139.
45. Hatlen B, Berge G, Odom J, Mundheim H, Ruyter B (2012) Growth performance, feed utilisation and fatty acid deposition in Atlantic salmon, Salmosalar L., fed graded levels of high-lipid/high-EPA Yarrowialipolytica biomass. Aquaculture 364:39-47.
46. Napier J, Sayanova O (2005) The production of a very-long-chain PUFA biosynthesis in transgenic plants: towards a sustainable source of fish oils. ProcNutrSoc 64:387-393.
47. Flachowsky G, Meyer U, Gruen M (2013) Plant and animal breeding as starting points for sustainable agriculture. In: Lichtfouse E (ed) Sustainable Agriculture Reviews 12. Springer, Netherlands, pp 201-224.
48. Alhazzaa R, Bridle A, Mori T, Barden A, Nichols P, et al. (2013) Echium oil is better than rapeseed oil in improving the response of barramundi to a disease challenge. Food Chem 141:1424-1432.
49. Martinez-Rubio L, Morais S, Evensen O, Wadsworth S, Vecino J, et al. (2013a) Effect of functional feeds on fatty acid and eicosanoid metabolism in liver and head kidney of Atlantic salmon (Salmosalar L.) with experimentally induced heart and skeletal muscle inflammation. Fish Shellfish Immunol 34:1533-1545.
50. Martinez-Rubio L, Wadsworth S, Vecino J, Bell G, Tocher D (2013b) Effect of dietary digestible energy content on expression of genes of lipid metabolism and LC-PUFA biosynthesis in liver of Atlantic salmon (Salmosalar L.). Aquaculture 384:94-103.
51. Ringo E, Olsen R, Gifstad T, Dalmo R, Amlund H, et al. (2010) Prebiotics in aquaculture- a review. AquacultNutr 16:117-136.
52. Oliva-Teles A (2012) Nutrition and health of aquaculture fish. J Fish Dis 35:83-108.
53. Meena D, Dad P, Kumar S, Mandal S, Prusty A, et al. (2013) Beta-glucan: an ideal immunostimulants in aquaculture (a review). Fish

Physiol. Biochem. 39: 431-457.

54. Mohapatra S, Chakraborty T, Prusty A, Das P, Paniprasad K, et al. (2012) Use of different microbial probiotics in the diet of rohu, Labeorohita fingerlings: effects on growth, nutrient digestibility and retention, digestive enzyme activities and intestinal microflora. AquacultNutr 18:1-11.
55. Franke A, Roth O, Clemmesen C (2013) Early stimulation of the immune system of an important aquaculture fish species: Probiotic application in European sea bass juveniles Fish & Shellfish Immunol 34:1707-1715.
56. Ruiz-Lopez N, Haslam R, Napier J, Sayanova O (2014) Successful high-level accumulation of fish oil omega-3 long-chain polyunsaturated fatty acids in a transgenic oilseed crop. The Plant J. 77: 198–208.
57. Thomassen M, Rein D, Berge G, Østbye T, Ruytera B (2012) High dietary EPA does not inhibit Δ5 and Δ6 desaturases in Atlantic salmon (Salmosalar L.) fed rapeseed oil diets. Aquaculture 361:78-85.
58. Deb R, Sajjanar B, Devi K, Reddy K, Prasad R, et al. (2013) Feeding animals with GM crops: Boon or bane? Indian J Biotech 12:311-322.
59. Ledford H. (2013) Transgenic salmon nears approval. Nature 497:17-18.
60. Devlin R, Biagi C, Yesaki T (2004) Growth, viability and genetic characteristics of GH transgenic coho salmon strains. Aquaculture 236:607-632.
61. Oakes J, Higgs D, Eales J, Devlin R (2007) Influence of ration level on the growth performance and body composition of non-transgenic and growth-hormone-transgenic coho salmon (Oncorhynchuskisutch). Aquaculture 265:309-324.
62. Tibbetts S, Wall C, Barbosa-Solomieu V, Bryenton M, Plouffe D, et al. (2013) Effects of combined 'all-fish' growth hormone transgenics and triploidy on growth and nutrient utilization of Atlantic salmon (Salmosalar L.) fed a practical grower diet of known composition. Aquaculture 406:141-152.
63. Leggatt R, Raven P, Mommsen T, Sakhrani D, Higgs D, et al.

(2009) Growth hormone transgenesis influences carbohydrate, lipid and protein metabolism capacity for energy production in coho salmon (Oncorhynchuskisutch). CompBiochemPhysiol B 154:121-133.

64. Xu Q, Feng C, Hori T, Plouffe D, Buchanan J, et al. (2013) Family-specific differences in growth rate and hepatic gene expression in juvenile triploid growth hormone (GH) transgenic Atlantic salmon (Salmosalar). CompBiochemPhysiol D 8:317–333.
65. Rasmussen R, Morrissey M (2007) Biotechnology in aquaculture: transgenics and polyploidy. Comp Rev Food Sci 6:2-16.
66. Diana J, Egna H, Chopin T, Peterson M, Cao L, et al. (2013) Responsible aquaculture in 2050: valuing local conditions and human innovations will be key to success. Bioscience 63:255-262.
67. Tacon A, Metian M (2013) Fish matters: Importance of aquatic foods in human nutrition and global food supply. Rev Fish Sci 21:22-38.
68. Mansfield B (2012) Is fish health food or poison? Farmed fish and the material production of un/healthy nature. Antipode 43:413-434.
69. Alvheim A, Torstensen B, Lin Y, Lillefosse H, Lock E, et al. (2013) Dietary linoleic acid elevates endogenous 2-arachidonolylglycerol and anandamide in Atlantic salmon (Salmosalar) and mice, and induces weight gain and inflammation in mice. Br J Nutr 109:1508-1517.
70. Nichols P, Glencross B, Petrie J, Singh S (2014) Readily available sources of long-chain omega-3 oils: Is farmed Australian seafood a better source of the good oil than wild-caught seafood? Nutrients 6:1063-1079.
71. Raatz S, Rosenberger T, Johnson L, Wolters W, Burr G, et al. (2013) Dose-dependent consumption of farmed Atlantic salmon (Salmosalar) increases plasma phospholipid n-3 fatty acids differentially. J AcadNutr Diet 113:282-287.
72. Torstensen B, Bell G, Rosenlund G, Henderson J, Graff I, et al. (2005) Tailoring of Atlantic salmon (Salmosalar) flesh lipid composition and sensory quality by replacing fish oil with a vegetable oil blend. J. Agric Food Chem 53:10166-10178.
73. Rudd M, Pelletier N, Tyedmyers P (2011) Preferences for health

and environmental attributes of farmed salmon amongst southern Ontario salmon consumers. Aquacult Econ Manag 15:18-45.

74. Foran J, Hites R, Carpenter D, Hamilton C, Matthews-Amos A, et al. (2004) A survey of metals in tissues of farmed Atlantic and wild Pacific salmon. Environ ToxicolChem 23:2108-2110.

75. Waagbo R, Berntssen M, Danielsen T, Helberg H, Kleppa A, et al. (2013) Feeding Atlantic salmon diets with plant ingredients during the seawater phase – a full-scale net production of marine protein with focus on biological performance, welfare, product quality and safety. AquacultNutr 19:598-613.

76. Bell G, Dick J, Strachan F, Guy D, Bertssen M, et al. (2012) Complete replacement of fish oil with a blend of vegetable oils affects dioxin, dioxin-like polychlorinated biphenyls (PCBs) and polybrominateddiphenyl ethers (PBDEs) in 3 Atlantic salmon (Salmosalar) families differing in flesh adiposity. Aquaculture 324:118-126.

77. Friesen E, Ikonomou M, Higgs D, Ang K, Dubetz C (2008) Use of terrestrial based lipids in aquaculture feeds and the effects on flesh organohalogen and fatty acid concentrations in farmed Atlantic salmon. Environ Sci Tech 42:3519-3523.

78. Farrell A, Friesen E, Higgs D, Ikonomou M (2010) Toward improved public confidence in farmed fish quality: a Canadian perspective on the consequences of diet selection. J World Aqua Soc 41:207–224.

79. Oliver J (2013) Food, water, soil, oil: Peak everything- Almost. In: Abstracts from Aquaculture Canada Conference, Guelph, Canada, June 1-4 2013.

80. Ahmadi P, Farahmanda H, Miandare H, Mirvaghefi A, Hoseinifar S (2014) The effects of dietary Immunogen® on innate immune response, immune related genes expression and disease resistance of rainbow trout (Oncorhynchusmykiss) Fish Shellfish Immunol 37:209-214.

81. Berge G, Hatlen B, Odom J, Ruyter B (2013) Physical treatment of high EPA Yarrowialipolitica biomass increases the availability of n-3 highly unsaturated fatty acids when fed to Atlantic salmon. AquacultNutr19:110-121.

82. Chang C, Huang S, Chen S, Chen S (2013) Innate immune responses

and efficacy of using mushroom beta-glucan mixture (MBG) on orange-spotted grouper, Epinepheluscoioides, aquaculture. Fish Shellfish Immunol 35:115-125.

83. Crampton V, Carr I (2012) Fish Forever. In: Spotlight 5 EWOS publication. EWOS, Norway
84. Dobsikova R, Blahova J, Mikulikova I, Modra H, Praskova E, et al. (2013) The effect of oyster mushroom β-1.3/1.6-D-glucan and oxytetracycline antibiotic on biometrical, haematological, biochemical, and immunological indices, and histopathological changes in common carp (Cyprinuscarpio L.). J Fish Shellfish Immunol 35:1813-1823.
85. Glencross B, Tocher D, Matthew C, Bell G (2014) Interactions between dietary docosahexaenoic acid and other long-chain polyunsaturated fatty acids on performance and fatty acid retention in post-smolt Atlantic salmon (Salmosalar). Fish PhysiolBiochem.
86. Guzman-Villanueva L, Ascencio-Valle F, Macias-Rodriguez M, Tovar-Ramirez D (2013) Effects of dietary β-1,3/1,6-glucan on the antioxidant and digestive enzyme activities of Pacific red snapper (Lutjanusperu) after exposure to lipopolysaccharides. Fish PhysiolBiochem.
87. Kuhlwein H, Merrifield D, Rawling M, Foey A, Davies S (2014) Effects of dietary β-(1,3)(1,6)-D-glucan supplementation on growth performance, intestinal morphology and haemato-immunological profile of mirror carp (Cyprinuscarpio L.). J AnimPhysiolAnimNutr 98:279-289.
88. Lokesh J, Fernandes J, Korsnes K, Bergh O, Brinchmann M, et al. (2012) Transcriptional regulation of cytokines in the intestine of Atlantic cod fed yeast derived mannan oligosaccharide or β-Glucan and challenged with Vibrio anguillarum. Fish Shellfish Immunol 33:626-631.
89. Morales-LangeB, BethkeJ, SchmittP, MercadoL(2014) Phenotypical parameters as a tool to evaluate the immunostimulatory effects of laminarin in Oncorhynchusmykiss. Aquacult Res.
90. Poley J, Purcell S, Igboeli O, Donkin A, Wotton H, et al. (2013) Combinatorial effects of administration of immunostimulatory compounds in feed and follow-up administration of triple-dose

SLICE® (emamectin benzoate) on Atlantic salmon, Salmosalar L., infection with Lepeophtheirussalmonis. J Fish Dis 36: 299-309.

91. Rozita K, Shila S, Mahdy C (2013) Effect of whole and cell wall of Saccharomyces cervisiae in immunity factors on rainbow trout (Oncorhynchusmykiss). American-Eurasian J Agric Environ Sci 13:633-638.
92. Sarker P, Bureau D, Hua K, Drew M, Forster I, et al. (2013) Sustainability issues related to feeding salmonids: a Canadian perspective. Rev Aqua 5:1-21.
93. Syakuri H, Jung-Schroers V, Adamek M, Brogden G, Irnazarow I, et al. (2014) Beta-glucan feeding differentiated the regulation of mRNA expression of claudin genes and prevented an intestinal inflammatory response post Aeromonashydrophila intubation in common carp, Cyprinuscarpio L. J Fish Dis 37:149-156.
94. Talpur A, Munir A, Mary A, Hashim R (2014) Aquaculture 426:14-20.
95. WHO (World Health Organization) (1996) Rome Declaration on World Food Security and World Food Summit Plan of Action. Food and Agricultural Organization of th e United Nations.

CHAPTER 10

EFFECTS OF SALINITY ON GROWTH, HORMONAL AND ENZYMATIC STATUS IN FISH: A REVIEW

Maysa Enayatmehr[1], Shahla Jamili[2]

[1] Phd student in Marine Biology, Department of Marine Biology, Science and Research Branch, Islamic Azad University, Tehran, Iran

[2] Iranian Fisheries Science Research Institute, Agricultural Research Education & Extension Organization

ABSTRACT

There has recently been an increasing interest in seafood products due to the growing awareness of their nutraceutical value. However, marine-based products are highly dependent on environmental conditions. Among other factors, many studies have reported an influence of water salinity on different aspect of fish physiology. In different studies, effect of different levels of marine salinity on growth factors, hormonal and enzymatic status, tissue histopathology and related gene expression have been investigated. Changes in growth rate, which depend on salinity, result from an action on: 1. standard metabolic rate; 2. food intake; 3. food conversion; and or 4. hormonal stimulation. Numerous studies reported the best range of salinity for better fish physiology performance but these domainsin different species and different places are different. In this review, we considered all of these factors and discuss the effects of marine salinity on growth factors in different species.

Keywords: Salinity, Fish, Growth, Enzymatic Status

INTRODUCTION

Due to especially physiological function in each species and also environmental factors, development and growth in fish is different. Fish physiology is depending on internal and external factors such as nervous, endocrinological, neuro endocrinological and also ecological factors respectively. Both of these factors control or synchronize many activities or functions, including growth capacity. In previous study ecological factors divided to two groups: 1- determining factors such as temperature, salinity and photoperiod which act directly through receptors to increase or decrease growth; and 2. limiting factors, which operate above ammoniaor below such as oxygen and a specific threshold or within a tolerance range such as pH.

Growth as a physiological function in fish is different from terrestrial animal. Growth pattern in terrestrial is discontinuous, this means that they just growth up to the limit age, but in fish iscontinuous, so that fish become larger the longer they live and they are much more dependent on external environmental conditions (Brett, 1979; Boeuf et al., 1999). Different external factors effect on growth in marine fish that temperature and environment salinity are most important of them (Brett, 1979). Many authors have demonstrated the influence of external salinity on growth capacities in fish (see Table1). It is important to other species that have migration between different locations with different salinity considered because the fish that mentioned in Table1 not influenced by salinity changes during their development and growth are rare. In other study it determined that many of fish prefer intermediary salinities especially in juvenile stage (Boeuf et al., 2001).So the question that can be asked is that, "how salinity influences growth?"

Table 1: Salinity and growth

Species	Tolerance	Best growth	Reference
Salmo salar (Atlantic salmon)	FW		Buckel et al. (1995)
	sW	17-19	Aristizabal Abud
Salmo salar	6 h SW	15	Woo and Kelly. (1992)
	10 d SW	8	Likongwe et al.
Oncorhynchus mykiss (rainbow trout)			Jonassen et al. (1997)
Scophthalmus maximus (turbot)		17.5	Chervinsky and Yashouv. (1971)
Limanda limanda (dab)		5-10	Suresh and Lin (1992)
Pleuronectes platessa (plaice)		10-15	Suresh and Lin (1992)
Anguilla anguilla (eel)		0	Suresh and Lin (1992)
Merlangius merlangus (whiting)		30-35	Suresh and Lin (1992)
Morone chrysops (White bass)		0-12	Heyward(1995)
Chanos chanos (Milkfish)		55	Swanson(1998)
Chanos chanos (Milkfish)		0	Alava(1998)

The influence of salinity on growth of different species of fish are indicated as follows: (positive +, or neutral =) tolerance (maximum salinity or range tested) , the best salinity conditions for growth, and the reference.

ACCLIMATION TO SALINITY AND ITS INFLUENCE ON GROWTH

Salinity is due to the presence of salts, dissolved in water, which represent 60 of the 92 'basic' chemical elements (Riley, 1965). Chloride Cl^- (560 mM) and sodium Na^+ (450 mM) are the most important in normal salinity sea water (SW, 35 ppt, 1050 mOsm l^{-1}). The most important aspectis that these elements are ionised in SW (19 g of Cl^- , 10.5 g of Na^+ l^{-1}). As a comparison,despitethe fact that it is extremely difficult tospecify what an 'average' composition of FW reallyis,Tchernia in 1969 tried to consider somedata and proposed 'general values' for FW (0.32ppt),35.1% of CO^{2-} , 20.4% of HCO_3^- , 12.1% ofSO_4^{2-} 11.7% of SiO_2 and 5.8% of Na^+ . Ifchlorideand sodium are predominant in SW, carbonatescarbonates and HCO_3^- are most abundant in FW. The composition of FW can vary a lot, but it is important to remember that FW is not deionized water! Many physiological experiments have maintained fish in deionised water and the fish obviously rapidly face physiological disturbance, often die and, if they survive, are unable to grow. Some species are known for their ability to acclimate to very different salinity media (Griffith, 1974;Suresh and Lin, 1992), including extreme environments (> 100 ppt). Killifish (Cyprinodontiformes) are particularly interesting because of their capacities to acclimate to such environments and represent nice models in biology. An extreme situation is the persistence of the Austrofundulus limnaeus population, because diapausing embryos become embedded in pond sediments (Podrabskyand Hand, 1999).

Osmosensitivity and salinity detection are of the highest physiological interest. Fish have prolactin (PRL) cells which are directly osmosensitive (Grau et al., 1994). They also possess chemoreceptors, situated in the pseudobranch (Laurent and Dunel-Erb, 1984), providing information on water salinity. These are connected to the central nervous system (CNS) and participate in triggering water drinking in SW. Teleost fish in SW drink, at least in part, to compensate for the water lost by diffusion (to balance osmotic water losses occurring due to the osmotic difference between their milieu int´erieur and the external medium), essentially through the gill. In summary, it appears that marine fish present higher developmental or growth rates at lower salinity and FW fish at higher salinity. Many species, often as juveniles, opt for intermediary salinity conditions of the brackish water (8-16 ppt) and grow optimally in estuaries and costal systems. It was indicated that the salinity tolerance during early developmental stages depends on how internal fluids perform in various extents (Holliday, 1969; Alderdice, 1988).Salinity effects have been studied extensively in marine fish embryos and larvae (Yang and Chen, 2006; Jørgensen and Hansen, 2010), but they are limited in freshwater species (Fashina-Bombata and Busari, 2003; Albert et al., 2004; Bonisławska, 2009). It has shown that some freshwater teleost eggs can be incubated and hatched in 5ppt salinity but fertility and eggs hatching rate of freshwater teleost decreases in saline water (Gbulubo and Erondu, 1998).The other report showed lower than 5ppt salinity tolerance in Zebrafish (Brachydanio rerio) (Sawant et al., 2001). Rockwell (1956) reported 70-90 % mortality in Pacific salmon, Oncorhynchus gorbuscha and O. keta 19 to 40 days after incubation period in 12ppt salinity but, salinity less than 6ppt has not very lethal effect in these species. Rubin (1994) Also showed, (For the salinity range found in the Baltic Sea), a small negative influence on egg-to-fry survival of sea trout eggs in 6ppt salinity (11% more compare to freshwater) but it caused a delay in hatching time and duration of the hatching period was increased. On the other hand some studies reported the salinity effects on hatching rate of freshwater teleost. For example Spined loach (Cobitis taenia) embryos developed successfully in the range of 0.12 to 4.80ppt salinity (Bohlen, 1999). But at 6.00ppt the hatching was strongly reduced, and development failed at or above 7.20 ppt. Froelich and Engelhardt (1996) reported that the low salinity incubation (as 2ppt) has control the fungus growth in koi carp (C. caprio) and Channel catfish (Ictalurus punctatus), also increases the hatching rate compare to control was reported but no effect on hatching time was reported in these species. Mortality occurs because of the embryo incapability to maintain osmotic pressure in normal rate in order to unbalanced ion gradient (Bunn et al., 2000). Holliday and Jones (1967) found that the egg salinity resistance was lowest in Blastula and gastrula in freshwater teleost eggs and Tylore (1971) found the most sensitivity just enclosing blastopore prior to hatch. However, there is limited information on how salinity affects developmental stages of sturgeons (Jian-Yi et al., 2006). Zotin (1965) defined five phases of water uptake by embryos of Beluga and sevruga during developmental stages. First between fertilization and gastrulation which differs as rapid uptake, the second stage continued up to yolk plug closure which ceases water uptake, third followed up to heart angle formation that water uptake was rapid. Forth related to heart pulsation beginnings and he reported some water lost in this stage and in fifth period to hatch that no water is absorbed. Khatooni et al (2013) was conduct a study investigate how salinity affects early developmental stages of the fertilized eggs and embryo of Persian sturgeon (A. persicus) during different developmental stages (Khatooni et al., 2013).

As result of Khatooni et al (2013) finding, the abnormalities were distinguished and classified in each of three developmental stages and the result showed in Fig. 1 to Fig.3. Result has shown that the effect of salinity on abnormality percent was different in each developmental stage (Table2). For example the abnormality percent increased significantly in gastrula stage only in 6 and 8ppt treatments. The abnormality rate in s- type heart formation stage was significantly higher in all treatments compare to control but it was still below 20%. The abnormality rate increased significantly in 4 and 6ppt treatments before hatch stage. Also our results showed that the salinity significantly affect abnormality percent during developmental stages of each treatment ($P< 0.05$). For example in 2ppt treatment the abnormality was increased significantly in s-type heart stage compare to gastrula stage but in 4ppt treatment the abnormality increased significantly in all three stages ($P< 0.05$). The significant decreased abnormality rate in 8ppt resulted from the high mortality of embryos in 8ppt treatment (Table2).

Table 2: The effects of different salinity treatments on abnormality percentage during three developmental stages of Persian sturgeon embryo

Treatments	0.5	2	4	6	8
Gastrula	2.56±0.6 a	4.33±1.0 a*	5.76±0.6 a	18.26±5.6 b	28.23±8.9 c
S- type heart formation	5.56±2.4 a	11.8±3.0 b	15.73±2.0 bc*	19.63±3.6 c	15.16±7.1 bc
Before hatch	8.16±3.8 a	14.66±4.5 ab	21.73±3.0 b**	30.33±5.4 c*	7.16±2.5 *

Three stages of Persian sturgeon embryonic developmental stages include in: Gastrula, S-type heart formation and before hatching (in developmental stage 35) in each treatment.* means significant difference between values of three stages in a treatment. Different superscript letters indicate significant difference between treatments in each stage ($P> 0.05$).

Hatching took place four days after fertilization (96h) and continued for three days more (7 d post-incubation) in all trials except 2ppt treatment which completed mostly in the day 6th.The hatchability of eggs incubated in 2ppt as well as control group ranged between 88.22 ±3.81% and 88.51±2.60%, respectively ($P< 0.05$). The hatching rate was recorded 84.19±3.2% in 4ppt treatment, while it was recorded 64.03%±1.8 in 6ppt treatments. Hatching did not occur in 8, 10 and 12ppt (Fig. 4). No Fungus was observed in salinities more than 2ppt.

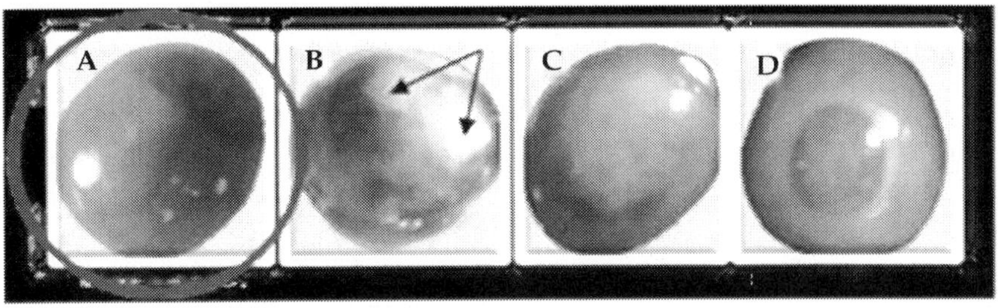

Figure 1: Different abnormalities samples observed during incubation period of gastrula stage in salinities treatment. A: normal embryo in gastrula stage. B: abnormalities in segmentation (arrows pointed the areas which the cells dose not made normal divisions or the cells has been damaged, this kind of abnormalities was mostly observed in salinities more than 4ppt salinity).C: the abnormal embryos with shortened yolk sac compare to normal stage(it was not numerous in salinities less than 6ppt). D: the developing retarded embryos compare to normal gastrula stage (this kind of abnormality was numerous in 8ppt treatment).

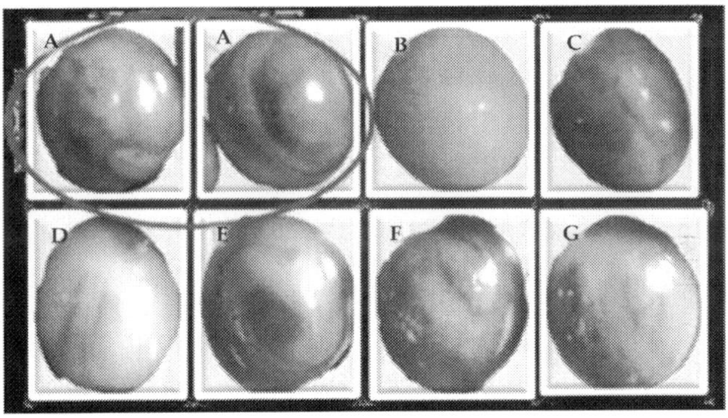

Figure 2: Different abnormalities samples observed during incubation period of s-type heart formation stage (35 embryonic developmental stages according to Dettlaff and Ginsburg, 1992). A: the normal embryo in s-type heart formation stage. B: retarded embryo. C: deficiency in notochord formation. D: deformity in head and tail formation, E: lack of head and fore part of body and deformed tail. E: deformity in head (lengthened than normal). G: deformed embryo with deficiency in back bone.

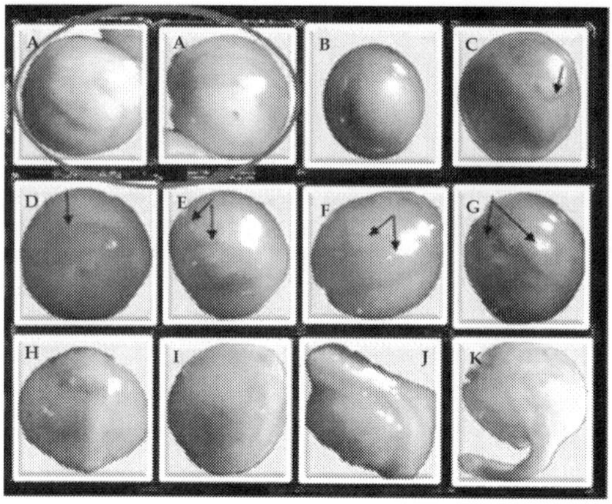

Figure 3: Different abnormalities samples observed during incubation period of before hatching stage. A: normal embryo in before hatch stage, B: Yolk sac Vol. decreasing more than normal mood in this stage, C: the arrow pointed to presence of abnormal assemblage of cells on the yolk sac (the larvae which hatched with this kind of abnormality does not have any problem in their activity but the assemblage become thinner than compare to control in salinities more than 4ppt), D: the abnormal embryo without head and the for head parts. E: the abnormal embryos which have two heads. F: the abnormal embryo with 2 spinal cords. G: the embryo with sever deformities on both head and tail. H: deformed retarded embryos which both head and tail parts are not form clearly. I: the retarded embryos which is not deformed yet in this stage. J and K: this samples are included the sever deformity in retarded embryos which was mostly seen in 8ppt salinity that no hatch was obtained.

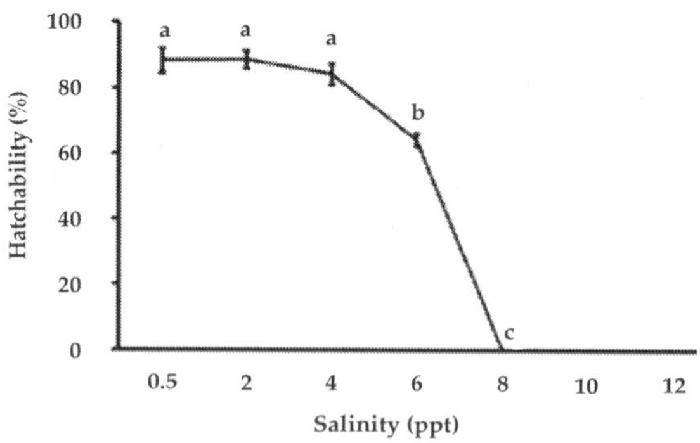

Figure 4: The egg hatchability percent incubated in different salinities. Different superscript letters indicate significant difference between treatments in each stage (P> 0.05).

Zotin (1965) reported that the water absorption will cease in gastrula stage and it will start to absorb rapidly in heart formation stage again in Beluga (Huso huso).So the increasing of abnormality and mortality rate in all brackish water treatments comparing to control in heart formation stage confirm the presence of the same mechanism in Persian sturgeon. However, Holliday and Jones (1967) found that the egg salinity resistance was lowest in blastula and gastrula in freshwater teleost eggs and Tylore (1971) found the most sensitivity just enclosing blastopore prior to hatch. Sawant, et al. (2001) also reported increasing salinity potential along with developmental stage in zebra fish in low salinity as 2ppt (from embryonic cleavage up to gastrula stage) and cytological examinations indicated that higher salinity mainly impaired the nuclear division of the embryonic cells. In Persian sturgeon it was shown that its embryos demonstrates a little adaptability during incubation period in higher salinity than freshwater(Sharif et al., 2015).It seems that there wouldn't be another ion regulatory mechanism (chloride cell) like some of tolerable teleost in early developmental stages (Kaneko et al., 2002). Decrease of hatching rate in salinities more than 6ppt and significant increase in mortality percent of Acipenser persicus embryo only in this treatment shows that the eggs could not tolerate salinities more than 6ppt during incubation stages. The same salinity potential was reported in some freshwater teleost eggs. For example Gbulubo and Erondu (1998) found the optimal salinity ranges for incubation 0.5 ppt and they found that hatching was significantly low above 5‰, no hatch was observed in 8ppt salinity in African catfish (Heterobranchus longifilis).

In summary, according to study that conducted by Mardaneh Khatooni et al., (2013), the mortality and abnormality of nonteleost species such as Persian Sturgeon embryos increased a long with increase in salinity but the embryos showed more tolerance during blastula andgastrula stages in lower salinity and larvae could nothatch in salinities more than 6 ppt. Results suggest that as the salinity increases more than 6ppt the embryonic development will be retarded or make it useless in Persian sturgeon. All in all, the salinity tolerable range during incubation period in Persian sturgeon is between 0.5- 6 ppt.

ENERGETIC COST

In previous study measurements of oxygen consumption in fish further showed a reduction in metabolic rate (2&28%) in isotonicsalinity relative to fresh water and seawater (Rao 1968; Sharif et al., 2015). These studies support the hypothesis that the energetic cost of ion regulation is lowest in an isotonic environment, where the ionic gradients between blood and water are minimal, and that this energy savings is substantial enough to increase growth. Other studies, however, fail to show isotonic salinity asthe point of maximal growth in salrnonids; growth rates were highest in fresh water in studies by Shaw et al. (1975), Clarke et al. (1981), McKay and Gjerde (1985).A theoretical estimate of the energetic cost of osmoregulationins almonids suggests that it would be very low, being less than 1 % of resting metabolism (Eddy1982). Accordingly, changes in metabolic rate of that magnitude would be very difficult to measure accurately. Studies such as that of Bullivant (196l), who reported no significant differences between the metabolic rates of yearling chinook salmon(Bncorhynchus tshatvytscha) reared in fresh water, half strength sea water, ad full-strength seawater, support those theoretical estimates.

Few studies have made simultaneous measurements on growth and metabolic rates of salmonids in relation to salinity. Here, we focus on a study thatMorgan et al. (1991) conducted. John et al. (1991) examined growth, metabolic rates, and ionic regulation in juvenile rainbow and steelhead trout and fall chinook salmon acclimated to a range of salinities, with one near isotonic. These species were chosen to investigate the possible effects of different life history patterns on growth and metabolic responses to environmental salinity. The steelhead trout is an anadromous fom of the rainbow trout, usually spending 2 or in fresh water before migrating to the sea (Scott and Crossman 1 973). Hall chinook salmon are also anadrorncsus, but spend a shorter period of time in fresh water than steelhead trout, migrating seaward within 3 months of emergence (Lister and Walker 1966). Their results showed that the mortality was less than 5% in all treatments except for chinook salmon fry reared in 28 ppt, which suffered a loss of 24 fish (Table 3). Themajorities of the mortalities were less than5 cm in fork length and weighed less than B .5 g.The mortality data indicate that the small fry stages (approximately 8.5-1 -0 g) of these species can adapt to salinities of up to 28 ppt when a gradual acclimation procedure is followed.

Growth rates declined with increasing salinity for all three species (Fig. 5 ; Table 3). Rainbow trout fry growth decreased. Growth of Chinook salmon fry was significantly higher in fresh water than all other treatments and showed a steady decline with increasing salinity to full strength seawater (fig. 6). Specific growth rate values for Chinook salmon fry in fresh water and seawater were similar to those reported by Clarke et al. (1983). Final condition factors varied among growth experiments, showing no consistent pattern between species (Table 3). Rainbow trout fry in fresh water and isotonic salinity had significantly higher final condition factors than the hypertonic treatment, but the condition factor for stelhead trout fry in fresh water wassignificantly lower than the other treatments. Differential growth rates between salinity treatments, however, resulted in size differences which masked the anticipated dehydration effect of salinity on moisture content (Table 3).

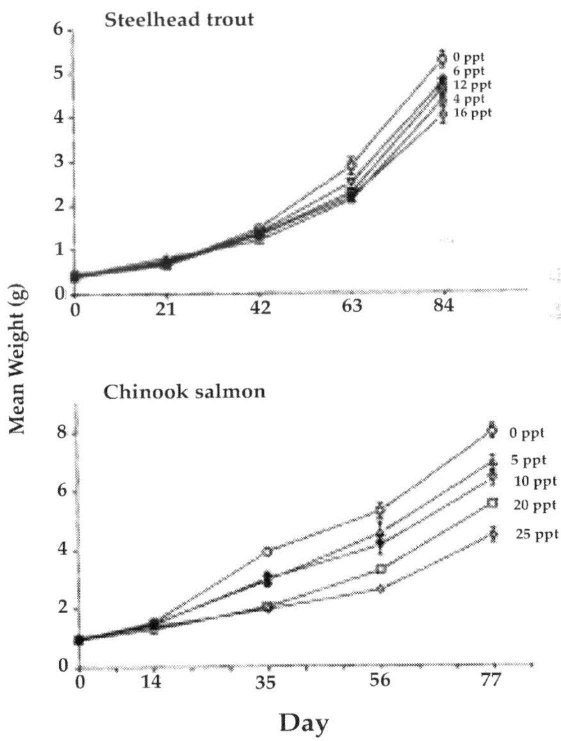

Figure 5: Growth cuwes for steelhead trout and chinook salmon fryreared in five different salinities. Absence sf standard emor bas indicatesthatthe SE was smaller than the symbol.

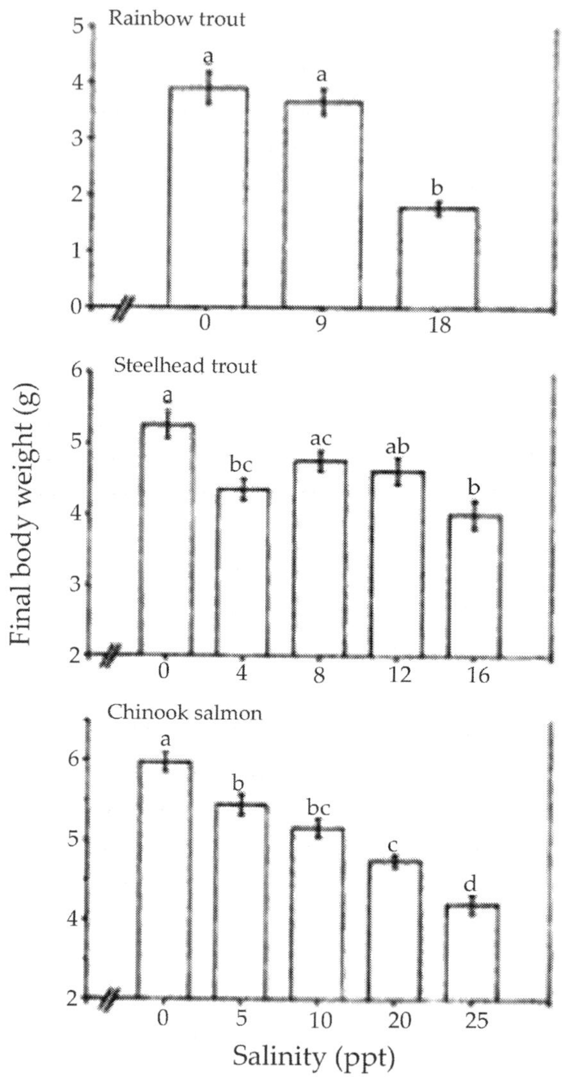

Figure 6: Effect sf salinity on growth of juvenile rainbow and steelheadtrout and chinook salmon (starting weights = 1.8, 0.4, and 1.0 g,respectively). Means (4 1 SE) with a common letter are not significantlydifferent by Tukey's test.

Table 3: Mortality specific growth rates, condition factors9 and moisture content of juvenile rainbowand steelhead tmut and chinook salmon reared invarious salinities. Mean values followed by the sameletter are not significantly different (P < 0.05) by Tukey9s test

Salinity treatment (ppt)	Percent mortality	Specific growth rate (%wt. d-1)	Final mean condition factior (±SE)	Final mean percent moisture content (±SE)
		Rainbow trout		
0	2	3.35	1.13 ±0.01 a	
9	0	3.25	1.17 ±0.01a	
18	4	1.57	1.07 ±0.01 b	
		Stellhead trout		
0	3	3.27	1.12 ±0.01 a	
4	0	2.64	1.15 ±0.01 b	
8	0	2.98	1.19 ±0.01 c	
12	0	3.11	1.17 ±0.01 bc	
16	0	2.75	1.17 ±0.01 bc	
		Stellhead trout		
0	0	2.62	1.15 ±0.01 a	78.0 ±0.02 a
5	0	2.44	1.06±0.01 a	78.1 ±0.02 ab
10	0	2.32	1.06±0.01 a	78.6 ±0.02 ab
20	0	2.30	1.05±0.01 a	79.1 ±0.02 b
28	24	1.84	1.05±0.01 b	81.6 ±0.02 c

As results, salinity had a negative effect on the growth of all three species, particularly at salinities above the isotonic level. These findings are in agreement with Shaw et al. (1975), Clarke et al. (1983)and McKay and Gjerde (1985), who also found that isotonic salinity was not the point of maximal growth in juvenile salmonids. It is possible, therefore, that confinement stress affected the study results. In the experiments conducted by Otto (1971), growth rates of coho salmon frywere actually higher in fresh water from June toSeptemberand increased in 18 ppt salinity only from October to February.This shift in salinityfor optimal growth during the fall is consistentwith other studies on the behaviaur andphysiology ofcoho salmon presmolts. As mentioned previously, activity of gill Na^+ ,K^+ -ATPase, the enzyme located in chloride cells which is indicative of ion excretion capacity, increased during mid-October incoho salmon underyearlings in fresh water. This resulted in improved growth and survival of the fish when they were transferred to seawater netpens during autumn compared with other times of the year (Haache et al. 1980). These results are all consistent with the observation that coho salmon are sometimes found in estuaries as underyearlings Tschapliwski 1987) and have therefore evolved physiological mechanisms to optimize growth in changing environments. Based on these data, we can say, even at the fry stage, amadrornous steelhead trout and chinook salmon adapt and grow better at hypertonic salinities than the freshwater freshwater resident rainbow trout.

John et al, (1976) found that oxygen consumption rates for all three species of sdmonids were not lowest in isotonic salinity; they were lowest in fresh water. The higher metabolic rates at higher water salinities apparently reflected a significant energetic cost, as growth rates declined with increasing salinity, and correlated very well with the changes in oxygen consumption.Assuming that these increased energy demands were related solely to ionic and osmoticregulation (= ionosmotic regulation), the metabolic rate data suggested that the energetic cost of ion-osmotic regulation for fry of these salmonid species was lowest in fresh water and increased to 12-1 8% in isotonic salinity.

Hematocrit values for rainbow trout fry increased with salinity and were significantly higher in 18 ppt than in fresh water. The measured values for both Na^+ and Cl^- in fresh water are in good agreement with normal values for all three species (Conte and Wagner 1965; Wagner et al. 1969). Na^+ and Cl^-, in rainbow and steelhead trout rose slightly when reared in salinities above the isotonic level, whileplasma ion concentrations in chinook salmon fry did not change across all salinities (the ionic status of mortalities in 28 ppt could last be determined). Despite these slight increases (8- 10%) in the trout, plasma ion concentrations for all three species were maintained within acceptable levels. To justify this phenomenon, some authors previously mentioned that the fish were regulating blood ions competently against concentration gradients (Table4). It is likely, therefore, that the observed growth and metabolic rates reflect steady state in fish with respect to ionic status. The increase in Ca^{2+} content of the external media at higher salinities may have assisted in regulating the passive movements of Na^+ and Cl^-, as Ca^{2+} is known to alter the permeability of gill membranes to these ions (Eddy 1975).Similar with these finding, Katuli et al., 2014 found the number of chloride cell (or mitochondria rich) along with rise of salinity, increased in Caspian Roach fingerling (Kauli et al., 2014).These cells through their ATPase activity can create more ion exchange in gill tissue and this followed by change in electrolytes values. The variable responses of hernatocrit to salinity observed in the present study have been reported previously (Bath and Eddy 1979) and may be explained by differential changes in red cell and plasma volumes which appear to be species specific.

Table 4: Chemical composition sf water samples collected from thesteelhead trout and chinook salmon fry rearing tanks (FW = freshwater;SW -- seawater)

Salinity treatment (ppt)	Na^+ (meq-E^{-1})	Cl^{-1} (meq.L^{-1}')	Ca^{2+} (meq.L^{-1}')	pH
		Stellhead trout		
0 (FW)	3	<1	>1	6.0
4	71	63	1	7.1
8	146	120	1	7.3
12	189	170	1	7.4
16	213	240	1	7.5
28 (SW)	386	450	6	7.7
		Chinook scklmoat		
0	3	<1	<1	6.0
5	81	71	1	7.1
10	163	156	3	7.3
20	273	288	5	7.5
28	931	456	7	7.6

In another study, external relation factors investigated. McCormick et al (1987).They concluded, inhibition of increases in gill Na+ ,K+-ATPase activity and salinity tolerance by exposure tocontinuous Bight did not preclude gradual adaptation of the L24 group to 3mo seawater. Although survival and capacity to maintain plasmaabsrnolxity after 5 mo in seawater was not significantly different from the other groups, the growth of the fish during this period was halved. This reduction in growth is due to salinity, since they found that poor growth does not result if L24 fish are maintained in freshwater. It seems likely that endogenous and environmental factors which affect changesin osmoregulatory physiology during the pan-smolt transformation will also affect growth in seawater(McCormick et al., 1987).

In summary, growth and metabolic rates of euryhaline fish species, such as rainbow trout, steelhead trout, and chinook salmon fry were optimum infresh water, their natural habitat at this life stage. Isotonic salinity did not provide metabolic or growth advantages in the present study, despite the hypothesis that it would provide the lowest energetic demands for ion-osmoregulation. Comparison with other studies indicated that optimal salinities for growth and metabolic rates were influenced by species, life stage, and season. Although the oxygen consumption data suggested that the energetic cost of ion-osmotic regulation increased with salinity, attempts to quantify this cost were probably affected by other metabolic processes which respond to changes in salinity.

As one contributing factor in salinity adaptation, the role of growth hormone (GH) and insulin-like growth factor 1 (IGF- 1) has been frequently emphasized in teleost fish such as S. trutta and rainbow trout, Oncorhynchus mykiss (Madsen and Bern 1992; Shimizu et al. 2007), and several non-salmonid species such as tilapia, Oreochromis mossambicus (Sakamoto et al. 1997; Riley et al. 2003), naked carp, Gymnocypris przewalskii (Cao et al. 2009), and zebrafish, Danio rerio (Almeida et al. 2013). However, little information is available on the specific functional roles and regulation of each GH hormone (i.e., GH1 and GH2) during growth and salinity adaptation. GH and its modulator, IGF-1, are the main components of the somatotropic axis, which is involved in a wide range of physiological mechanisms in mammals (Le Roith et al. 2001) and fish (Rousseau and Dufour 2007; Beckman 2011).Sharif et al., (2015) showed that the fish reared in brackish water exhibited a significantly higher SGR than did their counterparts in freshwater (Fig. 7). Also they found qPCR analysis revealed that the messenger RNA (mRNA) abundance of GH from the fish reared in brackish water exhibited an increasing trend over time and was significantly higher ($P< 0.05$) than that of their counterparts in freshwater after 5 d (Fig. 8A). The relative transcript abundance of IGF-1 revealed a trend comparable with that of GH, with a significant increase observed at 5 and 28 d in the fish kept in brackish water (Fig. 8B).In anadromous fish, the activation of GH/IGF-1 during seawateracclimation has frequently been reported. GH and IGF-1 are mitotic factors that trigger cell division and cell growth (Cao et al. 2009). It has been shown that one of the active sites for the action of GH/IGF-1 is the gill. Although the effects of brackish water on GH and IGF-1 gene expression and/orsomatic growth remain unknown, salmonid fish exhibit wide variation in the growth response to a salinity challenge depending on the stage of the life cycle (Shrimpton et al. 2005; Bystriansk yet al. 2006). For example, transferring young salmon before smoltification to sea water produces severe somatic growth retardation and abnormal development, most likely because of down regulation of hepatic GH receptor and low levels of plasma IGF-1, whereas post-smolt salmon respond positively to the salinity challenge and show a high growth rate and large body size (Duan 1997; Handel and and Stefansson2002). In their study, a significant increase in SGR was observed in brackish water. This phenotypic trait was supported by a significant increase in the GH and IGF-1 transcript levels. It is unclear whether the increase in IGF-1 mRNA levels is a result of increases in GH or represents a direct effect of salinity on IGF-1 gene expression (Sharif et al., 2015).

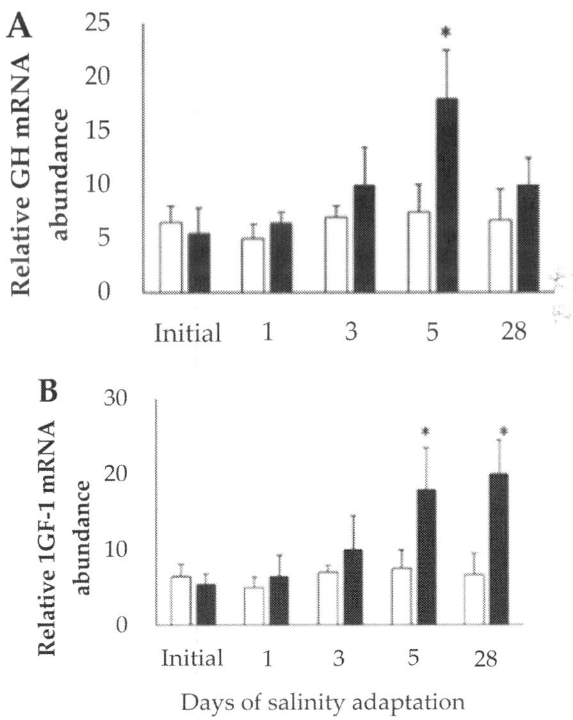

Figure 7: A-B. Transcript abundance of growth hormone (A)and insulin-like growth factor 1 (B) of Caspian trout, Salmo trutta caspius, after 1, 3, 5and 28 d of adaptationin brackish water. Open bars represent freshwater trials, and closed bars represent brackish water trials. Dataare shown

asthe mean±SEM. The asterisks indicate significant differences between freshwater and brackish water trials.

In conclusion, GH and IGF-1 genes are highly conserved in teleost fish. Brackish water adaptation of juvenile teleost fish such as Caspian trout in the early stage of smoltificationis associated with the stimulation of somatic growth, and it is possible that this effectis mediated through the GH/IGF-1 axis. Therefore, it appears that the early stage of smoltificationis the most suitable stage for introducing juvenile Caspian trout to Caspian Sea water. This result is in agreement with the findings of a recent study (Toorchi et al. 2012). Note also that the mineral composition and osmolarity range of brackish water relative to growth and smoltification stages need to be examined in future research to ensure the optimum conditions for the culture of other teleost species.

REFERENCES

1. Alava, V.R., 1998. Effect of salinity, dietary lipid source and level on growth of milkfish ŽChanos chanos. fry. Aquaculture 167, 229236.
2. Albert, A., Vetema, M., Saat, T., 2004. Effects of salinity on the development of Peipsi whitefish Coregonus lavaretus maraenoides Poljakow embryos. Annales Zoologici Fennici.. 41, 85-88.
3. Almeida, D. V., C. D. G. Martins, M. D. Figueiredo, C. F. C. Lanes, A. Bianchini, and L. F. Marins. 2013. Growth hormone transgenesis affects osmoregulation and energymetabolism in zebrafish, Danio rerio. Transgenic Research 22:75–88.
4. Aristizabal Abud, E.O., 1992. Effects of salinity and weight on routine metabolism in the juvenile croaker, Micropogonias furnieri ŽDesmarest 1823.. J. Fish Biol. 40, 471472.
5. BATHR, . N., AND h;. B. EDDY. 1979. %a18a nd water balance in rainbow trout QSalmo gairdneri) rapidly transferred from fresh water to seawater. J. Exp. Biol. 83: 193-202.
6. Beckman, B. R. 2011. Perspectives on concordant and discordant relations between insulin-like growth factor 1 (IGF-1) and growth in fishes. General and Comparative Endocrinology 170:233–252.
7. Boeuf, G., Boujard, D., Person-le Ruyet, J., 1999. Control of the somatic growth in turbot. J. Fish Biol. 55 ŽSuppl. A., 128_147.

8. Boeuf, G., & Payan, P. (2001). How should salinity influence fish growth?. Comparative Biochemistry and Physiology Part C: Toxicology & Pharmacology, 130(4), 411-423.
9. Bohlen, J., 1999. Influence of salinity on early development in the spined loach. Journal of Fish Biology. 55(1), 189–198.
10. Bonisławska, M., 2009. Effects of salinity on the duration and course of embryogenesis in sea trout (Salmo trutta L). Electronic Journal of Polish Agricultura .12(4), 7.
11. Brett, J.R., 1979. Environmental factors and growth. In:Hoar, W.S., Randall, D.J., Brett, J.R. ŽEds.., Fish Physiology, 8. Academic Press, New York, pp. 599675.
12. Buckel, J.A., Steinberg, N.D., Conover, D.O., 1995. Effects of temperature, salinity, and fish size on growth and consumption of juvenile bluefish. J. Fish Biol. 47, 696706.
13. Bunn, N.A., Fox, G.J., Webb, T., 2000. A literature review of studies on fish egg mortality: implications for the estimation of spawning stock biomass by the annual egg production method. Science Series Technical Report. Centre for Environment, Fisheries and Aquaculture Science: Lowestoft. 111, 37pp.
14. Bystriansky, J. S., J. G. Richards, P. M. Schulte, and J. S. Ballantyne. 2006. Reciprocal expression of gill Na+/K+–ATPase alpha-subunit isoforms alpha 1a and alpha 1b during seawater acclimation of three salmonid fishes that vary in their salinity tolerance. The Journal of Experimental Biology 209:1848–1858.
15. Cao, Y. B., X. Q. Chen, S.Wang, X. C. Chen, Y. X.Wang, J. P. Chang, and J. Z. Du. 2009. Growth hormone and insulin-like growth factor of naked carp, Gymnocypris przewalskii, in Lake Qinghai: expression in different water environments. General and Comparative Endocrinology 161:400–406.
16. Chervinsky, J., Yashouv, A., 1971. Preliminary experiments on the growth of Tilapia aurea, Steindachner ŽPisces, Cichlidae. in seawater ponds. Badmigeh 23, 125129.
17. CLARKEW, . C. 1982. Evaluation of the seawater challenge test as an index of marine survival. Aquaculture 28: 177-1 83.
18. CONTEF,. P., AND H. H. WAGNER19 65. Development of osmotic and ionic regulation in juvenile steelked trout (Saimo gairdneri).

Comp. Biochem. Physiol. 14: 603420.

19. Duan, C. M. 1997. The insulin-like growth factor system and its biological actions in fish. American Zoologist 37:491–503.
20. Fashina-Bombata, H.A., Busari, A.N., 2003. Influence of salinity on the developmental stage of African catfish Heterobranchus longifilis. Aquaculture. 224, 213-222.
21. Froelich, S.L., Engelhardt, T., 1996. Comparative effects of formalin and salt treatments on hatch rate of koi carp eggs. Progressive Fish-Cul- turist, 58, 209-211.
22. Gbulubo A.J., Erondu, E.S., 1998. Salinity influence on the early stages of the African catfish. Aquaculture International. 6, 369–379.
23. Gbulubo A.J., Erondu, E.S., 1998. Salinity influence on the early stages of the African catfish. Aquaculture International. 6, 369–379.
24. Grau, E.G., Richman, N.H.III, Borski, R.J., 1994. Osmoreception and a simple endocrine reflex of the prolactin cell of the tilapia Oreochromis mossambicus, Perspect. Comp. Endocrinol., 251256.
25. Griffith, R.W., 1974. Environment and salinity tolerance in the Genus Fundulus, Copeia, 319331.
26. Handeland, S. O. and S. O. Stefansson. 2002. Effects of salinity acclimation on pre-smolt growth, smolting and post-smolt performance in off-season Atlantic salmon smolts, Salmo salar L. Aquaculture 209:125–137.
27. Heyward, L.D., Smith, T.I.J., Jenkins, W.E., 1995. Survival and growth of white bass Moronechrysops reared at different salinities. J. World Aquacult. Soc. 26, 475479.
28. Holliday, F.G.T., 1969. Effects of salinity on the eggs and larvae of teleosts. In Fish Physiology (Hoar,W.S., Randall, D.S. eds) vol. 1, Academic Press, New York, p, 293–311.
29. Holliday, F.G.T., Jones, P.M., 1967. Some effects of salinity on the eggs and larvae of the plaice (Pleuronectes platessa). Journal of the Marine Biological Association of the United Kingdom., 47, 39- 48.
30. Jian-Yi, L., Qi-Wei, W., Xi-Hua, C., De-Guo, Y., 2006. The effects of physicochemical property of water on the oxygen consumption rate of Embryo and Larva of Chinese Sturgeon (Acipenser sinensis). Journal of Applied Ichthyology. 22 (Supplements1), 244–247.

31. Jørgensen, T.A., Hansen, B.W., 2010. High salinity tolerance in eggs and fry of a brackish Esox lucius population. Fisheries Management and Ecology. 17, 554–560.

32. Kaneko, T., Shiraishi, K., Katoh, F., Hasegawa, S., Hirio, J., 2002. Chloride cells during early life stages of fish and their functional differentiation. Fisheries Science. 68, 1-9.

33. Katuli, K. K., Amiri, B. M., Massarsky, A., & Yelghi, S. (2014). Impact of a short-term diazinon exposure on the osmoregulation potentiality of Caspian roach (Rutilus rutilus) fingerlings. Chemosphere, 108, 396- 404.

34. Khatooni, M. M., Amiri, B. M., Mirvaghefi, A., & Asgari, R. (2013). Effect of different levels of salinity on the viability of Persian sturgeon, Acipenser persicus, embryos during different developmental stages. Scientific Journal of Animal Science, 2(5), 116-123.

35. Laurent, P., Dunel-Erb, S., 1984. The peudobranch: morphology and function. In: Hoar, W.S. ŽEd.., Fish Physiology, 10B. Academic Press, New York, pp. 285320.

36. Likongwe, J.S., Stecko, T.D., Stauffer, J.R., Careline Jr., R.F., 1996. Combined effects of water temperature and salinity on growth and feed utilization of juvenile Nile tilapia Oreochromis niloticus ŽLinneaus.. Aquaculture 146, 3746.

37. LISTERD, . B. , AND C. E. WALKER1.9 66. The effect of flow control on freshwater survivd of chum, coho and chinook salmon in the Big Qualicum River. Can. Fish Cult. 37: 3-25.

38. Madsen, S.S., Bern, H.A., 1992. Antagonism of prolactin and growth hormone: impact on seawater adaptation in two salmonids, Salmo trutta and Oncorhynchus mykiss. Zool. Sci. 9, 775784.

39. MCKAYL, ., AND B. GJERDE1.9 85. The effect s f salinity on growth of rainbow trout. Aquaculture 49: 325-33 1.

40. Morgan, J. D., & Iwama, G. K. (1991). Effects of salinity on growth, metabolism, and ion regulation in juvenile rainbow and steelhead trout (Oncorhynchus mykiss) and fall chinook salmon (Oncorhynchus tshawytscha). Canadian Journal of Fisheries and Aquatic Sciences, 48(11), 2083-2094.

41. One, R. G. 1971. Effects of salinity on the survival and growth of pre-smolt coho salmon (Oncorhynchus kisutch). J. Fish. Res. Board Can. 28: 343-349.
42. Podrabsky, J.E., Hand, S.C., 1999. The bioenergetics of embryonic diapause in an annual killifish, Austrofundulus limnaeus. J. Exp. Biol. 202, 25672580.
43. Rao, G., 1968. Oxygen consumption of rainbow trout ŽSalmo gairdneri. in relation to activity and salinity. Can. J. Zool. 46, 781786.
44. Riley, J.P., 1965. Analytical chemistry of sea water. In: Riley, J.P., Skirrow, G. ŽEds.., Chemical Oceanography. Academic Press, London, pp. 295424.
45. Rockwell, J., 1956. Some effects of sea water and temperature on the embryos of Pacific salmon, Oncorhynchus gorbuscha and O. keta. In: Two Devices to Assess Incubation Survival and Emergence of Salmonid Fry in an Estuary Streambed (Scrivener, J.C. Ed), North American. Journal of Fish. Manage.8, 248- 258.
46. Rousseau, K. and S. Dufour. 2007. Comparative aspects of GH and metabolic regulation in lower vertebrates. Neuroendocrinology 86:165–174.
47. Rubin, J.F., 1994. Survival and development of sea trout, Salmo trutta (L.), eggs in Baltic Sea saltwater. Fisheries Research. 20, 1-12.
48. Sakamoto, T. and T. Hirano. 1993. Expression of insulin-like growth factor-1 gene in osmoregulatory organs during seawater adaptation of the salmonid fish - possible mode of osmoregulatory action of growthhormone. Proceedings of the National Academy of Sciences of the United States of America 90:1912– 1916.
49. Sawant, S. M., Zhang, S., Li, L., 2001. Effect of salinity on development of zebrafish, Brachydanio rerio. Current Science. 81, 1347-1350.
50. Sawant, S. M., Zhang, S., Li, L., 2001. Effect of salinity on development of zebrafish, Brachydanio rerio. Current Science. 81, 1347-1350.
51. SCOTT,W .B ., AND E. J. CROSSMAN19. 73. Freshwater fishes of Canada. Bull. Fish. Res. Board Can. 184: 966 p.

52. SHAWW, . M. , W. L. SAUNDERSA,N D HI. C. HALL.1 975. Environmental salinity: its failure to influence growth of Atlantic salmon (Sdrno salar) pm. 9. Pish. Res. Board Can. 32: 1821-1824.

53. Sharif, S., Shoae, A., Mojazi Amiri, B., & Farahmand, H. (2015). The Effects of Brackish Water on Growth Hormone/Insulin-like Growth Factor-1 Gene Expression of the Caspian Trout, Salmo trutta caspius (Kessler, 1877), During the Early Stage of Smoltification. Journal of the World Aquaculture Society, 46(2), 201-209.

54. Shrimpton, J. M., D. A. Patterson, J. G. Richards, S. J. Cooke, P. M. Schulte, S. G. Hinch, and A. P. Farrell. 2005. Lonoregulatory changes in different populations of maturing sockeye salmon, Oncorhynchus nerka, during ocean and river migration. The Journal of Experimental Biology 208:4069–4078.

55. Suresh, A.V., Lin, C.K., 1992. Tilapia culture in saline waters: a review. Aquaculture 106, 201226.

56. Suresh, A.V., Lin, C.K., 1992. Tilapia culture in saline waters: a review. Aquaculture 106, 201226.

57. Swanson, C., 1998. Interactive effects of salinity on metabolic rate, activity, growth and osmoregulation in the euryhaline milkfish ŽChanos chanos.. J. Exp. Biol. 201, 33553366.

58. Toorchi, M., A. Bani, and H. Alizadehsabet. 2012. Effects of salinity on osmoregulation and plasma cortisol levels of juvenile Caspian trout, Salmo trutta caspius, (Kessler, 1877). Journal of Applied Ichthyology 28:130–134.

59. WAGNERH, . M., F. P. CQMTEA, ND J. L. FESSLER1. 959. Development of osmotic. And ionic regulation in two races of chinook salmon (Oncorhynchus tskawyrscha). Comp. Biochern. Physiol. 29: 325-34 1.

60. Yang, Z., and Y. Chen., 2006. Salinity tolerance of embryos of obscure puffer Takifugu obscures. Aquaculture. 253, 393–397.

61. Zotin, A.I., 1965. The uptake and movement of water in embryos. Symposia Society for Experimental Biology. 19, 365-384.

CHAPTER 11

REPRODUCTIVE ENDOCRINOLOGY OF FEMALE CRUSTACEANS: PERSPECTIVE AND PROSPECTIVE

CH. Swetha [1], S.B. Sainath [1], P. Ramachandra Reddy [2] and P. Sreenivasula Reddy[3]

[1]Department of Biotechnology, Sri Venkateswara University, Tirupati-517 502, AP, India

[2]Department of Biochemistry, Yogi Vemana University, Kadapa-516 003, AP, India

[3]Department of Zoology, Sri Venkateswara University, Tirupati-517 502, AP, India

ABSTRACT

Reproduction in crustaceans is a highly complex process that requires precise coordination of external and internal factors to be successful. The field of crustacean reproductive endocrinology has moved from the classical approach of endocrinological techniques such as extirpation and additive methods, to the modern era of advanced biochemical, immunological, molecular biology and recombinant DNA technology methods. During last two decades, extensive endeavor has been captivating on the crustacean endocrines regulating reproduction, to authenticate their roles in the regulation of reproduction as well as their biochemical and molecular mechanism of action. In the present chapter, we have tried to recapitulate recent developments in molecular advances taken place in the reproduction regulation of crustacea via hormones, opioids, neurotransmitters and other molecules. In addition to providing a review of the scientific literature, we have also included our perspectives.

Keywords: Reproduction; Eyestalk neuropeptides; Neurotransmitters; Methyl farnesoate; Ecdysteroids

INTRODUCTION

Aquaculture is a rapidly growing global industry, comprising cultivation of various freshwater and marine species of finfish, shellfish (crustaceans), molluscs, and ornamental fish. Cultivation of crustacean species play very important role in the world aquaculture industry. It has the capacity to produce large amount of proteinaceous food, to meet the food demand for ever growing human population. Marine shrimps, prawns, lobsters and crabs are valuable sources of food, and thus of considerable economic importance to fishing industries throughout the world. Foods produced in crustacean aquaculture are often more expensive than other sources of animal protein. Perhaps, the most highly priced decapods are the true lobsters (*Homarus americanus*), regarded as the delicacies both in North America and in Europe. Developing of culture with inferior quality seed (susceptible to bacterial and viral infections) in the pond causes major damage to the aquaculture industry. Limited availability of quality and healthy seed constitutes the largest ultimate cause of destruction of aquaculture industry in recent past. In many respects, aquaculture is still in its infancy in the coastal region, and there is a pressing need to formulate strategies aimed at improving the income and assuring the availability of affordable protein to coastal communities. To accomplish this, it is necessary to improve the costal aquaculture by producing high quality seed from commercially important crustaceans. Fortunately due to a resurgent interest in invertebrate models for comparative studies and their commercial applications in the aquaculture industry, crustacean reproductive physiology is drawing renewed attention.

In the hatcheries, classical eyestalk ablation technique is employed to induce the reproduction in brood stock crustaceans. This method leads to production of inferior quality seed and often causes death in brood stock. To overcome this problem several research groups are working in many directions to induce successful reproduction especially on vitellogenin regulation in crustaceans throughout the world. One of the best ways to induce successful reproduction in crustaceans without damaging the brood stock and with high quality seed production is manipulation of endocrines at molecular level.

The regulation of reproduction in crustaceans is highly diverse [1] and is known to be controlled primarily by eyestalk neuropeptides, biogenic amines, opioids, gonad stimulating hormone, methyl farnesoate, ecdysteroids and vertebrate type steroids. Considerable extension has taken place in the crustacean reproduction with respect to the molecular approaches of endocrine hormones and other factors. In order to provide an overview of the state-of-the-art with regard to crustacean reproductive physiology, this chapter will consist of a brief description of the female reproductive system and its function, and the major advances in our understanding of regulation of reproduction. This chapter is not intended to be comprehensive. The reader is directed to several comprehensive reviews that discuss some aspects of the material covered in this review [2-7].

Description of Female Reproductive System

Most crustaceans have separate sexes, and these can be easily distinguished by observing the appendages on the abdomen called pleopods in shrimps or by observing the brood sac in crabs. The reproductive system of female crustaceans consists of the paired ovaries, oviducts, gonophores and an external sperm reception area. The decapod ovary is a paired and elongated tubular structure, whose transverse sections are identical from one location to other. The anteriolateral extensions lie dorsal to the hepatopancreas and are connected by a commissure located on the posterior edge of the stomach. A seminal receptacle is attached to each of the oviduct leading from the two branches at the anterior end of the pericardium and terminating in a medioventral gonopore.

Identification of Reproductive Stages

The maturation of the ovary can be grossly observed as it undergoes a series of changes in colour and increases in size during development. Different methods were employed by various workers in the identification of reproductive stages. In many studies morphological changes (colour) and gonad index were used as a marker for gonad maturation. In the freshwater crab, *Oziotelphusa senex senex,* the immature ovaries and adult previtellogenic ovaries are small, translucent to opaque white. With the onset of primary vitellogenesis, the ovary becomes a pale yellow (vitellogenic stage I). During secondary vitellogenesis, the ovary becomes orange (vitellogenic stage II) and it then becomes brown to dark brown prior to spawning (vitellogenic stage III). Maturation of the ovary in the crab *Oziotelphusa senex senex* also includes an increase in the size of the ovary as the oocytes proliferate and increase in diameter, due to yolk deposition.

In addition to the above, the histological examination of gonad at different stages of reproduction was used as an index for reproduction in crustaceans. Histological examination of ovary helps not only to determine the size of oocytes (oocyte diameter), but also to examine the yolk (vitellogenin) deposition in oocytes. Based on the cytological studies, vitellogenesis or yolk synthesis has been divided into two phases, primary and secondary. The onset of primary or endogenous vitellogenesis begins with the accumulation of proteinaceous and glycoproteinaceous granules on the rough endoplasmic rerticulum, where they are used to form yolk globules. Recently, determination of hemolymph and ovarian vitellogenin levels were used as an index for ovarian maturation [8].

Regulation of Reproduction

The regulation of reproduction in crustaceans is highly diverse, and mediated by both external and internal factors.

External Factors

Reproduction in crustaceans is cyclical in relation to seasonal changes. Environmental factors, such as temperature, photoperiod, food availability and salinity, all influence reproductive activity [9]. Photoperiod regulates the timing of ovarian growth. In crayfish *Orconectes virilis*, increased light hours accelerated the ovarian cycle of maturation, whereas in freshwater crab, *Oziotelphusa senex senex*, decreased light hours stimulated ovarian growth. The effects of temperature on ovarian growth and egg hatching have been observed in a number of crustaceans. A warmer water temperature induces egg laying in *Orconectes*[10] and increases the frequency of egg hatching in *Artemia salina* [11]. Li C et al., [12] also observed an increase in total fecundity and reproduction frequency in *Pseudodiaptomus dubia* as the water temperature increases, with the maximum at 30°C. Food abundance is essential for normal oocyte development, providing materials to synthesize yolk proteins and lipids. Starvation induced oocyte resorption and stimulated oocyte lipid conversion to polysaccharides, which are then released into the hemolymph.

Endogenous Factors

Gonad maturation in female crustaceans is regulated by both neuroendocrine and non-neuroendocrine secretions. A number of hormones from neuroendocrine organs play an important role in controlling gonad maturation [13-17]. Gonad maturation in crustaceans appears to be principally regulated by two antagonistic neuropeptides: GIH (also called vitellogenesis inhibiting hormone, VIH, in females) synthesized and secreted from the X-organ–sinus gland (XO-SG) complex of the eyestalk and gonad stimulating hormone (GSH), thought to be produced by the brain and thoracic ganglion. Besides these, methyl farnesoate and ecdysteroids secreted by mandibular organs and Y-organs respectively, are also involved in the regulation of female reproduction.

Role of eyestalks in regulation of reproduction: Since last century, the classical eyestalk extirpation experiments are in practice and revealed the involvement of eyestalk neuropeptides in the regulation of reproduction in several crustaceans. Panouse [18] was the first to demonstrate the existence of ovarian maturation inhibitory factor in the eyestalks of *Leander* *serratus*. This was further confirmed in *Orconectes virulis* [19] and *Uca pugilator* [20]. Accelerated ovarian development after eyestalk ablation was reported in the crayfish *Procambarus clarkii* [21]. Eyestalk ablation (unilateral and bilateral) induced precocious ovarian maturation in the Indian white shrimp *Penaeus indicus* [22]. Okumura and Aida [23] have studied the effect of bilateral eyestalk ablation on ovarian development in the freshwater prawn *Macrobrachium rosenbergii* and observed the shortened reproductive cycles in the destalked female prawns. In the kuruma prawn *Marsupenaeus japonicus*, Okumura and Sakiyama [24] have demonstrated the induced ovarian development (measured gonado somatic index as a reproductive end point) after eyestalk ablation in non-reproductive periods.

In several crustaceans, eyestalk ablation also resulted in premature yolk deposition in the ovary, both during the non-breeding and breeding seasons [25,26]. In the crabs *Paratelphusa hydrodromous* [25] and *Scylla serrata* [26] induction of pre-pubertal stages was observed after eyestalk ablation. Bilateral eyestalk ablation caused an increase in gonadosomatic index in immature prawns (*M. japonicus*) and also increased vitellogenin m-RNA levels in the ovary [27]. In the immature female kuruma prawn *M. japonicus* bilateral eyestalk ablation induced ovarian development and increased ovarian weight, hemolymph vitellogenin levels and vitellogenin mRNA levels in the ovary significantly [28]. The above studies clearly indicate that eyestalk ablation accelerates the process of ovarian maturation in crustaceans.

The medulla terminalis ganglionic X-organ-sinus gland complex is the major endocrine control center, analogous to vertebrate hypothalamo-neuro-hypophyseal complex, and located bilaterally in the eyestalks of most stalk-eyed crustaceans [3]. The X-organ is made with a cluster of neurosecretory somata. The axons of this X-organ cells form a major nerve in each eyestalk that converges upon a lateral hemolymph sinus where their dilated axon terminals form a discrete storage and releasing site i.e., sinus gland [29]. In crustaceans, hormones produced by the X-organ are stored in the sinus gland, to be secreted under the appropriate stimuli. The major peptide hormones secreted by the sinus gland are the following: MIH (molt inhibiting hormone), VIH (vitellogenesis inhibiting hormone), MOIH (mandibular organ inhibiting hormone), and CHH (crustacean hyperglycemic hormone); collectively called CHH-family peptides. MIH, VIH and MOIH target other endocrine glands, while CHH has somatic tissues as target. All these peptide hormones possess over-lapping biological activities [30- 33]. In addition to CHH-family peptides, the chromatophore regulating peptides viz. pigment dispersing hormone (PDH) and red pigment concentrating hormone (RPCH), and the neurotransmitters like enkephalins, serotonin, melatonin and dopamine were also produced in and released from the crustacean eyestalks. The involvement of eyestalk peptides and other molecules secreted by eyestalk is summarized below:

Vitellogenesis/Gonad-inhibiting hormone (VIH/GIH): The principal reproduction inhibitory principle present in the eyestalk is named as gonad-inhibiting hormone or vitellogenisis-inhibiting hormone (GIH/VIH). GIH was first demonstrated in the female prawn Palaemon (Leander) serratus, where removal of the eyestalk during sexual inactivity led to rapid increase in ovarian size and precocious egg deposition, apparently because of the removal of GIH [18]. Since then, several workers demonstrated the VIH-induced inhibition of ovarian growth in many crustaceans [34-40].

The VIH peptide has been characterized both structurally and functionally in limited species of crustaceans. In *H. americanus*Soyez *et al.*, [41] elucidated the primary structure of two VIH isoforms. The VIH isolated from the eyestalks of *Procambarus bouvieri* is characterized by incubating the peptide with ovarian fragments *in vitro* and observed the reduced levels of vitellogenin transcripts in the ovarian fragments [42]. The gene for VIH has been cloned from several crustacean species including *Homarus gammarus* [36], the American lobster *H. americanus* [43], the Norway lobster Nephrops norvegicus [35], the giant river prawn *Macrobrachium rosenbergii* [44], the shrimp *Rimicaris kairei* [45] and the woodlouse *Armadillidium vulgare* [46]. Molecular analysis of VIHs isolated from females of a few crustacean species shows that they consist of signal peptide (20-31 amino acid residues) and mature peptide (77-83 amino acid residues). They also show a considerable degree of sequence similarity with MIH, including the conservation of six cysteine residues at the same virtual locations forming three intra-molecular disulfide bonds (1-5; 2-4; 3-6 fashion), which aid to maintain the VIH tertiary structure [47-49].

Ollivaux et al., [36] cloned and sequenced the cDNA of VIH precursor from X-organ RNA [36]. Edomi et al., [35] synthesized a recombinant protein encoding GIH of the Norway lobster *Nephrops norvegicus*. Using this recombinant protein, they also localized the VIH synthesizing cells in the eyestalks of the *N. norvegicus*. Ohira *et al.,* [37], expressed recombinant VIH in bacterial system, then purified and allowed to refold. The refolded VIHs activity was determined using *in vitro* incubation of r*Hoa*-VIH with *M. japonicus* ovarian fragments. The r*Hoa*-VIH having amidated c-terminus showed significant reduction in the vitellogenin transcript levels in the ovary.

With the advances taking place in technology, the crustacean molecular endocrine research is also moving forward. The RNAi, is a post-transcription gene-silencing process in which double strand RNA triggers sequence-specific suppression of its cognate m-RNA is a powerful tool to reveal the gene function. In this direction, most recently the silencing of VIH has been reported in *Penaeus monodon* [39]. In this study, they observed the decreased VIH levels along with increased expression of vitellogenesis in the ovary by the functional knockdown of *P. monodon* VIH due to the double stranded *Pem*-VIH m-RNA injections *in vivo*.

Though the VIH is known to be involved in the regulation of vitellogenesis, the physiological mechanism of action of VIH-induced inhibition requires closed examination. Among the possibilities open to investigation are the following: (i) VIH may act directly on oocytes by inhibiting uptake of Vg or synthesis of yolk protein; this might be studied by *in vitro* culture of oocytes with or without Vg and a range of concentrations of pure VIH and (ii) by determination of GIH/VIH titer in hemolymph and mRNA levels during different reproductive stages. Another possibility is VIH may inhibit the release of a gonad stimulating hormone (GSH) from thoracic ganglia (TG) or brain of the central nervous system or MF from mandibular organ or ecdysteroid from Y-organ; determination of titers of GSH, MF and ecdysteroid will throw light on site of action of VIH. Clearly, more work is needed in this area.

- ***Crustacean hyperglycemic hormone (CHH):*** The most abundant neurohormone stored in sinus glands of crustaceans is CHH which is principally involved in the regulation of glycemia [50,51]. Besides regulating carbohydrate metabolism, CHH is also involved in the regulation of onset of vitellogenesis and thereby reproduction. Van Herp [52] in his book chapter explained the involvement of CHH in the regulation of vitellogenesis.

De Kleijn et al. [43,34] have demonstrated the involvement of CHH-A and CHH-B of *H. americanus* in the regulation of onset of vitellogenesis. They found that *Hoa*CHH-A is responsible for triggering the onset of vitellogenesis and *Hoa*CHH-B is involved in the stimulation of oocyte development before spawning. Recently, Gu *et al.* [53] also demonstrated the role of CHH-A and CHH-B of *Metapenaeus ensis* in the regulation of vitellogenesis in the female shrimps and found that high levels of *Me*CHH-A are needed for the initial gonadal development (vitellogenesis stage I) and a high level of *Me*CHH-B is required for gonadal maturation during middle (vitellogenesis stage II) and late vitellogenesis (stage III). In contrast, it was also demonstrated that the CHH of *Penaeus japonicus* induced inhibition of protein and mRNA synthesis in *in vitro* incubated ovarian fragments of the marine shrimp *P. semisulcatus* [54].

It is also possible that different isomorphs of the CHH have different receptors in different tissues and exhibit different functions. In support of this, CHH receptors were identified on the Y-organ membrane preparations suggesting the role of CHH in the regulation of ecdysteroid synthesis [55-57]. Webster [58] has also demonstrated the presence of CHH receptors on the oocyte membranes of the crabs Carcinus maenas and Cancer pagurus. Zarubin et al. [57] investigated the effect of recombinant CHH on ecdysteroidogenesis in YOs of European green crab Carcinus maenas and observed 51% reduction in ecdysteroid levels. The above studies confirm that CHH isomorphs may have different functions in different tissues. However, the mechanism of action of CHH in reproduction regulation in crustaceans remains to be elucidated. In addition to CHH family peptides, several additional non-neuroendocrine products also play direct or indirect roles in regulating the reproductive process. For example, MF and ecdysteroids stimulate reproduction in crustaceans and are regulated by mandibular organ inhibiting hormone and molt inhibiting hormone respectively secreted from eyestalks [13-15].

Mandibular organ-inhibiting hormone (MOIH): It is well established that methyl farnesoate (MF), a sesquiterpenoid secreted from the mandibular organs of crustaceans is involved in the regulation of crustacean reproduction. Byard [59] and Waddy *et al.,*[60] observed an increase in mandibular organ size during ovarian maturation. The implantation of mandibular organ into juvenile crab *Libinia emarginata* caused pre-mature vitellogenesis [61]. The secretory levels of MF by mandibular organs are high during the oocyte development [62]. Nagaraju *et al.*, [64] observed the hypertrophy of mandibular organ coincides with the rise in the ovarian index during ovarian maturation. The above studies suggest the involvement of mandibular organ and its secretory product (MF) in the induction of ovarian maturation in crustaceans.

It is also well known that mandibular organ is under the negative control of eyestalk neuropeptide mandibular organ-inhibiting hormone (MOIH) [64,65]. In the lobster *H. americanus* unilateral and bilateral eyestalk ablation elevated MF levels (54.1 and 106.9 ng/ gland respectively) in MO [66]. In the same lobster sinus gland extract injected into the bilateral eyestalk ablated animals caused significant reduction in hemolymph MF levels in a dose-dependent manner [66]. The eyestalk peptide induced inhibition of MF secretion by mandibular organs was demonstrated *in vitro* in some crustacean species, viz., in the crabs *C. pagurus*[64] and *O. senex senex* [8,67-69], and in the lobster *H. americanus* [60]. The effect of MOIH on mandibular organ was evaluated in the crayfish *P. clarkii* by measuring the activity of farnesoic acid O-methyl transferase, which mediates the final step of methyl farnesoate synthesis [70]. Chaves [70] also observed 20-100 fold increase in methyl transferase activity 8-12 days following eyestalk ablation in *P. clarkii.*

Using two-step reverse phase HPLC, Liu and Laufer [71] isolated and purified three peptides with MOIH activity from the spider crab *L. emarginata*. These three peptides having 72-76 amino acid residues are exhibited similarity with each other in terms of amino acid composition. All three peptides significantly inhibited the mandibular organ secretory activity. MOIH peptide and MOIH gene were isolated and characterized from *L. emarginata* which showed significant sequence similarity with other sinus gland neuropeptides and genes [72,73]. Tang *et al.,* [74] isolated and cloned two isoforms of MOIH cDNAs called MOIH-1 and MOIH-2 (differing by just one amino acid) from sinus glands of the edible crab, *C. pagurus*. Though the information with reference to MOIH is less, it is clear from the above studies that MOIH inhibits the methyl farnesoate secretion from mandibular organ thereby inhibits reproduction. However, much focus is needed on the role of MOIH in regulating reproduction in crustaceans.

Molt-inhibiting hormone (MIH): The MIH is mainly involved in the inhibition of ecdysteroid synthesis from Y-organs. There are very few reports on MIH regulated reproduction in crustaceans. Tiu and Chan [75] reported the inhibitory action of MIH on vitellogenesis in the shrimp *M. ensis*. They incubated recombinant *Me*MIH-B with hepatopancreas and ovary explants and observed the up-regulation of vitellogenin gene in both the explants in a dose-dependent manner. The function of MIH in the regulation of vitellogenesis is predicted to be through tissue specific receptors with different kinetics and signal transduction. Recently Nilli *et al*., [76] characterized MIH specific binding proteins (50 kDa protein) in the hepatopancreas of vitellogenic female *Callinectes sapidus*. They also conducted *in vitro* experiments by incubating hepatopancreas fragments with MIH and observed increased levels of cAMP. Based on the results Nilli *et al.,* [76] suggested that MIH employ cAMP as a secondary messenger in hepatopancreas. Nilli et al., [77] also studied the dynamics of MIH during vitellogenesis in the mature female crab *Callinectus sapidus*. They found that the MIH titers were four times higher at middle vitellogenesis than at previtellogenesis and suggested that the action of MIH on molting and reproduction in mature female *C. sapidus* is antagonistic, i.e., inhibition of molting and stimulation of vitellogenesis. Since molting and reproduction are sequential in prawns and shrimps and antagonistic in crabs, studies related to dynamics and physiological effects of MIH in the regulation of molting and reproduction will be very much essential to manipulate crustacean reproduction.

Red pigment concentrating hormone (RPCH): The red pigment concentrating hormone (RPCH), a neuro-endocrine peptide synthesized in and secreted from X-organ-sinus gland complex of eyestalks in crustaceans. The RPCH not only involved in the regulation of pigmentation, also involved in the regulation of ovarian maturation [78,79]. Kulakarni *et al,* [78] have reported the oocytes that had been incubated with thoracic ganglion and RPCH had entered into the midvitellogenic stage in *in vitro*. Sarojini *et al*. [79] demonstrated the *in* *vivo* and *in vitro* effect of RPCH on ovarian maturation in *P. clarkii*. They observed that the increase in oocyte diameter and mean ovarian indices in an in vivo, and significant ovarian maturation in an in vitro, where the ovaries incubated with thoracic ganglion alone (without eyestalk extract). It is very clear from the above studies that the RPCH induces the ovarian development by releasing the gonad-stimulating hormone from thoracic ganglia.

Other eyestalk factors involved in the regulation of reproduction: In addition to eyestalk neuropeptide hormones, some other factors present in the sinus gland have been reported to be stimulating the vitellogenesis [80-86].

Neurotransmitters: The role of neurotransmitters in the regulation of crustacean reproduction is well documented by Fingerman [87]. The major neurotransmitters identified in crustaceans are serotonin, dopamine, melatonin and octopamine (Figure 1). The neurotransmitter serotonin (5-hydroxytryptamine; 5-HT) is produced and released from eyestalks is involved in the induction of reproduction in crustaceans. 5-HT stimulates the release of gonad-stimulating hormone (GSH) from the brain and thoracic ganglion, where the GSH synthesized [81,88]. Fingerman *et al.,* [88] have measured the levels of 5-HT in the eyestalks and brain tissues of the fiddler crab *U. pugilator*. They also observed, 5-HT triggers the release of RPCH from eyestalks. 5-HT induced ovarian maturation was observed in *P. clarkii* [89]. The treatment of 5-HT caused significant increase in the ovarian index in the eyestalk-ablated crab *Paratelphusa hydrodromus* when compared with the eyestalk ablated saline injected crabs [90]. Recently Vaca and Alfaro [91] and Kumlu [92] have demonstrated 5-HT-induced ovarian maturation in the penaeid shrimp *P. semisulcatus* and found that the action of serotonin on ovarian maturation and spawning is significant but less effective than eyestalk ablation. The above results suggest that 5-HT induced reproduction is mediated either by release of GSH from brain and thoracic ganglion or by stimulating the release of RPCH from eyestalk neural tissue, there by GSH in crustaceans. It remains for future study to determine which model is the correct one.

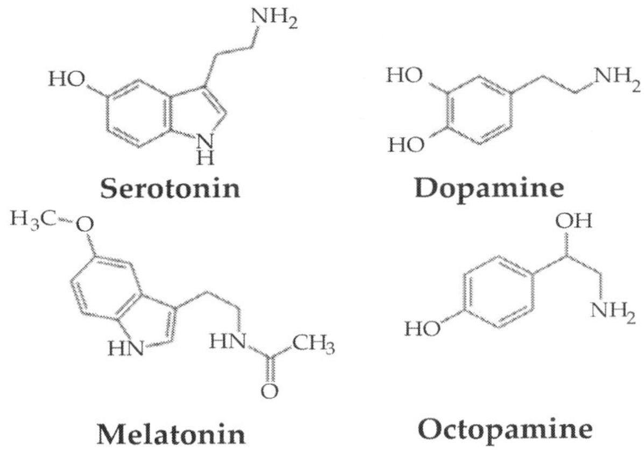

Figure 1: Structure of serotonin, dopamine, melatonin and octopamine.

Dopamine was identified in the eyestalks of several crustaceans employing the fluorescence method of Falck *et al.* [93]. Dopamine when injected into female *Procambarus clarkii inhibited* ovarian maturation. Cahansky *et al.* [94] studied the role of spiperone (dopaminergic antagonist) on ovaries of *Cherax quadricarinatus*. They also observed significant stimulation in the ovarian development in the presence of thoracic ganglia *in vitro*. The inhibitory action of dopamine could have been due to (a) Inhibition of GSH release; (b) Stimulation of release of GSH antagonist, GIH; (or)(c) Both (a) and (b).

Octopamine, another biogenic amine was identified in central nervous system of *H. americanus* using thin layer chromatography. The role of octopamine in reproduction appears to be atleast in part due to stimulation of contraction of the ovarian walls, liberating oocytes.

Melatonin (N-acetyl-5-methoxy trypatamine) is an evolutionary conserved molecule involved in the transduction of photoperiodic information from environment to organisms and popularly known as 'chemical of darkness' [95]. The presence of melatonin in invertebrates including crustaceans is well documented but its role in regulating their physiology is not clear [96]. Melatonin in crustaceans was first discovered in the visual system of crayfish, *P. clarkii* using spectrophotoflurometry [97]. The role of melatonin in the regulation of crustacean reproductive physiology is poorly understood. Recently, it was reported that injection of melatonin into female intact crabs, O. senex senex induced ovarian maturation [98] as evidenced by increase in the weight of the ovary, oocyte size and increase in ovarian vitellogenin levels over controls. From the results it is clear that melatonin is also involved in the stimulation of ovarian growth and vitellogenesis and regulation of melatonin synthesis can be used to manipulate crustacean reproduction. The relative roles and interaction of these biogenic amines in the regulation of reproduction remain to be elucidated.

Opioids: Opioids are the small peptides containing five amino acids and their significance is well known in vertebrates [82]. The search for the presence of endogenous opioids in invertebrates initially met with some uncertainty. However, using techniques like immunohistochemistry, radioimmunoassay (RIA) and high performance liquid chromatography (HPLC), opioids were discovered in several invertebrates, including crustaceans.

The potential role of opioids in the regulation of crustacean reproduction was well-documented [85]. The putative involvement of endogenous opioid system in the regulation of reproduction in the crab *U. pugilator* was investigated *in vivo* [99]. Injection of methionineenkephalin caused significant dose dependent inhibition of ovarian maturation, but the naloxone-injected crabs produced dose-dependent ovarian maturation. Sarojini *et al.* [100] incubated ovarian explants of *P. clarkii* with methionine-enkephalin and thoracic ganglia and found reduced ovarian development in a dose dependent manner as compared with explants incubated with thoracic ganglia alone. Reddy [84] reported the possible involvement of endogenous opioids in the regulation of reproduction in the prawn *P. indicus*, where injection of leucine-enkephalin caused significant increase in ovarian index and oocyte diameter in a dose-dependent manner. Increased ovarian growth has been observed in the ovarian explants of *C. quadricarinatus* incubated with naloxone, an enkephalonergic angonist [94]. It is clear from the literature that the opioid peptides are involved in the regulation of crustacean reproduction. Additional research is needed to clarify the molecular mechanism of action of opioid peptides in regulating crustacean reproduction.

Fushi tarazu-factor (FTZ-F1): The recently discovered another eyestalk factor which is responsible for the promotion of reproduction in crustaceans is fushi tarazu-factor (FTZ-F_1). FTZ-F_1 titer is more in late vitellogenic ovary as well as in early nauplius larvae of the shrimp *M. ensis* [83]. Since the levels of FTZ-F_1 is more during vitellogenic stage and larval stage it was concluded that FTZ-F_1 play a crucial role in embryogenesis and larval development of the shrimp.

Extra-Eyestalk Hormones and their Role in the Regulation of Reproduction

Gonad-stimulating hormone (GSH): The brain and thoracic ganglia have been identified as sources of a neurohormone called gonad stimulating hormone (GSH). Although GSH has not yet been isolated and purified to electrophoretic homogeneity, several *in vivo* and *in vitro* studies demonstrated the stimulating effects of the brain and thoracic ganglia extract on ovarian growth [13].

The involvement of GSH in the regulation of reproduction was demonstrated in different crustacean species. Gomez [25] implanted the brain tissues into the crab *P. hydrodromus* and observed accelerated oocyte development. The brain tissue induced ovarian growth was also observed in *Parapenaeopsis hardwickii, Macrobrachium kistnesis, Paratya compressa and Penaeus vannamei* [101-104]. Otsu [105] observed the factors that are present in thoracic ganglion are involved in the stimulation or acceleration of oocytes of sexually quiescent female crab *P. dehaani*. Brain and thoracic ganglia induced ovarian maturation were demonstrated in the red swamp crayfish *P. clarkii* in *in vitro* [89,100]. Thoracic ganglion induced reproduction is also reported in *C. quadricarinatus* [11].

GSH is a peptide regulates the onset of vitellogenesis by direct action on vitellogenin synthesizing sites [89,101]. Recently Medasani *et al.* [106] studied the role of heavy metals such as copper and cadmium on GSH thereby gonad maturation. They incubated ovarian explants along with thoracic ganglia and heavy metals, and found no change in the ovarian growth. Based on the results, they hypothesized that the copper and cadmium are having inhibitory effect on GSH release thereby on reproduction. Though there are reports stating the presence of GSH, but up to now it has not been characterized and it should be one important area of crustacean endocrinology for future research.

Methyl farnesoate (MF): Recently, the role of MF, a sesquiterpene secreted by the mandibular organs in the regulation of reproduction is well acknowledged [3,6,15,107]. MF is structurally similar to the insect JH III (Figure 2), only differing in the presence of an epoxide moiety at the terminal end. MF is negatively regulated by mandibular organ inhibiting hormone produced by the eyestalk [66,68,108-110]. The levels of MF were determined in the hemolymph of the mud crab *S. serrata* [111], American lobster *H. americanus* [62], crayfish *P. clarkii* [112] and giant fresh water prawn *M. rosenbergii* [113,114]. MF binding proteins which transport MF from synthesizing site to the target site was also characterized in *H. americanus* [115] and *L. emarginata* [116]. The highest MF levels in the individuals with an active vitellogenic stages, indicating the role of MF in the regulation of reproduction. Whether these alterations in the MF titers are due to changes in the rate of MF secretion by the MO or to some other mechanism, such as differential degradation, remains to be elucidated.

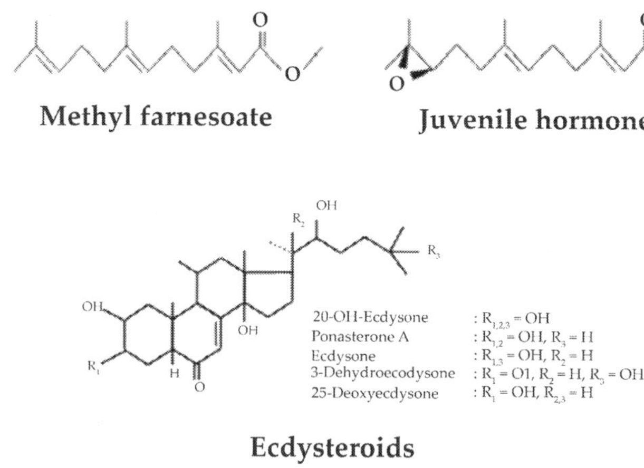

Figure 2: Structure of ecdysteroids, methyl farnesoate and insect juvenile hormone III.

The ecdysiotropic nature of MF was described in the crab *Cancer magister* [117]. Tamone and Chang [117] have demonstrated the effect of methyl farnesoate on ecdysteroid secretion from crab *C. magister* Y-organ in *in vitro*. They found that Y-organs when incubated in the presence of MF secreted significantly more ecdysteroids into the medium after 24 hours incubation and also find the magnitude of this stimulation increased with higher concentrations of MF and increasing incubation times.

There is evidence that MF may have functions in addition to acting as an ecdysiotropin. It may influence reproduction [118-121] and larval development [62]. Tsukimura and Kamemoto [122] have studied the effect of juvenile hormone III and MF on ovarian development in the shrimp *P. vannamei.* Laufer *et al.* [121] observed accelerated ovarian maturation in the red swamp crayfish *P. clarkii* after the administration of MF and concluded that exogenous MF can stimulate ovarian maturation in *P. clarkii.* Ovarian index and oocyte diameter were significantly increased in MF injected *P. indicus* and the increase of the ovary was 4 fold than those of controls [123]. They also observed that the ovaries of MF received animals, entered into the late vitellogenic stage (dark brown colour ovary). Reddy and Ramamurthi [124] observed that the ovaries of the crab, *O. senex senex* injected with MF had entered into the late vitellogenic stage (bright orange colour ovary) but the ovaries of initial control and concurrent control crabs were in the early vitellogenic stage (white colour ovary). Administration of MF, incorporated in food, resulted in the stimulation of ovarian growth in crayfish [121]. Similarly, pellets enriched with JHIII stimulated ovarian growth in crabs [125]. In vitro stimulation of ovarian growth by MF has also been reported for both penaeid shrimp [122] and crayfish [126].

In order to know more about the MF mediated reproductive regulation in crustaceans, the researchers are trying to exploit the intermediates of MF biosynthetic pathway. In the female crab *Charybdis feriatus* farnesoic acid, an intermediate of MF biosynthetic pathway induced vitellogenin gene expression [127,128]. Mak et al. [127] incubated hepatopancreas explants with farnesoic acid, MF and JH III and observed upregulation of Vg gene expression in hepatopancreas of farnesoic acid incubated explants. Whereas incubation of hepatopancreas explants with MF and JH caused insignificant change in Vg gene expression. The data suggests farnesoic acid may be the potential candidate to induce reproduction in crustaceans, which needs more focus in future.

***Ecdysteroids**:* The Y-organs are paired non-neural endocrine organs, homologous to prothoracic glands of insects are the sole source of molting hormones, secreted as a precursor, ecdysone, into the hemolymph, to be converted into the physiologically active hormone, 20-hydroxyecdysone by a 20-hydroxylase activity present in the epidermis and other organs and tissues of crustaceans [4]. The structure of 20-hydroxyecdysone (although originally called crustecdysone or ecdysterone) was elucidated as 2ß, 3ß, 14α, 20R, 22R, 25-hexahydroxy- 5ß-cholest-7-en-6-one; Figure 2). Ecdysteroids stimulates ovarian growth, especially in those crustaceans where reproduction occurs just after molting. In other crustaceans also, ecdysteroids are involved in the regulation of reproduction, but it is restricted to the first, i.e., proliferative stage of ovarian growth [129]. The primary function of ecdysteroid in crustaceans is to induce precocious molting. Recent studies indicate ecdysteroids also play an important role in the regulation of crustacean reproduction [130,131].

Charniaux-Cotton and Touir [132] observed the failure of folliculogenesis which is fundamental prerequisite for vitellogenin uptake by oocytes during secondary vitellogenesis by Y-organectomy in the shrimp *Lysmata seticaudata*. Removal of Y-organ also resulted in decreased vitellogenin synthesis and ovarian growth in the amphipod *Orchestia gammarella* [133]. A close correlation between the hemolymph ecdysteroid levels and ovarian maturation stages was observed in the prawn *Macrobrachium nipponense* [134] High concentrations of hemolymph ecdysteroids were also observed in reproducing females of terrestrial isopod *Oniscus asellus* [135]. However, situation is different in Brachyuran crab species where the entire reproductive cycle is completed in the intermolt period, when the ecdysteroid levels at very low [136]. The occurrence of different types of ecdysteroids in large amounts within the ovary is another evidence, to confirm the role of ecdysteroids in the regulation of reproduction in many crustacean species [130,131]. Despite the fact that extensive work has been carried out to establish the relationship between ecdysteroids and reproduction, the functional role of ecdysteroids in the control of reproduction is controversial and the mechanism of action on vitellogenesis is yet to be determined.

***Vertebrate-type steroids*:** Crustacean gonads have been shown to possess steroids more typically acknowledged with vertebrates and the enzymatic capacity to synthesize vertebrate sex steroids [137-139]. Progesterone, 17α-hydroxyprogesterone, estradiol (17β-estradiol), and testosterone are the vertebrate-type steroids discovered from several crustaceans (Figure 3). Progesterone and pregnenolone have been identified in the gonads and hemolymph of *Astacus leptodactylus* and *H. americanus*[140,141]. Fairs and colleagues [142] found high titers of estrogens during vitellogenic stages, suggesting a possible role in the stimulation of vitellogenesis.

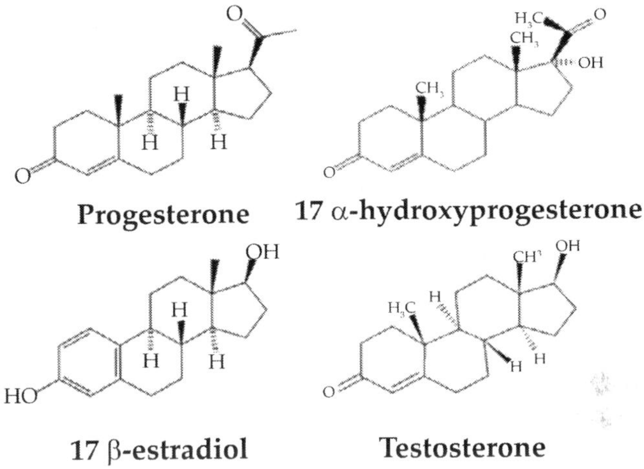

Figure 3: Structure of some vertebrate-type steroids.

A positive relationship between Vg levels in hemolymph and circulatory levels of both progesterone and 17β-estradiol has been observed in shrimps [143], prawns [144] and crabs [145]. Fluctuating levels of estradiol and progesterone in the ovary and hemolymph at different vitellogenic stages of the crab *Scylla serrata* were also reported [146]. In contrast, Okumura and Sakiyama [24] observed a negative relationship between ovarian maturation and hemolymph levels of steroids in *M. japonicus*. Injection of progesterone induced ovarian development in the shrimp *P. hardwickii* [147]. Progesterone and estradiol apparently stimulated Vg gene expression in both hepatopancreas and ovary explants of M. ensis [148]. Administration of 17α-hydroxyprogesterone stimulated ovarian growth and vitellogenesis in the kuruma prawn *M. japonicus* [149]. 17β-Estradiol stimulated vitellogenesis in ovary explants *in vitro* [144] and *in vivo* in crayfish [150]. It is also known that injection of progesterone and 17α-hydroxyprogesterone are able to induce ovarian maturation in the shrimp *M. ensis* [151] and in the crab *O. senex senex* [8] respectively. 17α-hydroxyprogesterone administered both by injection and by incorporation into the food, increased gonadosomatic index in prereproductive and post-reproductive periods of female estuarine crab *Chasmagnathus granulate* [125]. They also observed increased uptake of the amino acid L-Leucine by the ovaries incubated with 17α-hydroxyprogesterone *in vitro*.

Vertebrate-type steroids are also found in the shrimp *P. monodon* and the levels of some of these steroids (progesterone and 17β-estradiol) are high in the ovaries at the mature stage of the animal [143], suggesting the role of steroids in crustacean reproduction. In the red mud crab *S. serrata* the levels of 17β-estradiol and progesterone increased steeply in the tissues at the onset of vitellogenesis (vitellogenesis stage I) [146]. Rodriguez *et al.* [126] also reported the stimulatory role of steroids in the regulation of reproduction in the crayfish *P. clarkii.*

Some studies also showed circulating levels of progesterone and estradiol-like steroids that correlated with the vitellogenesis cycles of several crustaceans [143,145]. Both *in vivo* [125,152] and *in vitro* [122,125] stimulation of ovarian growth by 17α-hydroxyprogesterone have been reported. The precise role of steroid hormones in crustaceans is still unknown; however, stimulation of vitellogenin production within the ovary itself or in extra-ovarian sites has been suggested [144,152,153].

At present, it is not known whether crustaceans obtain the vertebrate type steroids from the environment i.e., via feed or synthesize internally. Additional studies are necessary to examine the mode and action of steroid hormones. Steroid hormones have to bind to nuclear receptors in order to generate physiological action. Progesterone and estradiol receptors have been detected in the hepatopancreas and gonad of the freshwater crayfish *Austropotamobius pallipes* using immunodiagnostic kits [154]. Though classical vertebrate steroid hormones and their receptors were identified from limited number of crustaceans, the manner by which they exert their effect on crustacean reproduction is less clear than steroid-induced responses in mammals. However, the general scheme seems remarkably similar in all organisms.

Involvement of other vertebrate hormones: Besides the presence and effects of vertebrate steroid hormones, the effects of other classical vertebrate gonadotropic hormones (follicle stimulating hormone, FSH, and luteinizing hormone, LH) were also studied in crustaceans. Zukowska-Arendarczyk [155] reported that administration of hypophysis gonadotropins induced ovarian maturation in Crangon crangon. However, to date no structurally similar peptides have been isolated from any invertebrate. Injection of human chorionic gonadotropin (hCG) stimulated maturation and spawning in shrimp and prawn [156,157]. Laufer and Landau [158] also observed ovarian maturation in *P. indicus* fed with food containing hCG. hCG administration also stimulated vitellogenin synthesis in the isopod, *Idotea balthica* [159], in the prawn *Crangon crangon* [160], and in the sand shrimp, *C. crangon* [158]. A molecule very similar to hCG has been identified in the prawn Penaeus serratus with the aid of a radioimmunoassay using hCG antibodies [161]. Further studies are required to fully understand the effects of mammalian gonadotropins on regulation of reproduction in crustaceans.

INTERACTION OF ENVIRONMENTAL AND INTERNAL FACTORS IN THE REGULATION OF REPRODUCTION

Crustacean reproduction is controlled by hormones of diversified chemical nature (peptide, steroid, terpenoid, biogenic amines, prostaglandins etc.) which aid in evaluating environmental conditions to determine the best time to reproduce. Environmental variables such as salinity, temperature, photoperiod and availability of food play vital roles in regulating crustacean physiology, including reproduction [13,16,162,163]. Lovett *et al.* [164] observed 5- to 10 fold increase in hemolymph MF level in the crab *C. maenas* when the crabs transferred from isosmotic sea water to 5 ppt seawater. Similar results were also observed in crabs when they were treated with low saline water at 11°C [165]. In contrast, exposure of another euryhaline crab, *C. sapidus*, to low salinity sea water (15 ppt) did not increase its hemolymph levels of MF [166]. Significant increase in hemolymph MF level was observed in the crab *C. maenas* after exposure to elevated temperature (32°C) and anoxia (0.25 ppm O_2). Olmstead and LeBlanc [167] observed that environmental factors can cause aberrant sex determination via perturbations in MF signaling in *D. magna*. Neuroendocrine coordination of the appropriate environmental factor is essential to ensure that seasonal reproductive events takes place in the appropriate temporal sequence. Information about the environmental and endogenous factors that control crustacean reproduction is required to enhance profitable aquaculture programs.

CONCLUSIONS AND PROSPECTIVES

The neuroendocrine and non-neuroendocrine regulation of reproduction in decapods has been the basis of study and a topic of debate for the past six decades because of interest in aquacultural activities. Various X-organ borne neurohormones are suggested to affect several physiological events, including the activity of peripheral endocrine glands, production of vitellogenin, the precursor of the main yolk protein, and gonad development. Due to multitropic nature and existence of different isomorphs of eyestalk peptides, it is also possible that other members of CHH-family (excluding VIH) may have influence on the reproduction. CHH receptors were detected on the oocyte membranes of *C. maenas* and *C. pagurus* [58]. CHH mediated reproduction was studied in *H. americanus* [34,43], *P. semisulcatus* [54] and *M. ensis* [53]. The reproductive inhibition was also observed by the eyestalk peptide MOIH in *C. pagurus* [64]. It affects the reproduction by inhibiting the synthesis and secretion of MF by mandibular organs in the crab *O. senex senex* [8]. The possible explanation for isoforms with multiple activities in CHH-family peptides is that they may be originated from single ancestral gene during evolution. This is in support with the highest % of structural similarity and amino acid homologies between CHH-family peptides isolated from intra and inter species. Furthermore, detection of specific target structures by means of identification and colocalization of hormone receptors has not been carried out. In addition to CHH-family peptides, RPCH belongs to chromatophores of eyestalk has stimulatory activity on ovary by releasing the GSH from the brain and thoracic ganglion [79]. The other eyestalk factors, like 5-HT [89], DA [168], Opioids [84,85] and FTZ-F_1 [83] also influences the reproduction in crustaceans.

GSH which possesses antagonistic activity to VIH is involved in the stimulation of reproduciton [100,104,105] and it is not yet characterized. In addition to the above hormones, the terpenoid, methyl farnesoate [118,169,170] and the steroid molting hormone, 20-hydroxyecdysone [130,131,171], are suggested to be involved in the expression of vitellogenin. More recently the farnesoic acid [127] and vertebrate-type steroids [8,125,146] have attracted the attention of crustacean reproductive endocrinologists. Though the hormones and neuromodulators involved in the regulation of vitellogenesis are well recognized, the cross-talk between them as well as overlapping and multiple functions of hormones, remain unclear.

Little is known about the role of eicosanoids in the regulation of reproduction in crustaceans, although considerable evidence has been collected during the last three decades pertaining to their synthesis and mode of action. In many cases, the evidence for an eicosanoid function is based on treatment of animals with a single compound and observation of the response. At this level of observation, it remains to be established that eicosanoids are physiologically involved. Eicosanoids are derived from fatty acids acquired through diet. It is well known that eicosanoids are produced to act at local, tissue, or cellular level in mammals. They play a very important role in the regulation of reproduction throughout animal kingdom. Martin-Creuzburg *et al.* [172] reported that population growth of *D. magna* is improved more by the addition of sterols and eicosapentaenoic acid (EPA) together than by adding either nutrient separately, which implies a synergistic effect of sterol and EPA on reproduction. Further, they also found in *D. magna* that somatic growth is primarily constrained by the availability of sterols, and reproduction is primarily constrained by the availability of EPA. Current evidence suggests that the effect of nonsteroidal antiinflammatory drug ibuprofen in *D. magna* is interruption of eicosanoid biosynthesis, thereby reduces fecundity [173], where both prostanoids and lipoxygenase products appear to be important agents in oogenesis and embryogenesis [174]. One can assume that as details of eicosanoid action in crustaceans become known they will contribute greatly to our understanding of crustacean reproduction. Our earlier work suggested that administration of PGE2 induced ovarian growth and ovarian vitellogenin levels in fresh water crab, *Oziotelphusa* [67]. Studies on decapod eicosanoids may generate information on fundamental aspects of eicosanoid biosynthesis and mechanisms of action, and may yield information of practical importance in hatchery industry. Young scientists are entering this emerging field, and these investigators will surely make truly exciting discoveries in the future.

The field of crustacean reproductive endocrinology is entering an exciting period with the introduction of modern techniques which permit detailed studies on the complexity of endocrine control of reproduction. Modern methodologies such as immunohistochemistry and molecular and biochemical methods have to be extended to crustaceans to elucidate the cellular dynamics of the neuroendocrine cell systems synthesizing VIH, MIH, MOIH and CHH. Using polyclonal/monoclonal antibodies, some studies have already been described the quantification methods for GIH [175] and CHH [176]. Based on the titers of circulatory hormones it might be possible to develop simple immunodiognostic kits which can be used in hatcheries for determining the conditions of the cultured females and even to manipulate their reproductive cycle by controlling external conditions such as temperature and photoperiod.

Several crustacean hormones exhibit multifunctional activities. Members of CHH family of neuropeptides not only present from embryos to adults and a single peptide can have multiple biological functions. Ecdysteroids may serve as morphogens during embryonic development, function as molting hormones during larval to adult life and also act as gonadotropins in adults. Methyl farnesoate also function as developmental hormone in larvae and as a gonadotropin in adults. These examples illustrate the complexity of crustacean endocrine system that several chemically diverse hormones can mediate different functions at different life stages. Understanding the cross-talk of these molecules in the internal milieu may provide rewarding insights into the regulation of crustacean reproduction. However, as crustacean endocrinologists probe deeper and deeper into the hormonal regulation of reproduction, they eventually will have to return to the whole animal to examine the ultimate stimulus to reproduction, namely, how the animal translates environmental cues viz., temperature, salinity, photoperiod, availability of food etc. into information that is utilized in the regulation of endocrine axis? Several workers acknowledged the importance of temperature, photoperiod and availability of food for breeding. This and other hypothetical and epithetical aspects of crosstalk between different internal molecules are summarized in Figure 4.

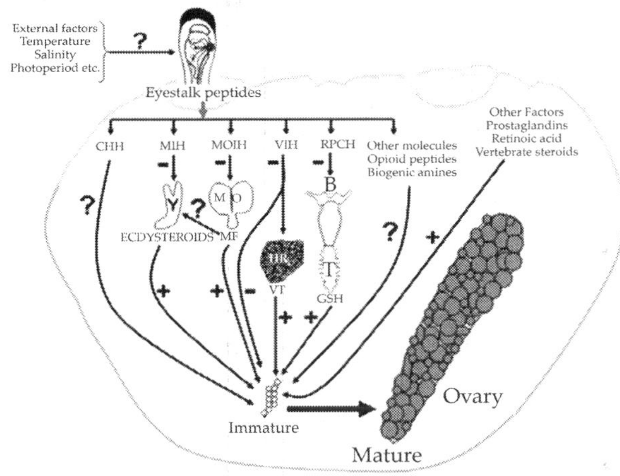

CHH: crustacean hyperglycemic hormone; MIH: molt-inhibiting hormone; VIH: vitellogenesis-inhibiting hormone; MOIH: mandibular organ-inhibiting hormone; RPCH: red-pigment concentrating hormone; GSH: gonad-stimulating hormone; MF: methyl farnesoate; B: brain; T: thoracic ganglia: Y: Y-organ; MO: mandibular organ; HP: hepatopancreas; VT: Vitellogenesis. + : stimulation; - : inhibition; ? = not established

Figure 4: Diagrammatic representation of external and internal factors regulating crustacean reproduction.

ACKNOWLEDGEMENTS

This review manuscript is dedicated to Prof. R. Ramamurthi, a committed teacher and advisor, and a brilliant scientist who introduced us to crustacean endocrinology. His suggestions helped us a lot to improve the manuscript. The work related to the crab *Oziotelphusa* and described in this paper was funded in part by grants from the Department of Science and Technology and Department of Biotechnology, New Delhi to PSR. Valuable contributions of colleagues and the students and postdoctoral researchers from PSR's laboratory are specially acknowledged.

REFERENCES

1. Chang E, Sagi A (2008) Male reproductive hormones. Reproductive Biology of Crustaceans. In: Mente E, (Ed.), Science Publishers, Enfield, NH. Pg: 299-318.
2. Adiyodi KG (1985) Reproduction and its control. In: Liss DE,

Mantel LH, (Eds.) The Biology of crustacean. Academi Press, New York. 9: 146-215.

3. Reddy PS, Ramamurthi R (1999) Recent Trends in Crustacean Endocrine Research. PINSA 65: 15-32.
4. Huberman A (2000) Shrimp endocrinology: A review. Aquaculture 191: 191-208
5. Tsukimura B (2001) Crustacean vitellogenesis: Its role in oocyte development. Amer Zool 41: 465-476.
6. Wilder MN, Subramoniam T, Aida K (2002) Yolk proteins of crustacea In: Raikhel AS, Sappington TW (Eds.) Reproductive Biology of Invertebrates. Vol XII, Part A: Progress in Vitellogenesis. Science Publishers, Inc., Plymouth, UK. Pg: 131-173.
7. Diwan A (2005) Current progress in shrimp endocrinology-A review. Indian J Exp Biol 43: 209-223.
8. Reddy PR, Reddy PS (2006) Isolation of peptide hormones with pleiotropic activities in the freshwater crab, *Oziotelphusa senex senex*. Aquaculture 259: 424-431.
9. Adiyodi KG, Adiyodi RG (1970) Endocrine control of reproduction in decapod Crustacea. Biol Rev Camb Philos Soc 45: 121-165.
10. Aiken DE (1969) Ovarian maturation and egg laying in the crayfish *Orconectes virilis*: influence of temperature and photoperiod. Can J Zool 47: 931-935.
11. Von Hentig R (1971) Influence of Salinity and Temperature on Development, Growth, Reproduction and Energy Budget of *Artemia Salina*. Marine Biology 9: 145-182.
12. Li C, Luo X, Huang X, Gu B (2009) Influences of temperature on development and survival, reproduction and growth of a calanoid copepod (*Pseudodiaptomus dubia*). ScientificWorldJournal 9: 866-879.
13. Fingerman M (1997) Crustacean endocrinology: a retrospective, prospective, and introspective analysis. Physiol Zool 70: 257-269.
14. Chang E, Chang S, Mulder E (2001) Hormones in the lives of crustaceans: an overview. Amer Zool 41: 1090-1097.
15. Nagaraju GPC (2007) Is methyl farnesoate a crustacean hormone? Aquaculture 272: 39-54.

16. Mazurová E, Hilscherová K, Triebskorn R, Köhler H, Marsálek B, et al. (2008) Endocrine regulation of the reproduction in crustaceans: identification of potential targets for toxicants and environmental contaminants. Biologia 63: 139-150.
17. Raviv S, Parnes S, Sagi A (2008) Coordination of reproduction and molt in decapods. In: Mente E, (Ed.) Reproductive Biology of Crustaceans. Science Publishers, Enfield, NH. Pg: 365-390.
18. Panouse JM (1943) Influence de l'ablation de pedoncule ocularie sur la croissance de 'ovarie chez la crevette, *Leander serratus*. C R Acad Sci Paris 217: 553-555.
19. Gorell TA, Gilbert LI (1971) Protein and RNA synthesis in the premolt crayfish, *Orconectes virilis*. J Comp Physiol A Neuroethol Sens Neural Behav Physiol 73: 345-356.
20. Eastman-Reks S, Fingerman M (1984) Effects of neuroendocrine tissue and cyclic AMP on ovarian growth *in vivo* and *in vitro* in the fiddler crab, *Uca pugilator*. Comp Biochem Physiol 79: 679-684.
21. Chaves AR (2000) Effect of x-organ sinus gland extract on (35S) methionine incorporation to the ovary of the red swamp cray fish, *Procambarus clarkii*. Comp Biochem Physiol A Mol Integr Physiol 126: 407-413.
22. Mohamed KS, Diwan AD (1991) Neuroendocrine regulation of ovarian maturation in the Indian white prawn *Penaeus indicus* H. Milne Edwards. Aquaculture 98: 381-393.
23. Okumura T, Aida K, (2001) Effects of bilateral eyestalk ablation on molting and ovarian development in the giant freshwater prawn *Macrobrachium rosenbergii*. Fish Sci 67: 1125-1135.
24. Okumura T, Sakiyama K (2004) Hemolymph levels of vertebrate-type steroid hormones in female kuruma prawn *Marsupenaeus japonicus*(Crustacea: Decapoda: Penaeida) during natural reproductive cycle and induced ovarian development by eyestalk ablation. Fish Sci 70: 372-380.
25. Gomez R (1965) Acceleration of development of gonads by implantations of brain in the crab *Paratelphusa hydrodromous*. Naturwissenschaften 52: 216.
26. Ragneker PV, Deshmukh RD (1968) Effect of eyestalk removal on the ovarian growth of the marine crab, *Scylla serrata* (Forskal). J

Anim Morph Physiol 17: 84-88.

27. Tsutsui N, Kim YK, Jasmani S, Ohira T, Wilder MN, et al. (2005) The dynamics of vitellogenin gene expression differs between intact and eyestalk ablated kuruma prawn *Penaeus (Marsupeaneus) japonicus*. Fish Sci 71: 249-256.
28. Okumura T (2007) Effects of bilateral and unilateral eyestalk ablation on vitellogenin synthesis in immature female kuruma prawns, *Marsupenaeus japonicus*. Zool Sci 24: 233-240.
29. Spaziani E, Mattson MP, Rudolph PH, (1994) Regulation of crustacean molt-inhibiting hormone. In: Davey KG, Peter RE, Tobe SS (Eds.) Perspect Comp Endocrinol 243-250.
30. Keller R (1992) Crustacean neuropeptides: structures, functions and comparative aspects. Experientia 48: 439-448.
31. Webster SG (1998) Neuropeptides inhibiting growth and reproduction in crustaceans. In: Cost GM, Webster SG (Eds.) Recent advances in arthropod endocrinology. Cambridge University Press, Cambridge. Pg: 33-52.
32. Chan SM, Gu PL, Chu KH, Tobe SS (2003) Crustacean neuropeptide genes of the CHH/MIH/GIH family: implications from molecular studies. Gen Comp Endocrinol 134: 214-219.
33. Nagaraju GP (2011) Reproductive regulators in decapod crustaceans: an overview. J Exp Bio 214: 3-16.
34. De Kleijn DP, Janssen KP, Waddy SL, Hegeman R, Lie WY, et al. (1998) Expression of the crustacean hyperglycemic hormones and the gonad-inhibiting hormone during the reproductive cycle of the female American lobster *Homarus americanus*. J Endocrinol 156: 291-298.
35. Edomi P, Azzoni E, Mettulio R, Pandolfelli N, Ferrero EA, et al. (2002) Gonad inhibiting hormone (GIH) of the Norway lobster (*Nephrops norvegicus*): cDNA cloning, expression, recombinant protein production, and immunolocalization. Gene 284: 93-102
36. Ollivaux C, Vinh J, Soyez D, Toullec JY (2006) Crustacean hyperglycemic and vitellogenesis-inhibiting hormones in the lobster *Homarus gammarus*. FEBS J 273: 2151-2160.
37. Ohira T, Okumura T, Suzuki M, Yajima Y, Tsutsui N, et

al. (2006) Production and characterization of recombinant vitellogenesis0inhibiting hormone from the American lobster *Homarus americanus*. Peptides 27: 1251-1258

38. Tsutsui N, Ohira T, Kawazoe I, Takahashi A, Wilder MN (2007) Purification of sinus gland peptides having vitellogenesis-inhibiting activity from the whiteleg shrimp *Litopenaeus vannamei*. Mar Biotechnol 9: 360-369.
39. Treerattrakool S, Panyim S, Chan SM, Withyachumnarnkul B, Udomkit A (2008) Molecular characterization of gonad-inhibiting hormone of *Penaeus monodon* and elucidation of its inhibitory role in vitellogenin expression by RNA interference. FEBS J 275: 970-980.
40. Treerattrakool S, Panyim S, Udomkit A (2011) Induction of ovarian maturation and spawning in Penaeus monodon broodstock by double-stranded RNA. Mar Biotechnol 13: 163-169.
41. Soyez D, le Caer JP, Noel PY, Rossier J (1991) Primary structure of two isoforms of the vitellogenesis inhibiting hormone from the lobster *Homarus americanus*. Neuropeptides 20: 25-32.
42. Augilar MB, Quackenbush LS, Hunt DT, Shabanowitz J, Huberman A (2002) Identification, purification and initial characterization of the vetellogenesis-inhibiting hormone from the Mexican crayfish *Procambarus bouvieri* (Ortmann). Comp Biochem Physiol B 102: 491-498.
43. De Kleijn DP, De Leeuw EP, Van Den Burg BC, Martens GJ, Van Herp F (1995) Cloning and expression of two mRNAs encoding structurally different crustacean hyperglycemic hormone precursors in the lobster *Homarus americanus*. Biochem Biophys Acta 1260: 62-66.
44. Yang WJ, Ranga Rao K (2001) Cloning of precursors for two MIH/VIH-related peptides in the prawn, *Macrobrachium rosenbergii*. Biochem Biophys Res Commun 289: 407-413.
45. Qian YQ, Dai L, Yang JS, Yang F, Chen DF, et al. (2009) CHH family peptides from an 'eyeless' deep-sea hydrothermal vent shrimp, *Rimicaris kairei*: characterization and sequence analysis. Comp Biochem Physiol B Biochem Mol Bio154: 37-47.
46. Grève P, Sorokine O, Berges T, Lacombe C, Van Dorsselaer

A, et al. (1999) Isolation and amino acide sequence of a peptide with vitellogenesis inhibiting activity from the terrestrial isopod Armadillidium vulgare (Crustacea). Gen Comp Endocrinol 115: 406-414.

47. Udomkit A, Chooluck S, Sonthayanon B, Panyim S (2000) Molecular cloning of a cDNA encoding a member of CHH/MIH/GIH family from *Penaeus monodon* and analysis of its gene structure. J Exp Mar Biol Ecol 224: 145-156.

48. Soyez D, Van Deijnen J, Martin M (2005) Isolation and characterization of a vitellogenesis-inhibiting factor from sinus glands of the lobster, *Homarus americanus*. J Exp Zool 244: 479-484.

49. Nagaraju GP, Kumari NS, Prasad GL, Rajitha B, Meenu M, et al. (2009) Structural prediction and analysis of VIH-related peptides from selected crustacean species. Bioinformation 4: 6-11.

50. Keller R, Sedlmeier D, (1988) A metabolic hormone in crustaceans: The hyperglycemic neuropeptide. In: Laufer H, Downer RGH (Eds.), Invertebrate endocrinology. Endocrinology of selected invertebrate types. Alan R. Liss., Inc., New York. 2: 315-326.

51. Lugo JM, Morera Y, Rodríguez T, Huberman A, Ramos L, et al. (2006) Molecular cloning and characterzation of the crustacean hyperglycemic hormone cDNA from *Litopenaeus schmitti*; functional analysis by double-stranded RNA interference technique. FEBS J 273: 5669-5677.

52. Van Herp, F., 1998. Molecular, cytological and physiological aspects of the crustacean hyperglycemic hormone family. In: Cost, E.M., Webster, S.G. (Eds.), Recent Advances in Arthropod Endocrinology. Cambridge University Press, Cambridge, UK 65: 53-70.

53. Gu PL, Yu KL, Chan SM (2000) Molecular characterization of an additional shrimp hyperglycemic hormone: cDNA cloning, gene organization, expression and biological assay of recombinant proteins. FEBS Lett 472: 122-128.

54. Khayat M, Yang W, Aida K, Nagasawa H, Tietz A, et al. (1998) Hyperglycemic hormone inhibits protein and m-RNA synthesis *in vitro* - incubated ovarian fragments of the marine shrimp, *Penaeus semisulcatus*. Gen Comp Endocrinol 110: 307-318

55. Webster SG, Keller R (1986) Purification, characterisation and amino acid composition of the putative moult-inhibiting hormone (MIH) of*Carcinus maenas* (Crustacea, Decapoda). J Comp Physiol 156: 617- 624.

56. Chung JS, Webster SG (2003) Moult cycle-related changes in biological activity of moult-inhibiting hormone (MIH) and crustacean hyperglycaemic hormone (CHH) in the crab, *Carcinus maenas*. From target to transcript. Eur J Biochem 270: 3280-3288.

57. Zarubin TP, Chang ES, Mykles DL (2009) Expression of recombinant eyestalk crustacean hyperglycemic hormone from the tropical land crab, *Gecarcinus lateralis*, that inhibits Y-organ ecdysteroidogenesis *in vitro*. Mol Biol Rep 36: 1231-1237

58. Webster SG (1993) High-affinity binding of putative moult-inhibiting hormone MIH and crustacean hyperglycaemic hormone CHH to membrane bound receptors on the Y-organ of the shore crab *Carcinus maenas*. Proc R Soc London 251: 53-59.

59. Byard EH (1975) The female specific protein and reproduction in the lobster, *Homarus americanus*. PhD thesis, University of Western Ontario, London, Ontario.

60. Waddy SL, Aiken DE, De Kleijn DPV (1995) Control of reproduction. In: Factor, J.R. (Ed.), Biology of the lobster, Homarus americanus. Academic press, San Diego, USA. 217-266.

61. Hinsch GW (1980) Effect of mandibular organ implants upon the spider crab ovary. Trans Amer Micros Soc 99: 317-322

62. Borst DW, Laufer H, Landau M, Change ES, Hertz WA, et al. (1987) Methyl farnesoate and its role in crustacean reproduction and development. Insect Biochem 17: 1123-1127.

63. Nagaraju GP, Reddy PR, Reddy PS (2004) Mandibular organ: it's relation to body weight, sex, molt and reproduction in the crab, *Oziotelphusa senex senex* Fabricius (1791). Aquaculture 232: 603-612.

64. Wainwright G, Webster SG, Wilkinson MC, Chung JS, Rees HH (1996) Structure and significance of mandibular organ-inhibiting hormone in the crab, *Cancer pagurus*. Involvement in multihormonal regulation of growth and reproduction. J Biol Chem 271: 12749-12754.

65. Reddy PR, Kiranmayi P, Thanuja Kumari K, Reddy PS (2006) 17α-Hydroxyprogesterone induced ovarian growth and vitellogenesis in the freshwater rice field crab *Oziotelphusa senex senex*. Aquaculture 254: 768-775.

66. Tsukimura B, Borst DW (1992) Regulation of methyl farnesoate in the hemolymph and mandibular organ of the lobster, *Homarus americanus*. Gen Com Endocrinol 2: 296-303.

67. Reddy PS, Reddy PR, Nagaraju GP (2004) The synthesis and effects of prostaglandins on the ovary of the crab *Oziotelphusa senex senex*. Gen Comp Endocrinol 135: 35-41.

68. Nagaraju GP, Prasad GL, Reddy PS (2005) Isolation and characterization of mandibular organ-inhibiting hormone from the eyestalks of freshwater crab *Oziotelphusa senex senex*. Int J Appl Sci Engineering 3: 61-68.

69. Nagaraju GPC, Reddy PR, Reddy PS (2006) *In vitro* methyl farnesoate secretion by mandibular organs isolated from different molt and reproductive stages of the crab *Oziotelphusa senex senex*. Fish Sci 72: 410-414.

70. Chaves AR (2001) Effects of sinus gland extract on mandibular organ size and methyl farnesoate synthesis in the crawfish. Comp Biochem Physiol A Mol Integr Physiol 128: 327-333.

71. Liu L, Laufer H (1996) Isolation and characterization of sinus gland neuropeptides with both MOIH and hyperglycemic effects from the spider crab *Libinia emarginata*. Arch Insect Biochem Physiol 32: 375-385.

72. Liu L, Laufer H, Wang Y, Hayes T (1997) A neurohormone regulating both methyl farnesoate synthesis and glucose metabolism in a crustacean. Biochem Biophys Res Commun 237: 694-701.

73. Liu L, Laufer H, Gogarten PJ, Wang M (1997) cDNA cloning of a mandibular organ inhibiting hormone from the spider crab *Libinia emarginata*. Invert Neurosci 3: 199-204.

74. Tang C, Weiqun LU, Wainwright G, Webster SG, Rees HH, et al. (1999) Molecular characterization and expression of mandibular organ inhibiting hormone a recently discovered neuropeptide involved in the regulation of growth and reproduction in the crab *Cancer pagurus*. Biochem J 343: 355-360.

75. Tiu SH, Chan SM (2007) The use of recombinant protein and RNA interference approaches to study the reproductive functions of a gonad-stimulating hormone from the shrimp *Metapenaeus ensis*. FEBS J 274: 4385-4395.

76. Zmora N, Sagi A, Zohar Y, Chung JS (2009) Molt-inhibiting hormone stimulates vitellogenesis at advanced ovarian developmental stages in the female blue crab, *Callinectes sapidus* 2: novel specific binding sites in hepatopancreas and cAMP as a second messenger. Saline Systems5: 6.

77. Zmora N, Trant J, Zohar Y, Chung JS (2009) Molt-inhibiting hormone stimulates vitellogenesis at advanced ovarian developmental stages in the female blue crab, *Callinectes sapidus* 1: an ovarian stage dependent involvement.Saline Systems 5: 7.

78. Kulkarni GK, Glade L, Fingerman M (1991) Oogenesis and effects of neuroendocrine tissues on *in vitro* synthesis of protein by the ovary of the red swamp crayfish *Procambarus clarkii* (Girard). J Crust Biol 11: 513-522.

79. Sarojini R, Nagabhushanam R, Fingerman M (1995) A neurotransmitter role for red-pigment-concentrating hormone in ovarian maturation in the red swamp crayfish *Procambarus clarkii*. J Exp Biol 198: 1253-1257.

80. Charniaux-Cotton H (1985) Vitellogenesis and Its Control in Malacostracan Crustace a. Amer Zool 25: 197-206.

81. Kulkarni GK, Nagabhushanam R, Amaldoss G, Jaiswal RG, Fingerman M (1991) 5-hydroxytryptamine stimulation of the ovary in the crayfish. *Procambarus clarkii*. Amer Zool 31: I I5A.

82. Nagabhushanam R, Sarojini R, Reddy PS, Devi M, Fingerman M (1995) Opioid peptides in invertebrates: localization, distribution and possible functional roles. Curr Sci 69: 659-671.

83. Chan SM, Chan KM (1999) Characterization of the shrimp eyestalk cDNA encoding a novel fushi tarazu-factor 1 (FTZ-F1). FEBS Lett 454: 109-114.

84. Reddy PS (2000) Involvement of opioid peptides in the regulation of reproduction in the prawn *Penaeus indicus*. Naturwissenschaften87: 535-538.

85. Kishori B, Reddy PS (2000) Antagonistic effects of opioid

peptides in the regulation of ovarian growth of the Indian rice field crab, *Oziotelphusa senex senex* Fabricius. Invertebr Reprod Develop 37: 107-111.

86. Kishori B, Reddy PS (2004) Influence of leucine-enkephalin on moulting and vitellogenesis in the freshwater crab, *Oziotelphusa senex senex* (Fabricius, 1791)(Decapoda, Brachyura). Crustaceana 76: 1281-1290.
87. Fingerman M (1997) Roles of neurotransmitters in regulating reproductive hormone release and gonadal maturation in decapod crustaceans. Invert Reprod Dev 31: 47-54.
88. Fingerman M, Julian WE, Spirtes MA, Kostrzewa RM (1974) The presence of 5-hydroxytryptamine in the eyestalks and brain of the fiddler crab Uca pugilaror. its quantitative modification by pharmacological agents, and possible role as a neurotransmitter in controlling the release of red pigment-dispersing hormone. Corny Gen Pharmac 5: 299-303.
89. Sarojini R, Nagabhushanam R, Fingerman M (1995) Mode of action of the neurotransmitter 5-hydroxytryptamine in stimulating ovarian maturation in the red swamp crayfish, *Procambarus clarkii*: An in vivo and in vitro study. J Exp Zool 271: 395 - 400.
90. Ragunathan MG, Arivazhagan A (1999) Influence of eyestalk ablation and 5-hydroxytryptamine on the gonadal development of a female crab, *Paratelphusa hydrodromous* (Herbst). Curr Sci 76: 583-587.
91. Vaca AA, Alfaro J (2000) Ovarian maturation and spawning the white shrimp, *Penaeus vannamei*, by serotonin Injection. Aquaculture 182: 373-385.
92. Kumlu M (2005) Gonadal maturation and spawning in *Penaeus semisulcatus* de Hann, 1844 by hormone injection. Turk J Zool 29: 193-199.
93. Falck B, Hillarp NA, Thieme G, Torp A (1982) Fluorescence of catechol amines and related compounds condensed with formaldehyde. Brain Res Bull 9: xi-xv.
94. Cahansky AV, Medesani DA, Chaulet A, Rodríguez EM (2011) *In vitro* effects of both dopaminergic and enkephalinergic antagonists on the ovarian growth of *Cherax quadricarinatus* (Decapoda,

Parastacidae), at different periods of the reproductive cycle. Comp Biochem Physiol A Mol Integr Physiol 158: 126-131.

95. Hardeland R (1997) New actions of melatonin and their relevanc to biometeorology. Int J Biometeorol 41: 47-57.

96. Vivien-Roels B, Pévet P (1993) Melatonin: presence and formation in invertebrates. Experentia 49: 642-647.

97. Balzer I, Espinola IR, Fuentes-Pard B (1997) Daily variations of immunoreactive melatonin in the visual system of crayfish. Biol Cell 89: 539-543.

98. Sainath SB, Reddy PS (2010) Effect of selected biogenic amines (dopamine, serotonin and melatonin on ovarian maturation in the fresh water edible crab, *Oziotelphusa senex senex*. Aquaculture 313: 144-148.

99. Fingerman M, Nagabhushanam R, Sarojini R, Reddy PS (1994) Biogenic amines in crustaceans: identification, localization and roles. J Crust Biol 14: 413-437.

100. Sarojini R, Nagabhushanam R, Fingerman M (1997) An *in vitro* study of the inhibitory action of methionine enkephalin on ovarian maturation in the red swamp crayfish, *Procambarus clarkii*. Comp Biochem Physiol 115: 149-153.

101. Kulkarni GK, Nagabhushanam R, Joshi PK (1981) Neuroendocrine regulation of reproduction in the marine female prawn, *Parapenaeopsis hardwickii* (Miers 1878). Ind J Mar Sci 10: 350-352.

102. Sarojini R, Mirajkar MS, Nagabhushanam R (1983) Bihormonal control of oogenesis in the freshwater prawn, *Macrobrachium kistnensis*. Acta Physiol Hung 61: 5-10.

103. Takayanagi H, Yamamoto Y, Takeda N (1986) An ovary-stimulating factor in the shrimp, *Paratya compressa*. J Exp Zool 240: 203-209.

104. Yano I, Wyban JA (1993) Induced ovarian maturation of *Penaeus vannamei* by injection of lobster brain extract. Bull Natl Res Inst Aquacult 21: 1-7.

105. Otsu T (1960) Precocious development of the *ovaries* in the crab, *Potamon dehaani,* following implantation of the thoracic ganglion. Annot Zool Japan 33: 90-96.

106. Medesani DA, López Greco LS, Rodríguez EM (2004) Interference of cadmium and copper with the endocrine control of ovarian growth, in the estuarine crab *Chasmagnathus granulata*. Aquat Toxicol 69: 165-174.

107. Laufer H, Biggers WJ (2001) Unifying concepts learned from methyl farnesoate for invertebrate reproduction and post-embryonic development. Amer Zool 41: 442-457.

108.

109. Borst DW, Ogan J, Tsukimura B, Claerhout T, Holford K (2001) Regulation of the crustacean mandibular organ. Amer Zool 41: 430-441.

110. Borst DW, Wainwright G, Rees HH (2002) *In vivo* regulation of the mandibular organ in the edible crab, *Cancer pagurus*. Proc Biol Sci 269: 483-490.

111. Nagaraju GP, Suraj N, Reddy PS (2003) Methyl farnesoate stimulates gonad development in *Macrobrachium malcolmsonii* (H. Milne Edwards) (Decapoda, Palaemonidae). Crustaceana 76: 1171-1178.

112. Tobe SS, Young DA, Khoo HW (1989) Production of methyl farnesoate by the mandibular organs of the mud crab, *Scylla serrata*: validation of a radiochemical assay. Gen Comp Endocrinol 73 : 342-353.

113. Landu M, Laufer H, Homola E, (1989) Control of methyl farnesoate synthesis in the mandibular organ of the crayfish *Procambarus clarkii*: evidence for peptide neurohormones with dual functions. Invertebr Reprod Dev 16: 165-168.

114. Sagi A, Homola E, Laufer H (1991) Methyl farnesoate in the prawn *Macrobrachium rosenbergii*: Synthesis by the mandibular organ *in vitro*, and titers in the hemolymph. Comp Biochem Physiol 99: 879-882.

115. Wilder MN, Okumura T, Suzuki Y, Fusetani N, Aida K (1994) Vitellogenin production induced by eyestalk ablation in juvenile giant freshwater prawn *Macrobrachium rosenbergii* and trial methyl farnesoate administration. Zool Sci 11: 45-53.

116. Prestwich GD, Bruce MJ, Ujváry I, Chang ES (1990) Binding proteins for methyl farnesoate in lobster tissues: detection by photoaffinity labeling. Gen Comp Endocrinol 80: 232-237.

117. Li H, Borst DW (1991) Characterization of a methyl farnesoate binding protein in hemolymph from *Libinia emarginata*. Gen Comp Endocrinol 81: 335-342.

118. Tamone SL, Chang ES (1993) Methyl farnesoate stimulates ecdysteroid secretion from crab Y-organs *in vitro*. Gen Comp Endocrinol 89: 425-432.

119. Laufer H, Borst D, Baker FC, Reuter CC, Tsai LW, et al. (1987) Identification of Juvenile hormone-like compound in a crustacean. Science 235: 202-205.

120. Homola E, Sagi A, Laufer H (1991) Relationship of claw form and exoskeleton condition to reproductive system size and methyl farnesoate in the male spider crab *Libinia emarginata*. Invert Reprod Dev 20: 2190-2225.

121. Sagi A, Homola E, Laufer H (1993) Distinct reproductive types of male spider crabs, *Libinia emarginata* differ in circulating and synthesizing methyl farnesoate. Biol Bull 185: 168-173.

122. Laufer H, Biggers WJ, Ahl JS (1998) Stimulation of ovarian maturation in the crayfish *Procambarus clarkii* by methyl farnesoate. Gen Comp Endocrinol 111: 113-118.

123. Tsukimura B, Kamemoto FI (1991) *In vitro* stimulation of oocytes by presumptive mandibular organ secretions in the shrimp, *Penaeus vannamei*. Aquaculture 92: 59-66.

124. Nagaraju GP, Ramamurthi R, Reddy PS (2002) Methyl farnesoate stimulates ovarian growth in *Penaeus indicus*. In: Harikumar VS (Ed.), Recent Trends in Biotechnology.Agrobios, India. 1: 85-89.

125. Reddy PS, Ramamurthy R (1998) Methyl farnesoate stimulates ovarian maturation in the freshwater crab *Oziotelphusa senex senex*. Curr Sci 74: 68-70.

126. Zapata V, Lopez GLS, Medasani D, Rodriguez M (2003) Ovarian growth in the crab *Chasmagnathus granulata* induced by hormones and neuroregulators throughout the year. *In vitro* and *in vivo* studies. Aquaculture 224: 339-352

127. Rodriguez EM, Medesani DA, Greco LS, Fingerman M (2002) Effects of some steroids and other compounds on ovarian growth of the red swamp crayfish, *Procambarus clarkii*, during early vitellogenesis. J Exp Zool 292: 82-87.

128. Mak AS, Choi CL, Tiu SH, Hui JH, He JG, et al. (2005) Vitellogenesis in the red crab *Charybdis feriatus*: hepatopancreas-specific expression and farnesoic acid stimulation of vitellogenin gene expression and farnesoic acid stimulation of vitellogenin gene expression. Mol Reprod Dev 70: 288-300.

129. Chan SM, Mak AS, Choi CL, Ma TH, Hui JH, et al. (2005) Vitellogenesis in the red crab, *Charybdis feriatus* contributions from small vitellogenin transcripts (*Cf*Vg) and farnesoic acid stimulation of *Cf*Vg gene expression. Ann N Y Acad Sci 1040: 74-79.

130. Adiyodi KG, Subramanian T (1983) Arthropoda Crustacea. In: Adiyodi KG, Adiyodi RG (Eds.). Reproductive biology of Invertebrates. Wiley Chichester. 1: 443-495.

131. Chang ES (1997) Chemistry of crustacean hormones that regulate growth and reproduction. In: Fingerman M, Nagabhushanam R, Thompson MF (Eds.), Recent Advances in Marine Biotechnology. Oxford and IBH Publishing, Co., New Delhi. 163-178.

132. Charniaux-Cotton H, Toui, A (1973) Contrale de la prgvitellogenese et de la vitellogenese chez la crevette hermaphrodite *Lysmata seticaudata* Risso. C R Acad Sci Ser D 276: 2717-2720.

133. Subramoniam T (2000) Crustacean ecdysteroids in reproduction and embryogenesis. Comp Biochem Physiol125: 135-156.

134. Meusy JJ, Blanche, MF, Junera H (1977) Molting and vitellogenesis in the Crustacean Amphopod Orchestia gammarella Pallas. II. Synthesis of vitellogenin (female protein fraction of hemolymph) after the destruction of Y organs. Gen Comp Endocrinol 33: 35-40.

135. Okumura T, Han CH, Suzuki Y, Aida K, Hanyu I (1992) Changes in hemolymph vitellogenin and ecdysteroid levels during the reproductive and non-reproductive molt cycles in the freshwater prawn *Macrobrachium nipponense*. Zool Sci 9: 37-45.

136. Demeusy N (1962) Role de la gland de mue dans l'evolution ovarianne du crabe *Carcinus maenas* Linn Can Biol Mar 3: 37-56.

137. Steel CGH, Vafopoulou X (1998) Ecdysteroid titres in haemolymph and other tissues during moulting and reproduction in the terrestrial isopod, *Oniscus asellus* (L.). Invertebr Reprod Dev 34: 187-194.

138. Lafont R, Dauphin-Villemant C, Warren J, Rees H (2005) Ecdysteroid chemistry and biochemistry. Compr Mol Insect Sci 3 :

125-195.

139. Gunamalai V, Kirubagaran R, Subramoniam T (2006) Vertebrate steroids and the control of female reproduction in two decapod crustaceans, *Emerita asiatica* and *Macrobrachium rosenbergii*. Curr Sci 90: 119-123.

140. Brown M, Sieglaff D, Rees H (2009) Gonadal ecdysteroidogenesis in Arthropoda: occurrence and regulation. Annu Rev Entomol 54: 105-125

141. Burns BG, Sangalang GB, Freeman HC, McMenemy M (1984) Isolation and identification of testosterone from serum and testes of the American lobster (Homarus americanus). Gen Comp Endocrinol 54: 429-432.

142. Ollevier F, De Clerck D, Diederik H, De Loof A (1986) Identification of nonecdystroid steroids in hemolymph of both male and female *Astacus leptodactylus* (Crustacea) by gas chromatography-mass spectrometry. Gen Comp Endocrinol 61: 214-228.

143. Fairs N, Quinlan P, Goad L (1990) Changes in ovarian unconjugated and conjugated steroid titers during vitellogenesis in *Penaeus monodon*. Aquaculture 89: 83-99.

144. Quinitio ET, Hara A, Yamauchi K, Nakao S (1994) Changes in the steroid hormone and vitellogenin levels during the gametogenic cycle of the giant tiger shrimp, *Penaeus monodon*. Comp Biochem Physiol 109: 21-26.

145. Yano I, Fingerman M, Nagabhushanam R (2000) Endocrine control of reproductive maturation in penaeid shrimp. In: Fingerman M, Nagabhushanam R (Eds.), Recent Advances in Marine Biotechnology: Aquaculture, Part A: Seaweeds and Invertebrates. Science Publishers, Enfield, NH.. 161-176.

146. Shih J (1997) Sex steroid-like substances in the ovaries, hepatopancreases, and body fluid of female *Mictyris brevidactylus*. Zool Stud 36: 136-145.

147. Kulkarni G, Nagabhushanam R, Joshi P (1979) Effect of progesterone on ovarian maturation in a marine penaeid prawn *Parapenaeopsis hardwickii* (Miers, 1878). Ind J Exp Biol 17: 986-987.

148. Warrier SR, Tirumalai R, Subramoniam T (2001) Occurence of vertebrate steroids, estradiole 17β and progesterone in the

reproducing females of the mud crab *Scylla serrata*. Comp Biochem Physiol A Mol Integr Physiol 130: 283-294.

149. Yano I (1987) Maturation of kuruma prawns *Penaeus japonicus* cultured in earthen ponds. NOAA Tech Rep NMFS 47: 3-7.

150. Tiu SH, Hui JH, He JG, Tobe SS, Chan SM (2006) Characterization of vitellogenin in the shrimp *Metapenaeus ensis*: expression studies and hormonal regulation of *Me*Vg1 transcription *in vitro*. Mol Reprod Dev 73: 424-436.

151. Coccia E, Lisa E, Cristo C, Cosmo A, Paolucci M (2010) Effects of estradiol and progesterone on the reproduction of the freshwater crayfish *Cherax albidus*. Biol Bull 218: 36-47.

152. Yano I (1985) Induced ovarian maturation and spawning in greasyback shrimp *Metapenaeus ensis,* by progesterone. Aquaculture47: 223-229.

153. Rodriguez EM, Lopez Greco LS, Medesani DA, Laufer H, Fingerman M (2002) Effect of methyl farnesoate, alone and in combination with other hormones, on ovarian growth of the red swamp crayfish, *Procambarus clarkii*, during vitellogenesis. Gen Comp Endocrinol 125: 34-40.

154. Quackenbush L (1994) Lobster reproduction: a review. Crustaceana 67: 82-94.

155. Paolucci M, Di Cristo C, Di Cosmo A (2002) Immunological evidence for progesterone and estradiol receptors in the freshwater crayfish *Austropotamobius pallipes*. Mol Reprod Dev 63: 55-62

156. Zukowska-Arendarczyk M (1981) Effect of hypophyseal gonadotropins (FSH and LH) on the ovaries of the sand shrimp *Crangon crangon*(Crustacea: Decapoda). Mar Biol 63: 241-247.

157. Jayaprakas V, Sambhu C (1998) Growth characteristics of white prawn *Penaeus indicus* (Decapoda-Crustacea) under dietary administration of protein hormones. Ind J Mar Sci 27: 389-395.

158. Yano I (1993) Ultraintensive culture and maturation in captivity of penaeid shrimp. In: McVey JP (Ed.), CRC Handbook of Mariculture: Crustacean Aquaculture. CRC Press, Boca Raton. 2: 289-313.

159. Laufer H, Landau M (1991) Endocrine control of reproduction in shrimp and other crustacea. In:DeLoach P, Dougherty WJ, Davidson MA (Eds.) Frontiers in shrimp research. Elsevier Science Publications, Amsterdam. 65-81.

160. Souty C, Picaud JL (1984) Effect of injection of a human gonadotropin on the synthesis and liberation of vitellogenin by adipose tissue of the marine crustacean isopod Idotea balthica basteri Audouin. Gen Comp Endocrinol 54: 418-421.

161. Toullec JY, Van Wormhoudt A (1987) Variations quantitatives de peptides apparentés à l›hormone de croissance humaine chez *Palaemon serratus*. CR Acad Sci Paris 353: 265-269.

162. Bomirski A, Klek-Kawińska E (1976) Stimulation of oogenesis in the sand shrimp, *Crangon crangon*, by a human gonadotrophin. Gen Comp Endocrinol 30: 239-242.

163. Charmantier-Daures M, Charmantier G, Janssen KP, Aiken DE, vanHerp F (1994) Involvement of eyestalk factors in the neuroendocrine control of osmoregulation in adult American lobster *Homarus americanus*. Gen Comp Endocrinol 94: 281-293.

164. Spanings-Pierrot C, Soyez D, Van Herp F, Gompel M, Skaret G, et al. (2000) Involvement of crustacean hyperglycemic hormone in the control of gill ion transport in the crab *Pachygrapsus marmoratus*. Gen Comp Endocrinol 119: 340-350.

165. Lovett D, Tanner C, Glomski K, Ricart T, Borst DW (2006) The effect of seawater composition and osmolality on hemolymph levels of methyl farnesoate in the green crab *Carcinus maenas*. Comp Biochem Physiol A Mol Integr Physiol 143: 67-77.

166. Nagaraju GP, Borst DW (2008) Methyl farnesoate couples environmental changes to testicular development in the crab *Carcinus maenas*. J Exp Biol 211: 2773-2778.

167. Henry R, Borst DW (2005) Effects of eyestalk ablation on carbonic anhydrase activity in the euryhaline blue crab *Callinectes sapidus*: neuroendocrine control of enzyme expression. J Exp Zool A Comp Exp Biol 305: 23-31.

168. Olmstead A, LeBlanc G (2007) The environmental-endocrine basis of gynandromorphism (intersex) in a crustacean. Int J Biol Sci 3: 77-84.

169. Laufer H, Homola E, Landau M (1987) Control of methyl farnesoate synthesis in crustacean mandibular organs. Amer Zool 69: 27.

170. Sarojini R, Nagabhushanam R, Fingerman M (1995) *In vivo* effects of dopamine and dopaminergic antagonists on testicular maturation in the red swamp crayfish, *Procambarus clarkii*. Biol Bull 189: 340-346.

171. Reddy PR, Nagaraju GPC, Reddy PS (2004) Involvement of methyl farnesoate in the regulation of molting and reproduction in the freshwater crab *Oziotelphusa senex senex*. J Crust Biol 24: 511-515.

172. Gunamalai V, Kirubagaran R, Subramoniam T (2004) Hormonal coordination of molting and female reproduction by ecdysteroids in the mole crab *Emerita asiatica* (Milne Edwards). Gen Comp Endocrinol 138: 128-138.

173. Martin-Creuzburg D, Sperfeld E, Wacker A (2009) Colimitation of a freshwater herbivore by sterols and polyunsaturated fatty acids. Proc Biol Sci 276: 1805-1814.

174. Heckmann LH, Sibly RM, Connon R, Hooper HL, Hutchinson TH, et al. (2008) Systems biology meets stress ecology: linking molecular and organismal stress responses in *Daphnia magna*. Genome Biol 9: 40.

175. Medeiros MN, Mendonca LH, Hunter AL, Paiva-Silva GO, Mello FG, et al. (2004) The role of lipoxygenase products on the endocytosis of yolk proteins in insects: Participation of cAMP. Arch Insect Biochem Physiol 55: 178-187.

176. Meusy JJ, Martin G, Soyez D, van Deijnen JE, Gallo JM, (1987) Immunochemical and immunocytochemical studies of the crustacean vitellogenesis-inhibiting hormone (VIH). Gen Comp Endocrinol 67: 333-341.

CHAPTER 12

ENDOCRINE AND LOCAL IGF-I IN THE BONY FISH IMMUNE SYSTEM

Anne-Constance Franz[1], Oliver Faass[1,2], Bernd Köllner[3], Natallia Shved[1], Karl Link[1,4], Ayako Casanova[5], Michael Wenger[1,5], Helena D'Cotta[6], Jean-François Baroiller[6], Oliver Ullrich[1,2], Manfred Reinecke[1], and Elisabeth Eppler[1,2,7,8]

[1] Institute of Anatomy, University of Zurich, Winterthurerstrasse 190, Zurich 8057, Switzerland

[2] Study Group "Magdeburger Arbeitsgemeinschaft für Forschung unter Raumfahrt- und Schwerelosigkeitsbedingungen" (MARS), Otto-von Guericke-University Magdeburg, Universitätsplatz 2, Magdeburg 39106, Germany

[3] Friedrich-Loeffler-Institute, Federal Research Institute for Animal Health, Greifswald-Insel Riems 17493, Germany

[4] Institute of Evolutionary Medicine, University of Zurich, Winterthurerstrasse 190, Zurich 8057, Switzerland

[5] Institute for Fish and Wildlife Health, University of Bern, Länggassstrasse 122, Bern 3012, Switzerland

[6] Fish Physiology and Genomics, UMR ISEM—Institut des Sciences de l'Evolution de Montpellier, Département Conservation et Domestication, CIRAD, Campus International de Baillarguet, TA B-116/A, Montpellier cedex 5, Montpellier 34398, France

[7] Institute of Anatomy, University of Basel, Pestalozzistrasse 20, Basel 4056, Switzerland

[8] Institute of Anatomy II, University of Erlangen-Nürnberg, Universitätsstraße 19, Erlangen 91054, Germany

ABSTRACT

A role for GH and IGF-I in the modulation of the immune system has been under discussion for decades. Generally, GH is considered a stimulator of innate immune parameters in mammals and teleost fish. The stimulatory effects in humans as well as in bony fish often appear to be correlated with elevated endocrine IGF-I (liver-derived), which has also been shown to be suppressed during infection in some studies. Nevertheless, data are still fragmentary. Some studies point to an important role of GH and IGF-I particularly during immune organ development and constitution. Even less is known about the potential relevance of local (autocrine/paracrine) IGF-I within adult and developing immune organs, and the distinct localization of IGF-I in immune cells and tissues of mammals and fish has not been systematically defined. Thus far, IGF-I has been localized in different mammalian immune cell types, particularly macrophages and granulocytes, and in supporting cells, but not in T-lymphocytes. In the present study, we detected IGF-I in phagocytic cells isolated from rainbow trout head kidney and, in contrast to some findings in mammals, in T-cells of a channel catfish cell line. Thus, although numerous analogies among mammals and teleosts exist not only for the GH/IGF-system, but also for the immune system, there are differences that should be further investigated. For instance, it is unclear whether the primarily reported role of GH/IGF-I in the innate immune response is due to the lack of studies focusing on the adaptive immune system, or whether it truly preferentially concerns innate immune parameters. Infectious challenges in combination with GH/IGF-I manipulations are another important topic that has not been sufficiently addressed to date, particularly with respect to developmental and environmental influences on fish growth and health.

Keywords: GH, TNF-α, IGF, receptor, head kidney, spleen, lymphocytes, phagocytes, innate immune system, adaptive immune system

INTRODUCTION: THE GH/IGF-SYSTEM AND THE IMMUNE SYSTEM

In mammals and in bony fish, a role for the growth hormone (GH)/insulin-like growth factor (IGF)-system in the modulation of the immune system has been under debate for decades. In general, GH is considered a stimulator of the immune system in mammals [1,2,3,4] as well as in bony fish [5,6]. Generally, IGF-I is a potent growth factor that stimulates immune cell growth and differentiation and inhibits apoptosis [1,3].

In mammals and fish, (endocrine) IGF-I is mainly produced in the liver, with GH being the major stimulus for its synthesis and release into the circulation. There is general agreement that the serum IGF-I concentration influences GH release from the anterior pituitary via a negative feedback mechanism by specific inhibition of GH gene transcription and secretion in mammals [7,8,9] and teleosts [10,11,12,13,14]. While the GH/IGF-I axis is generally considered well-preserved throughout phylogeny [13,14,15,16], the role of IGF-II is still enigmatic in fish [11,15]. Furthermore, in bony fish a few years ago, a third IGF (IGF-3) has been postulated to be expressed exclusively in gonads [17]. In tilapia, IGF-3 expression was also detected by real-time PCR in other tissues including head kidney and spleen, but to a much lower extent than in the gonads [18]. The GH/IGF-system of bony fish further consists of multiple subtypes of GH receptors (GHR), IGF receptors and IGF binding proteins that arose through gene duplication events, which occurred during evolution of the teleost lineage (for further reading, see [19]).

In this review, we will mainly focus on the GH/IGF-I axis, particularly since most published data in the context of the fish (and mammalian) immune system concern the axis GH-GHR-IGF-I with some comparative aspects on mammals at the end of the respective chapters.

GH AND IGF-I ACTIONS IN BONY FISH IMMUNE RESPONSES

Indications exist that IGF-I plays important roles in cell proliferation, differentiation and functional modulation of the immune system of fish [5,6,20,21]. Studies so far have shown that in the piscine immune system, exogenous administration of GH enhanced innate immune activities of leukocytes [22] and phagocytes [23] and activated rainbow trout and gilthead sea bream phagocytes *in vitro* [24] and *in vivo*[25,26]. Furthermore, GH stimulated the respiratory burst of leukocytes in vibriosis-furunculosis-vaccinated rainbow trout [27] and the plasma lysozyme activity, which is a marker for the innate immune system, in rainbow trout and channel catfish [20,28]. Further, both GH and IGF-I stimulated the proliferation of gilthead sea bream head kidney leukocytes *in vitro* [21]. Thus far in two different fish species, GH treatment was combined with bacterial infection: in rainbow trout challenged with *Vibrio anguillarum*, the bactericidal activity was not enhanced by GH [26]. In another study in channel catfish infected with *Edwardsiella ictaluri*, the mortality was not positively influenced [28]. Nevertheless, the immune systems of these species investigated so far, strongly differ from each other, not only due to their different environments, which is cold water for rainbow trout and warm water for channel catfish. Thus, more studies considering different life traits, environments and pathogens are needed to gain further insights why positive influences of GH on different immune functions may not lead to an efficient immune response and an improved health status.

Innate immune functions are stimulated also by IGF-I: *in vitro* administration of salmon IGF-I, human IGF-I, and IGF-II increased superoxide production in zymosan-stimulated rainbow trout head kidney leukocytes, and salmon IGF-I had a significant effect on superoxide production, which was observed to be equipotent to that of salmon GH [29]. Two IGF type-I receptors (IGFR Ia and IGFR Ib) and GHR1 and GHR2 genes were detected in brain, pituitary, liver, and gills, as well as in head kidney leukocytes, except IGFR Ia gene expression was absent in head kidney leukocytes. In the same study, *in vivo* intraperitoneal injection of salmon IGF-I increased the plasma levels of lysozymes [29]. In summary, these data mainly point to the stimulation of innate immune reactions including cell expansion, particularly of phagocytes by both, GH and IGF-I.

Although, admittedly, the majority of published reports examined only innate immune parameters and did not consider adaptive immune parameters, it remains an open question whether the immune-modulating action of GH/IGF-I in fish indeed is restricted to the innate system or whether this current state of knowledge is due to an unbalanced data basis.

Thus far, only fragmentary data regarding the relevance of endocrine IGF-I (IGF-I serum levels) and/or paracrine/autocrine actions exists. The importance of systemic GH and IGF-I in the immune competence of fish, respectively, has been observed in the above mentioned channel catfish challenged with *Edwardsiella ictaluri* [28]. Here, lower IGF-I serum levels and down-regulated hepatic GHR expression were found. While treatment with recombinant bovine GH increased lysozyme activity, it did not elevate the IGF-I serum levels and GHR expression as compared to the control fish, which the authors attributed to the dietary regimen of this study. In addition, Bilodeau *et al.* [30] reported reduced mean plasma IGF-I levels in juvenile hybrid catfish after challenge with *Edwardsiella ictaluri*.

In light of the general importance of IGF-I during fish growth and organ development (e.g., [18,31]), not much is known on its role during development of the immune system. Thus, further studies should enlighten the potential interactions between GH/IGF and the immune systems, particularly during development.

First reports in human patients also point to a potential relevance of systemic IGF-I in immune competence: children who did not survive a septic shock possessed lower IGF-I serum levels than the survivors [32]. Furthermore, children during sepsis and septic shock also showed a suppression of IGF-I serum levels [33]. In both studies, high GH serum levels were measured, which were attributed to a stimulated GH release via the above mentioned feed-back mechanism or to increased endotoxin and cytokine load, respectively [32,33,34]. Furthermore, in HIV-infected patients, the IGF-I serum level gain was found to be a critical marker of the efficiency of GH treatment for the restoration of CD4+ T helper cells [35].

Thus, systemic elevation of GH and/or IGF-I levels seems to play a crucial role in the immune response of mammals and this may apply for bony fish as well as indicated from the few studies published so far.

However, local effects may exist that have not been sufficiently studied to date. Better knowledge in this field is necessary with respect to the increasing detection of the presence of hormones, their receptors and releasing hormones in immune cells themselves. This local presence of hormone systems within the immune system has recently been reviewed by Csaba [36], who highlighted the fact, that immune cells synthesize, store and secrete hormones identical to those of the endocrine glands, for example:

- pro-opio-melanocortin hormones;
- thyroid system hormones;
- GH and prolactin;
- melatonin, histamine, serotonin and catecholamines;
- gonadotropin-releasing hormones; and
- their respective receptors.

Both GH and prolactin, in addition to erythropoietin [37] and leptin, have been classified amongst the short-chain cytokines of the type 1 cytokine family, which supports the dual role of GH as a hormone and a cytokine (for further reading, see [38]).

In his review, Csaba [36] points out that the immune cells may be similar to the unicells (Tetrahymena), and thus retain the characteristics of this early phylogenetic level, while other cells during evolution accumulated together to form endocrine glands. As mobile cells, immune cells are able to transport the stored hormone to different locations (packed transport) or when attracted by local factors, accumulate in the region of the target, to synthesize and secrete hormones locally. The targeted packed transport is more economical than the hormone-release into the blood circulation by glandular endocrine organs.

However, while the distinct localization of IGF-I has been studied for many endocrine glands in mammals (e.g., [7,8,15]), as well as in fish (e.g., [13,14,15,16,31,39]), in the immune system, particularly in fish, it has not been systematically explored to date.

IMMUNE CELLS OF FISH AS SITE OF LOCAL IGF SYNTHESIS

Initial studies have detected IGF-I gene expression in immune organs of teleost fish and even found some correlation with infection or stress. However, the distinct localization at the cellular level has not been investigated in bony fish to date. Particularly since the head kidney is anatomically located nearby the excretory kidney, it has to be considered that it possesses a complex cellular composition. Different cell types express IGF-I as reported in tilapia, e.g., the cells of the *Oreochromis niloticus* renal tubules [31], or the *Oreochromis mossambicus* interrenal cells [40].

In addition, in the gilthead sea bream *Sparus aurata* and the rainbow trout *Oncorhynchus mykiss*, IGF-I gene was expressed in head kidney [41,42,43,44]. *In situ* hybridization of the tilapia head kidney revealed IGF-I mRNA in numerous immune cells [42,43,45]. In kidney and spleen of male and female yellow perch (*Perca flavescens*), a moderate expression of both IGF-I amplification products, IGF-Ia and IGF-Ib was detected [46]. The expression of both, IGF-I and IGF-II genes, has been described in adult tilapia spleen [42,43,45,47], gilthead sea bream and tilapia head kidney [41,42,43]. In red-banded sea bream head kidney and spleen, the expression levels were much lower for IGF-I than for IGF-II, which were similar to IGF-II expression levels in liver [48].

In order to further elucidate the distinct localization of IGF-I in leukocytes and to evaluate whether there exist analogies between higher and lower vertebrates, we investigated the presence of IGF-I in different leukocytes. For the detection of IGF-I mRNA (Figure 1a) and peptide (Figure 1b), we used T-cells from a defined Channel catfish (*Ictalurus punctatus*) T-cell line G14D established by Norman Miller, University of Mississippi Medical Center, Jackson, MS and co-workers [49]. Cells were purchased from ATCC (Cat.No: CRL-2759, [49]) and cultured according to protocol [50] in incomplete channel catfish medium supplemented with channel catfish serum kindly provided by Professor Miller. For real-time PCR (Figure 1a), primers and probes were created based on mRNA sequences of *I. punctatus* IGF-I (GenBank Access No: NM_001200295.1 [51]), and *O. mossambicus* 18S rRNA

(GenBank Access No: AF497908.1) and obtained from Microsynth as described for other species and tissues (e.g., [42,43,47,52,53]). For immunocytochemistry (Figure 1b), cells from the same T-cell line G14D were fixed and by cytospin transferred to slides for immunostaining as described previously [54] using antiserum against IGF-I as described [31,55].

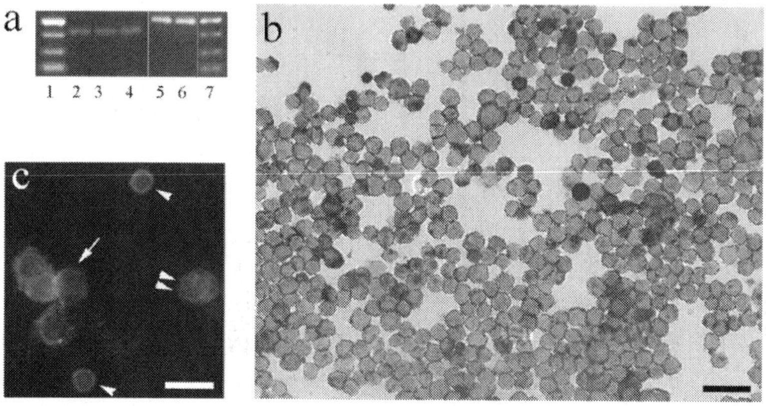

Figure 1: etection of IGF-I in channel catfish (I. punctatus) and rainbow trout (O. mykiss) leukocytes. (a,b) IGF-I mRNA and peptide in I. punctatus T-cell line G14D using (a) PCR and (b) immunocytochemistry. (a) Agarose gels of real-time PCR products confirming the presence of IGF-I mRNA (bands 2–4) at the expected size (73 bp) and reference gene 18S rRNA (bands 5–6) at the expected size (85 bp). Bands 1 and 7: molecular weight marker. (b) The majority of the T-cells show IGF-I immunoreactive material (brown color) within the cytoplasm. Nuclei counterstained with haematoxylin. Bar: 60 μm. (c) Head kidney leukocytes freshly isolated from rainbow trout were adjusted to 1 × 107 cells/mL, incubated with mouse mAb against rainbow trout granulocytes (mAbQ4E; [56]) visualized with a Texas red-coupled anti-mouse antibody as previously described [54]. Subsequently, cells were fixed for intracellular labeling for IGF-I using the same antiserum as in (b) followed by a FITC-coupled anti rabbit antibody for double immunofluorescence of IGF-I (green) in phagocytic cells (red, arrow). Note the small cells immunoreactive for IGF-I, which do not express the granulocyte marker (single arrowheads) and the large cell immunoreactive for the granulocyte marker, but without immunoreactivity for IGF-I (double arrow heads). Bar: 60 μm. (c)

Head kidney leukocytes freshly isolated from rainbow trout were adjusted to 1 x 10^7 cells/mL, incubated with mouse mAb against rainbow trout granulocytes (mAbQ4E; [56]) visualized with a Texas red-coupled anti-mouse antibody as previously described [54]. Subsequently, cells were fixed for intracellular labeling for IGF-I using the same antiserum as in (b) followed by a FITC-coupled anti rabbit antibody for double immunofluorescence of IGF-I (green) in phagocytic cells (red, arrow). Note the small cells immunoreactive for IGF-I, which do not express the granulocyte marker (single arrowheads) and the large cell immunoreactive for the granulocyte marker, but without immunoreactivity for IGF-I (double arrow heads). Bar: 30 pm.

In order to investigate another fish species important in local aquaculture, we further detected IGF-I in head kidney leukocytes (Figure 1c) freshly isolated from adult female rainbow trout (*Oncorhynchus mykiss*) reared at the facilities of the Institute for Fish and Wildlife Health, University of Bern as described elsewhere [54]. In brief, the fish were kept under flow-through conditions (0.5 L/min) in air-saturated tap water (temperature 15 ± 1 °C). The head kidney leukocytes were characterized by an immunocytochemistry protocol [54] using a myeloid lineage marker specific monoclonal antibody [56], which recognizes a cell surface marker expressed on rainbow trout immune leukocytes ranging from immature myeloblasts to phagocyting cells, which are capable of respiratory burst. Thus, it can be considered as a common lineage marker for the myeloid lineage [56].

While IGF-I could be detected in T-cells from both fish species investigated, in contrast, in mature mammalian T-cells, intracellular IGF-I has not been detected to date (e.g., in the thymus, IGF-I was not found in T-cells and local IGF-I supply seems to be derived from thymus epithelial cells [57]). Nevertheless, its role in mammalian T-cell development appears to be pronounced (e.g., [58]). Most recently, in humans, we described the local presence of IGF-I in neutrophils of the tumor microenvironment of a lymphoma [59], which to the best of our knowledge was the first detection of IGF-I in granulocytes. Since we proposed an important role for IGF-I in cell shaping and restitution, e.g., of the high endothelial venules of human lymph nodes, we assume that IGF-I in phagocytic cells might be involved in cell growth, repair and proliferation, particularly of phagocytes. The production and/or local presence of IGF-I in macrophages have been previously demonstrated for humans and mice (e.g., [60,61]).

REGULATION OF IGF-I IN FISH IMMUNE ORGANS AND CELLS BY GH AND OTHER PHYSIOLOGICAL FACTORS

There is evidence also for regulation of extrahepatic IGF-I and IGF-II by GH [45,53,62,63,64,65,66,67] in numerous tissues, but minimal data exists about the immune organs. The GHR has been cloned and localized, for instance in the spleen of salmon [68], and rainbow trout [69], in head kidney and spleen and on hematopoetic cells of sea bream [41,70], and in tilapia spleen and head kidney [42]. GH-binding sites were detected by receptor autography in splenic melanomacrophage centers of gilthead sea bream [71] and in sea bass by immunohistochemistry in immune cells, which were morphologically identified from head kidney smears [21]. Here, aside from the different precursors of the erythropoietic line, also members of the myelocytic line exhibited GH binding capacity, particularly prominent in myeloblasts and to an appreciable extent in myelocytes. While mature eosinophils did not show binding of GH, immature eosinophils were labeled mainly at the periphery. A light binding was also observed in immature neutrophils while basophils were not detected on the smears which can be explained by their similar scarcity as in humans. Monocytes/macrophages which were very sparsely detectable, exhibited a faintly positive reaction for GH binding. Large and small lymphocytes also showed appreciable staining while no binding was detected in thrombocytes [21], which have been recently demonstrated to be capable of acting as phagocytic immune cells in teleost fish [72]. Interestingly, in rainbow trout, erythroid cells and also a subpopulation of B-lymphocytes responded to erythropoietin, while granulocytes and thrombocytes did not [37]. Thus, more studies are needed on stimuli of immune cell lineage differentiation and proliferation.

In rainbow trout spleen, the IGF-I steady-state levels were unaffected by GH treatment and only increased over time [66]. In the tilapia *O. niloticus*, GH injections increased IGF-I mRNA in the head kidney, even more than in the liver, which is the main source of endocrine IGF-I. This elevation in the head kidney could be morphologically verified by *in situ* hybridization. Here again, IGF-I mRNA could be localized in numerous immune cells. IGF-II and GHR gene expressions were not affected by GH treatment [43]. In the spleen, also no change occurred in GHR mRNA, but both, IGF-I and IGF-II mRNAs were increased as could be localized by *in situ* hybridization where IGF-I mRNA was increased in immune cells.

However, the immunological potential of local IGF-I and potential stimulation by GH treatment remains to be investigated. Thus far, in the head kidney of *Enteromyxum leei*-infected gilthead sea bream (*Sparus aurata*), IGF-I gene expression declined, which was accompanied by a down-regulation also of the GHR expression. In contrast, the non-infected individuals showed increased IGF-I and GHR expression levels. From these results, the authors concluded that an increase in IGF-I and GHR expression might be advantageous for immune cells in response to the parasite [41].

The role of autocrine/paracrine IGF-I as related to endocrine IGF-I is also not that clear to date [53,73], and even less is known with respect to fish health. Disturbances of fish health, e.g., by stress and infection, might alter the GH/IGF-I axis at different levels and, thus, impact fish growth, however potential interactions among stress, health and growth are generally extremely complex and not much is known thus far as summarized by Beckman [73]. Thus far, in sea bream during vibriosis infection, the local IGF-I gene expression in liver and kidney was down-regulated [74], but whether this may be due to the general stress situation of the organism or to alterations of the GH/IGF-I axis or specifically induced by the infection remains to be clarified. For instance, also in gilthead sea bream exposed to confinement stress, IGF-I gene expression was declined in the liver [75] and rearing and handling conditions as well as many other factors impact the complex fish physiology, in particular the GH/IGF-I axis [73].

A potential crosstalk between IGF-I and other growth factors and cytokines is coming into focus for research in mammals (e.g., [76]), and this might be of particular interest also for fish research.

As mentioned above, human and murine myeloid cells, particularly macrophages, produce IGF-I in relatively large amounts [60,61], also stimulated by cytokines such as tumor necrosis factor (TNF)-α [77,78], which was also reported to induce a state of resistance against IGF-I

[79]. *Vice versa*, there exist indications that IGF-I counteracted sickness behavior and depression in mice induced by cytokines including TNF-α [80]. However, data are rudimentary and even contradictory: in GH-treated growth-deficient children altered metabolic parameters including TNF-α levels are reported e.g., [81], and modulation of cytokines by GH at high doses is suggested to be involved in adverse consequences in critically ill patients, such as septic shock and uncontrolled infections [82].

Thus, the idea of a bidirectional communication between neuroendocrine and immune system exists and demands a search for better knowledge also in fish [38]. While different cytokines have already been explored in fish (for reviews, see [83,84]), data on potential interactions with the GH/IGF-I axis are still very fragmentary and mainly deal with TNF-α: e.g., in GH-injected tilapia, TNF-α expression was elevated in the liver [43] and in a GH-RH-treated flounder embryonic cell line, TNF-α mRNA was also induced, while in a primary pituitary cell culture, TNF-α mRNA was decreased [85].

IGF-I AND TNF-α DURING PATHOGENIC INFECTION OF FISH

A possible role of IGF-I and TNF-α in infection and the capability of the host to cope with tissue damage has been previously demonstrated by treating developing rainbow trout, prior to infection with *Yersinia ruckeri*, with 17β-estradiol (E2). The effects of either E2-treatment or the bacterial infection or combination of both on the expression pattern of cytokines (*i.e.*, TNF-α) and/or growth factors (*i.e.*, IGF-I) was tissue and treatment specific [44]. The results indicate a stimulatory effect of infection on IGF-I and TNF-α gene expression in the spleen: IGF-I was elevated 6.802-fold ($p = 0.0001$) at Day 3 after infection with 10^6 colony-forming unit (CFU) of *Yersinia ruckeri* [86,87] and TNF-α gene expression 60.497-fold ($p= 0.001$) as compared with the uninfected control [44]. The infection reduced the survival rates of the rainbow trout to 61% [87]. In contrast, in the head kidney, three days after infection both, IGF-I and TNF-α were lowered, IGF-I to 0.093-fold ($p = 0.0001$), and TNF-α to 0.079-fold ($p = 0.0001$), respectively. IGF-I and TNF-α gene expressions were lower also prior to infection already after the developmental E2-treatment. Both treatments induced additive effects [44].

There exists evidence for interactions between the reproductive state and fish health, particularly the immune system (reviewed by [5,14,38,88]). The suppression of IGF-I by E2 in head kidney was consistent with previous findings by our group where IGF-I was still lower even 1 month after termination of the estrogen treatment [42]. Interestingly, both IGF-I and TNF-α reacted in parallel to both stimuli, E2 treatment and infection [44]. In addition, after E2-treatment of juvenile sea bass (*Dicentrarchus labrax*), suppressed TNF-α gene expression in head kidney leukocytes was reported [89].

In the spleen, we observed a stimulatory effect of infection with Yersinia on both, IGF-I and TNF-α gene expressions with a dose-dependent increase in the E2-treated groups. Particularly TNF-α expression levels were massively elevated by the infection (60.497-fold in the group without E2 pre-treatment, 322.568-fold in the low E2 pre-treatment, 5651.008-fold in the high E2 pre-treatment group) [44]. However, the lack of response of IGF-I gene expression to estrogen treatment was different from previous findings by our group where IGF-I was strongly suppressed even one month after cessation of treatment [42].

In conclusion, the results of the study point to tissue specific effects of either treatment. While in liver and head kidney, the effects were suppressive on both, IGF-I and TNF-α expression levels, in spleen, the gene expression levels were stimulated. Thus, the proposed high TNF-α expression as an indicator for the degree of infection in rainbow trout in another study [90] may be counteracted by the spleen with a parallel elevation of IGF-I, although to a lesser extent. This local elevation of IGF-I as described [44] may be supportive for survival in fish as previously suggested by other studies in humans and fish. Further studies are needed to explore the complex interactions between cytokines and growth factors (for reviews on research in fish see, e.g., [5,6,10,12,38,83,91]).

CONCLUSIONS

Numerous indications derived from studies in different fish species point to interactions between the GH/IGF and the immune system. Although most data support the current assumption of a general stimulatory action of GH and IGF-I on immune cell expansion and functional parameters, it remains an open question whether the immune-modulating action of GH/IGF-I in fish indeed is restricted to the innate system or whether this is due to the thus far available data basis. Our detection of IGF-I in channel catfish T-lymphocytes points to a role also in the adaptive immune system.

More studies are required to take into consideration different life traits, environments, and pathogens in order to better compare species differences. In particular, studies are needed to investigate why positive influences of GH on different immune functions do not generally lead to an efficient immune response and an improved health status. A better understanding of this phenomenon may be revealed by research into the role of the GH/IGF-I axis during development of fish immune organs and cell subtypes. Here too, more tools to characterize cell lineages and the increasing number of cytokines detected in bony fish will also contribute to a better understanding of the complex orchestration of cytokines and growth factors in the fish immune system.

ACKNOWLEDGMENTS

The authors are grateful to Norman W. Miller and his team, University of Mississippi Medical Center, Jackson, MS, USA for helpful support with the cell culture. The data are based on previous studies supported by the Swiss National Science Foundation and the Foundation at the University of Zurich/Baugarten Stiftung. Recent work is supported by the Prof. Dr. med. Karl und Rena Theiler-Haag Foundation. Special thank is due to Helmut Segner, Institute for Fish and Wildlife Health, University of Bern, Switzerland for critical reading of the manuscript and lively discussions.

AUTHOR CONTRIBUTIONS

All authors contributed substantially to the content of this review article. Anne-Constance Franz, Bernd Köllner, Natallia Shved, Karl Link, Ayako Casanova, Elisabeth Eppler generated the original data presented in Figure 1. Elisabeth Eppler wrote the paper.

REFERENCES

1. Clark R. The somatogenic hormones and insulin-like growth factor-1: Stimulators of lymphopoiesis and immune function. Endocr. Rev. 1997;18:157–179. doi: 10.1210/edrv.18.2.0296.
2. Hattori N. Expression, regulation and biological actions of growth hormone (GH) and ghrelin in the immune system. Growth Horm. IGF Res. 2009;19:187–197. doi: 10.1016/j.ghir.2008.12.001.
3. Smith T.J. Insulin-like growth factor-I regulation of immune function: A potential therapeutic target in autoimmune diseases? Pharmacol. Rev. 2010;62:199–236. doi: 10.1124/pr.109.002469.
4. Weigent D.A. Lymphocyte GH-axis hormones in immunity. Review. Cell Immunol. 2013;285:118–132. doi: 10.1016/j.cellimm.2013.10.003.
5. Segner H., Eppler E., Reinecke M. The impact of environmental hormonally active substances on the endocrine and immune system of fish. Review. In: Reinecke M., Zaccone G., Kapoor B.G., editors. Fish Endocrinology. Volume 2. Science Publishers; Enfield, NH, USA: 2006. pp. 809–865.
6. Yada T. Growth hormone and fish immune system. Gen. Comp. Endocrinol. 2007;152:353–358. doi: 10.1016/j.ygcen.2007.01.045.
7. Jones J.I., Clemmons D.R. Insulin-like growth factors and their binding proteins: Biological actions. Endocr. Rev. 1995;16:3–34.
8. Wallenius K., Sjögren K., Peng X.D., Park S., Wallenius V., Liu J.L., Umaerus M., Wennbo H., Isaksson O., Frohman L., et al. Liver-derived IGF-I regulates GH secretion at the pituitary level in mice. Endocrinology. 2001;142:4762–4770. doi: 10.1210/endo.142.11.8478.
9. Honda J., Manabe Y., Matsumura R., Takeuchi S., Takahashi S. IGF-I regulates pro-opiomelanocortin and GH gene expression in the mouse pituitary gland. J. Endocrinol. 2003;178:71–82. doi: 10.1677/joe.0.1780071.
10. Duan C. Nutritional and developmental regulation of insulin-like growth factors in fish. J. Nutr. 1998;128:306S–314S.
11. Reinecke M., Björnsson B.T., Dickhoff W.W., McCormick S.D.,

Navarro I., Power D.M., Gutiérrez J. Growth hormone and insulin-like growth factors in fish: Where we are and where to go. Gen. Comp. Endocrinol. 2005;142:20–24. doi: 10.1016/j.ygcen.2005.01.016.

12. Wood A.W., Duan C., Bern H.A. Insulin-like growth factor signaling in fish. Int. Rev. Cytol. 2005;243:215–285.
13. Reinecke M. Insulin and the insulin-like growth factors. In: Reinecke M., Zaccone G., Kapoor B.G., editors. Fish Endocrinology. Volume 1. Science Publishers; Enfield, NH, USA: 2006. pp. 3–13.
14. Reinecke M. Influences of the environment on the endocrine and paracrine fish growth hormone-insulin-like growth factor-I system. J. Fish Biol. 2010;76:1233–1254. doi: 10.1111/j.1095-8649.2010.02605.x.
15. Reinecke M., Collet C. The phylogeny of the insulin-like growth factors. Int. Rev. Cytol. 1998;183:1–94.
16. Reinecke M. Insulin-like growth factors and fish reproduction. Biol. Reprod. 2010;82:656–661. doi: 10.1095/biolreprod.109.080093.
17. Wang D.S., Jiao B., Hu C., Huang X., Liu Z., Cheng C.H. Discovery of a gonad-specific IGF subtype in teleost. Biochem. Biophys. Res. Comm. 2008;367:336–341. doi: 10.1016/j.bbrc.2007.12.136.
18. Berishvili G., Baroiller J.-F., Eppler E., Reinecke M. Insulin-like growth factor-3 (IGF-3) in male and female gonads of the tilapia: Development and regulation of gene expression by growth hormone (GH) and 17alpha-ethinylestradiol (EE2) Gen. Comp. Endocrinol. 2010;167:128–134. doi: 10.1016/j.ygcen.2010.01.023.
19. Reindl K.M., Sheridan M.A. Peripheral regulation of the growth hormone-insulin-like growth factor system in fish and other vertebrates. Comp. Biochem. Physiol. A. 2012;163:231–245. doi: 10.1016/j.cbpa.2012.08.003.
20. Yada T., Azuma T., Takagi Y. Stimulation of non-specific immune functions in seawater-acclimated rainbow trout, *Oncorhynchus mykiss*, with reference to the role of growth hormone. Comp. Biochem. Physiol. B. 2001;129:695–701. doi: 10.1016/S1096-4959(01)00370-0.
21. Calduch-Giner J.A., Sitjà-Bobadilla A., Alvarez-Pellitero P., Pérez-Sánchez J. Evidence for a direct action of GH on haemopoietic cells of a marine fish, the gilthead sea bream (*Sparus aurata*) J.

Endocrinol. 1995;146:459–467. doi: 10.1677/joe.0.1460459.

22. Kajita Y., Sakai M., Kobayashi M., Kawauchi H. Enhancement of non-specific cytotoxic activity of leucocytes in rainbow trout *Oncorhynchus mykiss* injected with growth hormone. Fish Shellfish Immunol. 1992;2:155–157. doi: 10.1016/S1050-4648(05)80044-0.
23. Sakai M., Kobayashi M., Kawauchi H. Enhancement of chemoluminiscent responses of phagocytic cells from rainbow trout, *Oncorhynchus mykiss*, by injection of growth hormone. Fish Shellfish Immunol. 1995;5:375–379. doi: 10.1006/fsim.1995.0035.
24. Sakai M., Kobayashi M., Kawauchi H. *In vitro* activation of fish phagocytic cells by GH, prolactin and somatolactin. J. Endocrinol. 1996;151:113–158. doi: 10.1677/joe.0.1510113.
25. Calduch-Giner J.A., Sitjà-Bobadilla A., Alvarez-Pellitero P., Pérez-Sánchez J. Growth hormone is a phagocyte-activating factor in the gilthead sea bream (*Sparus aurata*) Cell Tissue Res. 1997;287:535–540. doi: 10.1007/s004410050777.
26. Sakai M., Kajita Y., Kobayashi M., Kawauchi H. Immunostimulating effect of growth hormone: *In vivo* administration of growth hormone in rainbow trout enhances resistance to *Vibrio anguillarum*infection. Vet. Immunol. Immunopathol. 1997;57:147–152. doi: 10.1016/S0165-2427(96)05771-6.
27. Kitlen J.W., Hejbol E.K., Zinck T., Varming K., Byatt J.C., McLean E. Growth performance and respiratory burst activity in rainbow trout treated with growth hormone and vaccine. Fish Shellfish Immunol. 1997;7:297–304. doi: 10.1006/fsim.1997.0085.
28. Peterson B.C., Small B.C., Bilodeau L. Effects of GH on immune and endocrine responses of channel catfish challenged with *Edwardsiella ictaluri*. Comp. Biochem. Physiol. A. 2007;146:47–53. doi: 10.1016/j.cbpa.2006.08.042.
29. Yada T. Effects of insulin-like growth factor-I on non-specific immune functions in rainbow trout. Zool. Sci. 2009;26:338–343. doi: 10.2108/zsj.26.338.
30. Bilodeau A.L., Peterson B.C., Bosworth B.G. Response of toll-like receptors, lysozyme, and IGF-I in back-cross hybrid (F1 male (blue × channel) × female channel) catfish challenged with

virulent *Edwardsiella ictaluri*. Fish Shellfish Immunol. 2006;20:29–39. doi: 10.1016/j.fsi.2005.03.005.

31. Berishvili G., Shved N., Eppler E., Clota F., Baroiller J.F., Reinecke M. Organspecific expression of IGF-1 during early development of bony fish as revealed in the tilapia, *Oreochromis niloticus*, by *in situ*hybridization and immunohistochemistry: Indication for the particular importance of local IGF-I. Cell Tissue Res. 2006;325:287–301. doi: 10.1007/s00441-005-0133-9.
32. De Groof F., Joosten K.F., Janssen J.A., de Kleijn E.D., Hazelzet J.A., Hop W.C., Uitterlinden P., van Doorn J., Hokken-Koelega A.C. Acute stress response in children with meningococcal sepsis: Important differences in the growth hormone/insulin-like growth factor I axis between nonsurvivors and survivors. J. Clin. Endocrinol. Metab. 2002;87:3118–3124. doi: 10.1210/jcem.87.7.8605.
33. Onenli-Mungan N., Yildizdas D., Yapicioglu H., Topaloglu A.K., Yüksel B., Ozer G. Growth hormone and insulin-like growth factor 1 levels and their relation to survival in children with bacterial sepsis and septic shock. J. Paediatr. Child Health. 2004;40:221–226. doi: 10.1111/j.1440-1754.2004.00342.x.
34. Ohlsson C., Mohan S., Sjögren K., Tivesten A., Isgaard J., Isaksson O., Jansson J.O., Svensson J. The role of liver-derived insulin-like growth factor-I. Endocr. Rev. 2009;30:494–535. doi: 10.1210/er.2009-0010.
35. Napolitano L.A., Schmidt D., Gotway M.B., Ameli N., Filbert E.L., Ng M.M., Clor J.L., Epling L., Sinclair E., Baum P.D., et al. Growth hormone enhances thymic function in HIV-1-infected adults. J. Clin. Investig. 2008;118:1085–1098. doi: 10.1172/JCI32830.
36. Csaba G. Hormones in the immune system and their possible role. A critical review. Acta Microbiol. Immunol. Hung. 2014;61:241–260. doi: 10.1556/AMicr.61.2014.3.1.
37. Körbel S., Bittorf T., Siegl E., Köllner B. Recombinant human erythropoietin induces proliferation and Ca(2+)-influx in specific leukocyte subpopulations of rainbow trout (*Oncorhynchus mykiss*) blood and head kidney cells. J. Comp. Physiol. B. 2004;174:121–128. doi: 10.1007/s00360-003-0396-y.
38. Verburg-van Kemenade L., Stolte E.H., Metz J.R., Chadzinska M.

Neuroendocrine-immune interactions in teleost fish. In: Bernier N.J., van der Kraak G., Farrell A.P., Brauner C.J., editors. Fish Neuroendocrinology, Fish Physiology. Volume 28. Academic Press; Burlington, MA, USA: 2009. pp. 313–364.

39. Eppler E. The insulin-like growth factor I (IGF-I) within the bony fish pituitary: New morphofunctional and phylogenetic aspects. Open Neuroendocrinol. J. 2011;4:43–50. doi: 10.2174/1876528901104010043.

40. Reinecke M., Schmid A., Ermatinger R., Loffing-Cueni D. Insulin-like growth factor I in the teleost *Oreochromis mossambicus*, the tilapia: Gene sequence, tissue expression, and cellular localization. Endocrinology. 1997;138:3613–3619. doi: 10.1210/en.138.9.3613.

41. Sitjà-Bobadilla A., Calduch-Giner J., Saera-Vila A., Palenzuela O., Álvarez-Pellitero P., Pérez-Sanchez J. Chronic exposure to the parasite *Enteromyxum leei* (Myxozoa: Myxosporea) modulates the immune response and the expression of growth, redox and immune relevant genes in gilthead sea bream, *Sparus aurata* L. Fish Shellfish Immunol. 2008;24:610–619. doi: 10.1016/j.fsi.2008.01.014.

42. Shved N., Berishvili G., Häusermann E., D'Cotta H., Baroiller J.F., Eppler E. Challenge with EE2 during early development persistently impairs growth, differentiation and local expression of IGF-1 and-2 in immune organs of tilapia. Fish Shellfish Immunol. 2009;26:524–530. doi: 10.1016/j.fsi.2009.02.003.

43. Shved N., Berishvili G., Mazel P., Baroiller J.F., Eppler E. Growth hormone treatment acts on the endocrine and autocrine/paracrine GH/IGF-axis and on TNF-α expression in bony fish pituitary and immune organs. Fish Shellfish Immunol. 2011;31:944–952. doi: 10.1016/j.fsi.2011.08.012.

44. Wenger M., Shved N., Akgül G., Caelers A., Casanova A., Segner H., Eppler E. Developmental oestrogen exposure differentially modulates IGF-I and TNF-α expression levels in immune organs of *Yersinia ruckeri*-challenged young adult rainbow trout (*Oncorhynchus mykiss*) Gen. Comp. Endocrinol. 2014;205:168–175. doi: 10.1016/j.ygcen.2014.05.022.

45. Eppler E., Berishvili G., Mazel P., Caelers A., Hwang G., Maclean

N., Reinecke M. Distinct organ-specific up- and downregulation of IGF-I and IGF-II mRNA in various organs of a GH-overexpressing transgenic Nile tilapia. Transgenic Res. 2010;19:231–240. doi: 10.1007/s11248-009-9314-8.

46. Lynn S.G., Shepherd B.S. Molecular characterization and sex-specific tissue expression of prolactin, somatolactin and insulin-like growth factor-I in yellow perch (*Perca flavescens*) Comp. Biochem. Physiol. B. 2007;147:412–427. doi: 10.1016/j.cbpb.2007.02.005.
47. Caelers A., Berishvili G., Meli M.L., Eppler E., Reinecke M. Establishment of a real-time RT-PCR for the determination of absolute amounts of IGF-1 and IGF-2 gene expression in liver and extrahepatic sites of the tilapia. Gen. Comp. Endocrinol. 2004;137:196–204. doi: 10.1016/j.ygcen.2004.03.006.
48. Ponce M., Infante C., Funes V., Manchado M. Molecular characterization and gene expression analysis of insulin-like growth factors I and II in the redbanded seabream, *Pagrus auriga*: Transcriptional regulation by growth hormone. Comp. Biochem. Physiol. B. 2008;150:418–426. doi: 10.1016/j.cbpb.2008.04.013.
49. Hogan R.J., Waldbieser G., Goudie C.A., Antao A., Godwin U.B., Wilson M.R., Miller N.W., Clem L.W., McConnell T.J., Wolters W.R., et al. Molecular and immunological characterization of gynogenetic channel catfish (*Ictalurus punctatus*) Mar. Biotechnol. 1999;1:317–327. doi: 10.1007/PL00011781.
50. Zou J., Secombes C.J., Long S., Miller N., Clem L.W., Chinchar V.G. Molecular identification and expression analysis of tumor necrosis factor in channel catfish (*Ictalurus punctatus*) Dev. Comp. Immunol. 2003;27:845–858. doi: 10.1016/S0145-305X(03)00085-5.
51. Clay L.A., Wang S.Y., Wolters W.R., Peterson B.C., Waldbieser G.C. Molecular characterization of the insulin-like growth factor-I (IGF-1) gene in channel catfish (*Ictalurus punctatus*) Biochim. Biophys. Acta. 2005;1731:139–148. doi: 10.1016/j.bbaexp.2005.10.001.
52. A Marca Pereira M.L., Eppler E., Thorpe K.L., Wheeler J.R., Burkhardt-Holm P. Molecular and cellular effects of chemicals disrupting steroidogenesis during early ovarian development of brown trout (*Salmo trutta fario*) Environ. Toxicol. 2014;29:199–206. doi: 10.1002/tox.20786.

53. Eppler E., Caelers A., Shved N., Hwang G., Rahman A.M., Maclean N., Zapf J., Reinecke M. Insulin-like growth factor I (IGF-I) in a growth-enhanced transgenic (GH-overexpressing) bony fish, the tilapia (*Oreochromis niloticus*): Indication for a higher impact of autocrine/paracrine than of endocrine IGF-I. Transgenic Res. 2007;16:479–489. doi: 10.1007/s11248-007-9093-z.
54. Nakayama A., Riesen I., Köllner B., Eppler E., Segner H. Surface marker-defined head kidney granulocytes and B lymphocytes of rainbow trout express Benzo[a]pyrene-inducible cytochrome P4501A protein. Toxicol. Sci. 2008;103:86–96. doi: 10.1093/toxsci/kfn024.
55. Eppler E., Shved N., Moret O., Reinecke M. IGF-I is distinctly located in the bony fish pituitary as revealed for *Oreochromis niloticus*, the Nile tilapia, using real-time RT-PCR, *in situ* hybridisation and immunohistochemistry. Gen. Comp. Endocrinol. 2007;150:87–95. doi: 10.1016/j.ygcen.2006.07.013.
56. Kuroda A., Okamoto N., Fukuda H. Characterization of monoclonal antibodies against antigens shared with neutrophils and macrophages in rainbow trout (*Oncorhynchus mykiss*) Fish Pathol. 2000;35:205–213. doi: 10.3147/jsfp.35.205.
57. Marinova T.T., Kuerten S., Petrow D.B., Angelov D.N. Thymic epithelial cells of human patients affected by myasthenia gravis overexpress IGF-I immunoreactivity. APMIS. 2008;116:50–58. doi: 10.1111/j.1600-0463.2008.00666.x.
58. Law H.K., Tu W., Liu E., Lau Y.L. Insulin-like growth factor I promotes cord blood T cell maturation through monocytes and inhibits their apoptosis in part through interleukin-6. BMC Immunol. 2008;9:74. doi: 10.1186/1471-2172-9-74.
59. Eppler E., Janas E., Link K., Weidmann L., Bischofberger H., Wenger M., Tinguely M., Schraml P., Moch H., Fellbaum C. Insulin-like growth factor-I is expressed in classical and nodular lymphocyte-predominant Hodgkin's lymphoma tumour and microenvironmental cells. Cell Tissue Res. 2015;359:841–851. doi: 10.1007/s00441-014-2052-0.
60. Arkins S., Rebeiz N., Biragyn A., Reese D.L., Kelley K.W. Murine

macrophages express abundant insulin-like growth factor-I class I Ea and Eb transcripts. Endocrinology. 1993;133:334–343.

61. Oberlin D., Fellbaum C., Eppler E. Insulin-like growth factor I messenger RNA and protein are expressed in the human lymph node and distinctly confined to subtypes of macrophages, antigen-presenting cells, lymphocytes and endothelial cells. Immunology. 2009;128:342–350. doi: 10.1111/j.1365-2567.2009.03136.x.
62. Duguay S.J., Swanson P., Dickhoff W.W. Differential expression and hormonal regulation of alternatively spliced IGF-I mRNA transcripts in salmon. J. Mol. Endocrinol. 1994;12:25–37. doi: 10.1677/jme.0.0120025.
63. Duguay S.J., Lai-Zhang J., Steiner D.F., Funkenstein B., Chan S.J. Developmental and tissue regulated expression of IGF-I and IGF-II mRNAs in *Sparus aurata*. J. Mol. Endocrinol. 1996;16:123–132. doi: 10.1677/jme.0.0160123.
64. Otteson D.C., Cirenza P.F., Hitchcock P.F. Persistent neurogenesis in the teleost retina: Evidence for regulation by the growth-hormone/insulin-like growth factor-I axis. Mech. Dev. 2002;117:137–149. doi: 10.1016/S0925-4773(02)00188-0.
65. Vong Q.P., Chan K.M., Cheng C.H. Quantification of common carp (*Cyprinus carpio*) IGF-I and IGF-II mRNA by real-time PCR: Differential regulation of expression by GH. J. Endocrinol. 2003;178:513–521. doi: 10.1677/joe.0.1780513.
66. Biga P.R., Schelling G.T., Hardy R.W., Cain K.D., Overturf K., Ott T.L. The effects of recombinant bovine somatotropin (rbST) on tissue IGF-I, IGF-I receptor, and GH mRNA levels in rainbow trout, *Oncorhynchus mykiss*. Gen. Comp. Endocrinol. 2004;135:324–333. doi: 10.1016/j.ygcen.2003.10.014.
67. Gahr S.A., Vallejo R.L., Weber G.M., Shepherd B.S., Silverstein J.T., Rexroad C.E., III Effects of short-term growth hormone treatment on liver and muscle transcriptomes in rainbow trout (*Oncorhynchus mykiss*) Physiol. Genomics. 2008;32:380–392. doi: 10.1152/physiolgenomics.00142.2007.
68. Fukada H., Ozaki Y., Pierce A.L., Adachi S., Yamauchi K., Hara A., Swanson P., Dickhoff W.W. Salmon growth hormone

receptor: Molecular cloning, ligand specificity, and response to fasting. Gen. Comp. Endocrinol. 2004;139:61–71. doi: 10.1016/j.ygcen.2004.07.001.

69. Very N.M., Kittilson J.D., Norbeck L.A., Sheridan M.A. Isolation, characterization, and distribution of two cDNAs encoding for growth hormone receptor in rainbow trout (*Oncorhynchus mykiss*) Comp. Biochem. Physiol. B. 2005;140:615–628. doi: 10.1016/j.cbpc.2004.12.008.

70. Jiao B., Huang X., Chan C.B., Zhang L., Wang D., Cheng C.H. The co-existence of two growth hormone receptors in teleost fish and their differential signal transduction, tissue distribution and hormonal regulation of expression in seabream. J. Mol. Endocrinol. 2006;36:23–40. doi: 10.1677/jme.1.01945.

71. Muñoz-Cueto J.A., Martínez-Barbera J.P., Pendón C., Rodríguez R.B., Valdivia M.M., Sarasquete M.C. Autoradiographic localization of growth hormone receptors in tissues of the gilthead sea bream, *Sparus aurata*; Proceedings of the 17th Conference of European Comparative Endocrinologists, From Basic to Applied Aspects; Córdoba (Spain). 5–10 September 1994; Abstract 237.

72. Nagasawa T., Nakayasu C., Rieger A.M., Barreda D.R., Somamoto T., Nakao M. Phagocytosis by Thrombocytes is a Conserved Innate Immune Mechanism in Lower Vertebrates. Front. Immunol. 2014;5:445. doi: 10.3389/fimmu.2014.00445.

73. Beckman B.R. Perspectives on concordant and discordant relations between insulin-like growth factor 1 (IGF1) and growth in fishes. Gen. Comp. Endocrinol. 2011;170:233–252. doi: 10.1016/j.ygcen.2010.08.009.

74. Deane E.E., Woo N.Y. Modulation of beta-actin, insulin-like growth factor 1 and glucose-6-phosphate dehydrogenase gene expression during vibriosis of sea bream, *Sparus* (=*Rhabdosargus*) *sarba* Forsskål. J. Fish Dis. 2005;28:593–601. doi: 10.1111/j.1365-2761.2005.00664.x.

75. Saera-Vila A., Calduch-Giner J.A., Prunet P., Pérez-Sánchez J. Dynamics of liver GH/IGF axis and selected stress markers in juvenile gilthead sea bream (*Sparus aurata*) exposed to acute confinement: Differential stress response of growth hormone

receptors. Comp. Biochem. Physiol. A. 2009;154:197–203. doi: 10.1016/j.cbpa.2009.06.004.

76. Li S., Pinard M., Wang Y., Yang L., Lin R., Hiscott J., Su B., Brodt P. Crosstalk between the TNF and IGF pathways enhances NF-κB activation and signaling in cancer cells. Growth Horm IGF Res. 2015;25:253–261. doi: 10.1016/j.ghir.2015.07.008.

77. Noble P.W., Lake F.R., Henson P.M., Riches D.W.H. Hyaluronate activation of CD44 induces insulin-like gowth factor-1 expression by a tumor necrosis factor-alpha dependent mechanism in murine macrophages. J. Clin. Investig. 1993;91:2368–2377. doi: 10.1172/JCI116469.

78. Fournier T., Riches D.W., Winston B.W., Rose D.M., Young S.K., Noble P.W., Lake F.R., Henson P.M. Divergence in macrophage insulin-like growth factor-I (IGF-I) synthesis induced by TNF-alpha and prostaglandin E2. J. Immunol. 1995;155:2123–2133.

79. O'Connor J.C., Cusker R.H., Strle K., Johnson R.W., Dantzer R., Kelley K.W. Regulation of IGF-I function by proinflammatory cytokines: At the interface of immunology and endocrinology. Cell. Immunol. 2008;252:91–110. doi: 10.1016/j.cellimm.2007.09.010.

80. Palin K., Bluthé R.M., McCusker R.H., Moos F., Dantzer R., Kelley K.W. TNFalpha-induced sickness behavior in mice with functional 55 kD TNF receptors is blocked by central IGF-I. J. Neuroimmunol. 2007;187:55–60. doi: 10.1016/j.jneuroim.2007.04.011.

81. Meazza C., Elsedfy H.H., Pagani S., Bozzola E., El Kholy M., Bozzola M. Metabolic parameters and adipokine profile in growth hormone deficient (GHD) children before and after 12-month GH treatment. Horm. Metab. Res. 2014;46:219–223. doi: 10.1055/s-0033-1358730.

82. Uronen-Hansson H., Allen M.L., Lichtarowicz-Krynska E., Aynsley-Green A., Cole T.J., Höidén-Guthenberg I., Fryklund L., Klein N. Growth hormone enhances proinflammatory cytokine production by monocytes in whole blood. Growth Horm IGF Res. 2003;13:282–286. doi: 10.1016/S1096-6374(03)00034-0.

83. Secombes C.J., Wang T., Hong S., Peddie S., Crampe M., Laing K.J., Cunningham C., Zou J. Cytokines and innate immunity of

fish. Dev. Comp. Immunol. 2001;25:713–723. doi: 10.1016/S0145-305X(01)00032-5.

84. Goetz F.W., Planas J.V., MacKenzie S. Tumor necrosis factors. Dev. Comp. Immunol. 2004;28:487–497. doi: 10.1016/j.dci.2003.09.008.

85. Nam B.H., Moon J.Y., Kim Y.O., Kong H.J., Kim W.J., Kim K.K., Lee S.J. Molecular and functional analyses of growth hormone-releasing hormone (GHRH) from olive flounder (*Paralichthys olivaceus*) Comp. Biochem. Physiol. B. 2011;159:84–91. doi: 10.1016/j.cbpb.2011.02.006.

86. Wenger M., Sattler U., Goldschmidt-Clermont E., Segner H. 17Beta-estradiol affects the response of complement components and survival of rainbow trout (*Oncorhynchus mykiss*) challenged by bacterial infection. Fish Shellfish Immunol. 2011;31:90–97. doi: 10.1016/j.fsi.2011.04.007.

87. Wenger M., Krasnov A., Skugor S., Goldschmidt-Clermont E., Sattler U., Segner H. Estrogen modulates hepatic gene expression and survival of rainbow trout infected with pathogenic bacteria *Yersinia ruckeri*. Mar. Biotechnol. 2012;14:530–543. doi: 10.1007/s10126-012-9473-0.

88. Li M., Leatherland J.F. The interaction between maternal stress and the ontogeny of the innate immune system during teleost embryogenesis: Implications for aquaculture practice. J. Fish Biol. 2012;81:1793–1814. doi: 10.1111/j.1095-8649.2012.03447.x.

89. Seemann F., Knigge T., Rocher B., Minier C., Monsinjon T. 17β-estradiol induces changes in cytokine levels in head kidney and blood of juvenile sea bass (*Dicentrarchus labrax*, L., 1758) Mar. Environ. Res. 2013;87–88:44–51. doi: 10.1016/j.marenvres.2013.03.003.

90. Raida M.K., Holten-Andersen L., Buchmann K. Association between *Yersinia ruckeri* infection, cytokine expression and survival in rainbow trout (*Oncorhynchus mykiss*) Fish Shellfish Immunol. 2011;30:1257–1264. doi: 10.1016/j.fsi.2011.03.022.

91. Wang T., Secombes C.J. The cytokine networks of adaptive immunity in fish. Fish Shellfish Immunol. 2013;35:1703–1718.

CITATIONS

Chapter 7

R. Panicz, J. Sadowski, and R. Drozd, Genetic and structural characterization of the growth hormone gene and protein from tench, Tinca tinca, DOI 10.1007/s10695-012-9661-x.

Chapter 8

Lõhmus M, Sundström LF, Björklund M, Devlin RH (2010) Genotype-Temperature Interaction in the Regulation of Development, Growth, and Morphometrics in Wild-Type, and Growth-Hormone Transgenic Coho Salmon. PLoS ONE 5(4): e9980. https://doi.org/10.1371/journal.pone.0009980.

Chapter 9

Hixson SM (2014) Fish Nutrition and Current Issues in Aquaculture: The Balance in Providing Safe and Nutritious Seafood, in an Environmentally Sustainable Manner. J Aquac Res Development 5:234. doi:10.4172/2155-9546.1000234.

Chapter 10

Maysa Enayatmehr, Shahla Jamili, Effects of Salinity on Growth, Hormonal and Enzymatic Status in Fish: A Review, ISSN (Online): 2319-7064.

Chapter 11

Swetha CH, Sainath SB, Reddy PR, Reddy PS (2011) Reproductive Endocrinology of Female Crustaceans: Perspective and Prospective. J Marine Sci Res Development S3:001. doi: 10.4172/2155-9910.S3-001.

Chapter 12

Anne-Constance Franz, Oliver Faass, Bernd Köllner, Natallia Shved, Karl Link, Ayako Casanova, Michael Wenger, Helena D'Cotta, Jean-François Baroiller, Oliver Ullrich, Manfred Reinecke, and Elisabeth Eppler, Endocrine and Local IGF-I in the Bony Fish Immune System, doi: 10.3390/biology5010009.

INDEX